Two Manuals Combined into One Book

Military Mountaineering

And

FM 7-85 Ranger Unit Operations

Department of Defense

Pentagon
Publishing

PREFACE

Mountains exist in almost every country in the world and almost every war has included some type of mountain operations. This pattern will not change; therefore, soldiers will fight in mountainous terrain in future conflicts. Although mountain operations have not changed, several advancements in equipment and transportation have increased the soldiers' capabilities. The helicopter now allows access to terrain that was once unreachable or could be reached only by slow methodical climbing. Inclement weather, however, may place various restrictions on the capabilities of air assets available to a commander. The unit must then possess the necessary mountaineering skills to overcome adverse terrain to reach an objective.

This field manual details techniques soldiers and leaders must know to cope with mountainous terrain. These techniques are the foundation upon which the mountaineer must build. They must be applied to the various situations encountered to include river crossings, glaciers, snow-covered mountains, ice climbing, rock climbing, and urban vertical environments. The degree to which this training is applied must be varied to conform to known enemy doctrine, tactics, and actions. This FM also discusses basic and advanced techniques to include acclimatization, illness and injury, equipment, anchors, evacuation, movement on glaciers, and training.

This field manual is a training aid for use by *qualified* personnel in conjunction with FM 3-97.6, Mountain Operations, which is used for planning operations in mountainous terrain. Personnel using FM 3-97.61 should attend a recognized Department of Defense Mountain Warfare School for proper training. **Improper use of techniques and procedures by untrained personnel may result in serious injury or death.** Personnel should be certified as Level I, Basic Mountaineer; Level II, Assault Climber; or Level III, Mountain Leader before using FM 3-97.61 for training (see Appendix A).

The measurements in this manual are stated as they are used in training (either metric or standard). Appendix B contains a measurement conversion chart for your convenience.

The proponent of this publication is HQ TRADOC. Submit changes for improving this publication to doctrine@benning.army.mil or on DA Form 2028 (Recommended Changes to Publications and Blank Forms) and forward to the Commander, U.S. Army Infantry School, ATTN: ATSH-RBO, Fort Benning, GA 31905-5593.

Unless otherwise stated, whenever the masculine gender is used, both men and women are included.

FIELD MANUAL
No. 3-97.61

HEADQUARTERS
DEPARTMENT OF THE ARMY
WASHINGTON, DC, 26 August 2002

MILITARY MOUNTAINEERING

CONTENTS

Page

PREFACE vii

CHAPTER 1. MOUNTAIN TERRAIN, WEATHER, AND HAZARDS

Section I. Mountain Terrain 1-1
- 1-1 Definition 1-1
- 1-2 Composition 1-1
- 1-3. Rock and Slope Types 1-1
- 1-4. Rock Classifications 1-2
- 1-5. Mountain Building 1-4
- 1-6. Route Classification 1-5
- 1-7. Cross-Country Movement 1-9
- 1-8. Cover and Concealment 1-10
- 1-9. Observation 1-10
- 1-10. Fields of Fire 1-11

Section II. Mountain Weather 1-11
- 1-11. Considerations for Planning 1-11
- 1-12. Mountain Air 1-12
- 1-13. Weather Characteristics 1-12
- 1-14. Wind 1-13
- 1-15. Humidity 1-14
- 1-16. Cloud Formation 1-14
- 1-17. Types of Clouds 1-15
- 1-18. Fronts 1-23
- 1-19. Temperature 1-23
- 1-20. Weather Forecasting 1-24
- 1-21. Recording Data 1-25

Section III. Mountain Hazards 1-27
- 1-22. Subjective Hazards 1-27
- 1-23. Objective Hazards 1-27
- 1-24. Weather Hazards 1-28
- 1-25. Avalanche Hazards 1-29

*This publication supersedes TC 90-6-1, dated 26 April 1989.

		Page
CHAPTER 2.	**MOUNTAIN LIVING**	
Section I.	Survival	2-1
	2-1. Water Supply	2-1
	2-2. Nutrition	2-2
	2-3. Personal Hygiene and Sanitation	2-5
Section II.	Acclimatization and Conditioning	2-6
	2-4. Symptoms and Adjustments	2-6
	2-5. Physical and Psychological Conditioning	2-7
Section III.	Medical Considerations	2-9
	2-6. Illness and Injury	2-9
	2-7. Treatment and Evacuation	2-10
	2-8. Solar Injuries	2-10
	2-9. Cold-Weather Injuries	2-11
	2-10. Heat Injuries	2-20
	2-11. Acute Mountain Sickness	2-21
	2-12. Chronic Mountain Sickness	2-22
	2-13. Understanding High-Altitude Illnesses	2-22
	2-14. High-Altitude Pulmonary Edema	2-23
	2-15. High-Altitude Cerebral Edema	2-24
	2-16. Hydration in HAPE and HACE	2-25
CHAPTER 3.	**MOUNTAINEERING EQUIPMENT**	
Section I.	Equipment Description and Maintenance	3-1
	3-1. Footwear	3-1
	3-2. Clothing	3-2
	3-3. Climbing Software	3-8
	3-4. Climbing Hardware	3-11
	3-5. Snow and Ice Climbing Hardware	3-21
	3-6. Sustainability Equipment	3-26
Section II.	Equipment Packing	3-30
	3-7. Choice of Equipment	3-30
	3-8. Tips on Packing	3-33
CHAPTER 4.	**ROPE MANAGEMENT AND KNOTS**	
Section I.	Preparation, Care and Maintenance, Inspection, Terminology	4-1
	4-1. Preparation	4-1
	4-2. Care and Maintenance	4-1
	4-3. Inspection	4-3
	4-4. Terminology	4-3
Section II.	Coiling, Carrying, Throwing	4-4
	4-5. Coiling and Carrying the Rope	4-4
	4-6. Throwing the Rope	4-8

Page

Section III. Knots 4-8
4-7. Square Knot 4-9
4-8. Fisherman's Knot 4-9
4-9. Double Fisherman's Knot 4-10
4-10. Figure-Eight Bend 4-11
4-11. Water Knot 4-12
4-12. Bowline 4-13
4-13. Round Turn and Two Half Hitches 4-14
4-14. Figure-Eight Retrace (Rerouted Figure-Eight) 4-15
4-15. Clove Hitch 4-16
4-16. Wireman's Knot 4-17
4-17. Directional Figure-Eight 4-18
4-18. Bowline-on-a-Bight (Two-Loop Bowline) 4-19
4-19. Two-Loop Figure-Eight 4-20
4-20. Figure-Eight Loop (Figure-Eight-on-a-Bight) 4-21
4-21. Prusik Knot 4-22
4-22. Bachman Knot 4-23
4-23. Bowline-on-a-Coil 4-24
4-24. Three-Loop Bowline 4-25
4-25. Figure-Eight Slip Knot 4-26
4-26. Transport Knot (Overhand Slip Knot/Mule Knot) 4-27
4-27. Kleimhiest Knot 4-28
4-28. Frost Knot 4-29
4-29. Girth Hitch 4-30
4-30. Munter Hitch 4-30
4-31. Rappel Seat 4-31
4-32. Guarde Knot 4-32

CHAPTER 5. ANCHORS
Section I. Natural Anchors 5-1
5-1. Trees 5-1
5-2. Boulders 5-2
5-3. Chockstones 5-2
5-4. Rock Projections 5-3
5-5. Tunnels and Arches 5-4
5-6. Bushes and Shrubs 5-4
5-7. Slinging Techniques 5-4
Section II. Anchoring With the Rope 5-7
5-8. Rope Anchor 5-7
5-9. Tensionless Anchor 5-7
Section III. Artificial Anchors 5-8
5-10. Deadman 5-8
5-11. Pitons 5-9
5-12. Chocks 5-11

Page
5-13. Spring-Loaded Camming Device 5-13
5-14. Bolts 5-14
5-15. Equalizing Anchors 5-15

CHAPTER 6. CLIMBING
Section I. Climbing Fundamentals 6-1
6-1. Route Selection 6-1
6-2. Terrain Selection for Training 6-1
6-3. Preparation 6-2
6-4. Spotting 6-2
6-5. Climbing Technique 6-3
6-6. Safety Precautions 6-8
6-7. Margin of Safety 6-9
Section II. Use of Holds 6-10
6-8. Climbing With the Feet 6-10
6-9. Using the Hands 6-13
6-10. Combination Techniques 6-19
Section III. Roped Climbing 6-27
6-11. Tying-in to the Climbing Rope 6-27
6-12. Presewn Harnesses 6-28
6-13. Improvised Harnesses 6-32
Section IV. Belay Techniques 6-33
6-14. Procedure for Managing the Rope 6-34
6-15. Choosing a Belay Technique 6-39
6-16. Establishing a Belay 6-40
6-17. Setting Up a Belay 6-43
6-18. Top-Rope Belay 6-43
Section V. Climbing Commands 6-43
6-19. Verbal Commands 6-44
6-20. Rope Tug Commands 6-45
Section VI. Roped Climbing Methods 6-45
6-21. Top-Roped Climbing 6-45
6-22. Lead Climbing 6-45
6-23. Aid Climbing 6-54
6-24. Three-Man Climbing Team 6-57

CHAPTER 7. ROPE INSTALLATIONS
Section I. Fixed Rope 7-1
7-1. Installation 7-1
7-2. Utilization 7-1
7-3. Retrieval 7-2
7-4. Fixed Rope With Intermediate Anchors 7-3

			Page
Section	II.	Rappelling	7-5
	7-5.	Selection of a Rappel Point	7-6
	7-6.	Installation of the Rappel Point	7-6
	7-7.	Operation of the Rappel Point	7-6
	7-8.	Recovery of the Rappel Point	7-8
	7-9.	Types of Rappels	7-9
Section	III.	One-Rope Bridge	7-14
	7-10.	Site Selection	7-15
	7-11.	Installation Using Transport Tightening System	7-15
	7-12.	Installation Using Z-Pulley Tightening System	7-18
	7-13.	Utilization	7-19
	7-14.	Hauling Line	7-21
	7-15.	Retrieval	7-22
Section	IV.	Suspension Traverse	7-22
	7-16.	Site Selection	7-23
	7-17.	Installation	7-23
	7-18.	Retrieval	7-27
Section	V.	Vertical Hauling Line	7-27
	7-19.	Site Selection	7-27
	7-20.	Installation	7-28
	7-21.	Retrieval	7-29
Section	VI.	Simple Raising Systems	7-29
	7-22.	Z-Pulley System	7-29
	7-23.	U-Pulley System	7-31
CHAPTER 8.		**MOUNTAIN WALKING TECHNIQUES**	
	8-1.	Basic Principles	8-1
	8-2.	Techniques	8-2
	8-3.	Safety Considerations	8-5
	8-4.	Navigation	8-5
	8-5.	Route Planning	8-10
	8-6.	Route Selection	8-13
CHAPTER 9.		**MOUNTAIN STREAM CROSSING**	
	9-1.	Reconnaissance	9-1
	9-2.	Preparation of Troops and Equipment	9-3
	9-3.	Individual Crossings	9-3
	9-4.	Team Crossing	9-4
	9-5.	Rope Installations	9-5
	9-6.	Safety	9-8
	9-7.	Swimming	9-8

Page

CHAPTER 10. MOVEMENT OVER SNOW AND ICE
10-1. Movement Over Snow 10-1
10-2. Movement Over Ice 10-2
10-3. Use of Ice Ax and Crampons 10-2
10-4. Glissading 10-13
10-5. Snow and Ice Anchors 10-14
10-6. Roped Climbing on Ice and Snow 10-18
10-7. Movement on Glaciers 10-19
10-8. Glacier Bivouac Procedures 10-33

CHAPTER 11. MOUNTAIN RESCUE AND EVACUATION
11-1. Considerations 11-1
11-2. Planning Rescue Operations 11-2
11-3. Mass Casualties 11-3
11-4. Special Training 11-4
11-5. Preparation for Evacuation 11-4
11-6. Manual Carries 11-4
11-7. Litters 11-7
11-8. Rescue Systems 11-9
11-9. Low-Angle Evacuation 11-10
11-10. High-Angle Evacuation 11-12

APPENDIX A. LEVELS OF MILITARY MOUNTAINEERING A-1
APPENDIX B. MEASUREMENT CONVERSION FACTORS B-1
APPENDIX C. AVALANCHE SEARCH AND RESCUE TECHNIQUES C-1
GLOSSARY Glossary-1
REFERENCES References-1
INDEX Index-1

CHAPTER 1

MOUNTAIN TERRAIN, WEATHER, AND HAZARDS

Commanders must consider the effects terrain and weather will have on their operations, mainly on their troops and logistics efforts. Weather and terrain combine to challenge efforts in moving supplies to forward areas. Spring storms, which may deposit a foot of snow on dry roads, combined with unprepared vehicles create hazardous situations. Helicopters are a valuable asset for use in moving men and supplies, but commanders should not plan to use them as the only means of movement and resupply. Alternate methods must be planned due to the variability of weather. Units scheduled for deployment in mountainous terrain should become self-sufficient and train under various conditions. Commanders must be familiar with the restraints that the terrain can place on a unit.

Section I. MOUNTAIN TERRAIN

Operations in the mountains require soldiers to be physically fit and leaders to be experienced in operations in this terrain. Problems arise in moving men and transporting loads up and down steep and varied terrain in order to accomplish the mission. Chances for success in this environment are greater when a leader has experience operating under the same conditions as his men. Acclimatization, conditioning, and training are important factors in successful military mountaineering.

1-1. DEFINITION

Mountains are land forms that rise more than 500 meters above the surrounding plain and are characterized by steep slopes. Slopes commonly range from 4 to 45 degrees. Cliffs and precipices may be vertical or overhanging. Mountains may consist of an isolated peak, single ridges, glaciers, snowfields, compartments, or complex ranges extending for long distances and obstructing movement. Mountains usually favor the defense; however, attacks can succeed by using detailed planning, rehearsals, surprise, and well-led troops.

1-2. COMPOSITION

All mountains are made up of rocks and all rocks of minerals (compounds that cannot be broken down except by chemical action). Of the approximately 2,000 known minerals, seven rock-forming minerals make up most of the earth's crust: quartz and feldspar make up granite and sandstone; olivene and pyroxene give basalt its dark color; and amphibole and biotite (mica) are the black crystalline specks in granitic rocks. Except for calcite, found in limestone, they all contain silicon and are often referred to as silicates.

1-3. ROCK AND SLOPE TYPES

Different types of rock and different slopes present different hazards. The following paragraphs discuss the characteristics and hazards of the different rocks and slopes.

a. **Granite.** Granite produces fewer rockfalls, but jagged edges make pulling rope and raising equipment more difficult. Granite is abrasive and increases the danger of ropes or accessory cords being cut. Climbers must beware of large loose boulders. After a rain, granite dries quickly. Most climbing holds are found in cracks. Face climbing can be found, however, it cannot be protected.

b. **Chalk and Limestone.** Chalk and limestone are slippery when wet. Limestone is usually solid; however, conglomerate type stones may be loose. Limestone has pockets, face climbing, and cracks.

c. **Slate and Gneiss.** Slate and gneiss can be firm and or brittle in the same area (red coloring indicates brittle areas). Rockfall danger is high, and small rocks may break off when pulled or when pitons are emplaced.

d. **Sandstone.** Sandstone is usually soft causing handholds and footholds to break away under pressure. Chocks placed in sandstone may or may not hold. Sandstone should be allowed to dry for a couple of days after a rain before climbing on it—wet sandstone is extremely soft. Most climbs follow a crack. Face climbing is possible, but any outward pull will break off handholds and footholds, and it is usually difficult to protect.

e. **Grassy Slopes.** Penetrating roots and increased frost cracking cause a continuous loosening process. Grassy slopes are slippery after rain, new snow, and dew. After long, dry spells clumps of the slope tend to break away. Weight should be distributed evenly; for example, use flat hand push holds instead of finger pull holds.

f. **Firm Spring Snow (Firn Snow).** Stopping a slide on small, leftover snow patches in late spring can be difficult. Routes should be planned to avoid these dangers. Self-arrest should be practiced before encountering this situation. Beginning climbers should be secured with rope when climbing on this type surface. Climbers can glissade down firn snow if necessary. Firn snow is easier to ascend than walking up scree or talus.

g. **Talus.** Talus is rocks that are larger than a dinner plate, but smaller than boulders. They can be used as stepping-stones to ascend or descend a slope. However, if a talus rock slips away it can produce more injury than scree because of its size.

h. **Scree.** Scree is small rocks that are from pebble size to dinner plate size. Running down scree is an effective method of descending in a hurry. One can run at full stride without worry—the whole scree field is moving with you. Climbers must beware of larger rocks that may be solidly planted under the scree. Ascending scree is a tedious task. The scree does not provide a solid platform and will only slide under foot. If possible, avoid scree when ascending.

1-4. ROCK CLASSIFICATIONS

Rock is classified by origin and mineral composition.

a. **Igneous Rocks.** Deep within the earth's crust and mantle, internal heat, friction and radioactive decay creates magmas (melts of silicate minerals) that solidify into igneous rocks upon cooling. When the cooling occurs at depth, under pressure, and over time, the minerals in the magma crystallize slowly and develop well, making coarse-grained plutonic rock. The magma may move upward, propelled by its own lower density, either melting and combining with the overlying layers or forcing them aside. This results in an intrusive rock. If the melt erupts onto the surface it cools rapidly and the minerals form little or no crystal matrix, creating a volcanic or extrusive rock.

(1) ***Plutonic or Intrusive Rocks.*** Slow crystallization from deeply buried magmas generally means good climbing, since the minerals formed are relatively large and interwoven into a solid matrix. Weathering develops protrusions of resistant minerals, which makes for either a rough-surfaced rock with excellent friction, or, if the resistant

crystals are much larger than the surrounding matrix, a surface with numerous knobby holds. Pieces of foreign rock included in the plutonic body while it was rising and crystallizing, or clusters of segregated minerals, may weather differently than the main rock mass and form "chicken heads."

(a) Intrusions are named according to location and size. Large (100 square kilometers or larger) masses of plutonic rock are called "batholiths" and small ones "stocks." Most plutonic rock is in the granite family, differing only in the amounts of constituent minerals contained. A core of such batholiths is in every major mountain system in the world. In the Alps, Sierras, North Cascades, Rockies, Adirondacks, and most other ranges this core is at least partly exposed.

(b) Small plutonic intrusions are stocks, forced between sedimentary strata, and dikes, which cut across the strata. Many of these small intrusive bodies are quickly cooled and thus may look like extrusive rock.

(2) ***Volcanic or Extrusive Rocks.*** Explosive eruptions eject molten rock so quickly into the air that it hardens into loose aerated masses of fine crystals and uncrystallized glass (obsidian). When this ash consolidates while molten or after cooling, it is called "tuff," a weak rock that breaks down quickly and erodes easily. Quieter eruptions, where widespread lava flows from large fissures, produce basalt. Basaltic rocks are fine-grained and often sharp-edged.

(3) ***Jointing Rocks.*** In plutonic rocks, joints or cracks are caused by internal stresses such as contraction during cooling or expansion when overlying rock erodes or exfoliates. Some joints tend to follow a consistent pattern throughout an entire mountain and their existence can often be predicted. Therefore, when a ledge suddenly ends, the joint—and thus the ledge—may begin again around the corner. When molten rock extrudes onto the surface as a lava flow or intrudes into a cold surrounding mass as a dike or sill, the contraction from rapid cooling usually causes so much jointing that climbing can be extremely hazardous. Occasionally, this jointing is regular enough to create massed pillars with usable vertical cracks such as Devil's Tower in Wyoming.

b. **Sedimentary Rocks.** Sedimentary rocks are born high in the mountains, where erosion grinds down debris and moves it down to rivers for transportation to its final deposition in valleys, lakes, or oceans. As sediments accumulate, the bottom layers are solidified by pressure and by mineral cements precipitated from percolating groundwater. Gravel and boulders are transformed into conglomerates; sandy beaches into sandstone; beds of mud into mudstone or shale; and shell beds and coral reefs into limestone or dolomite.

(1) Though in general sedimentary rocks are much more friable than those cooled from molten magmas, pressure and cementing often produce solid rocks. In fact, by sealing up internal cracks cementing can result in flawless surfaces, especially in limestone.

(2) Most high mountain ranges have some sedimentary peaks. Ancient seafloor limestone can be found on the summits of the Himalayas and the Alps. The Canadian Rockies are almost exclusively limestone. With the exception of the Dolomites, in general sedimentary rocks do not offer high-angle climbing comparable to that of granite.

c. **Metamorphic Rocks.** These are igneous or sedimentary rocks that have been altered physically and or chemically by the tremendous heat and pressures within the earth. After sediments are solidified, high heat and pressure can cause their minerals to recrystallize. The bedding planes (strata) may also be distorted by folding and squeezing. Shale changes to slate or schist, sandstone and conglomerate into quartzite, and limestone to marble. These changes may be minimal, only slightly altering the sediments, or extensive enough to produce gneiss, which is almost indistinguishable from igneous rock.

(1) Metamorphic rocks may have not only joints and bedding, but cleavage or foliation, a series of thinly spaced cracks caused by the pressures of folding. Because of this cleavage, lower grades of metamorphic rocks may be completely unsuitable for climbing because the rock is too rotten for safe movement.

(2) Higher degrees of metamorphism or metamorphism of the right rocks provide a solid climbing surface. The Shawangunks of New York are an excellent example of high-grade conglomerate quartzite, which offers world class climbing. The center of the Green Mountain anticline contains heavily metamorphosed schist, which also provides solid climbing.

1-5. MOUNTAIN BUILDING

The two primary mechanisms for mountain-building are volcanic and tectonic activity. Volcanoes are constructed from lava and ash, which begin within the earth as magma. Tectonic activity causes plates to collide, heaving up fold mountains, and to pull apart and crack, forming fault-block mountain ranges.

a. **Plate Tectonics.** The massive slabs composing the outer layer are called tectonic plates. These plates are made up of portions of lighter, granitic continental crust, and heavier, basaltic oceanic crust attached to slabs of the rigid upper mantle. Floating slowly over the more malleable asthenosphere, their movement relative to each other creates earthquakes, volcanoes, ocean trenches, and mountain ridge systems.

b. **Mountain Structure.** The different horizontal and vertical stresses that create mountains usually produce complex patterns. Each type of stress produces a typical structure, and most mountains can be described in terms of these structures.

(1) ***Dome Mountains.*** A simple upward bulge of the crust forms dome mountains such as the Ozarks of Arkansas and Missouri, New York's Adirondacks, the Olympics of Washington, and the High Uintahs of Utah. They are usually the result of the upward movement of magma and the folding of the rock layers overhead. Erosion may strip away the overlying layers, exposing the central igneous core.

(2) ***Fault-Block Mountains.*** Faulting, or cracking of the crust into large chunks, often accompanies upwarp, which results in fault-block mountains. Many forms are created by the motion of these chunks along these faults.

(a) The ranges of the desert country of California, Nevada, and Utah provide the clearest display of faulting. The breakage extends to the surface and often during earthquakes—caused by slippage between the blocks—fresh scarps many feet high develop.

(b) Sometimes a block is faulted on both sides and rises or falls as a unit. More often, however, it is faulted on one side only. The Tetons of Wyoming and the Sierra Nevada display this—along the single zone of faults the range throws up impressive steep scarps, while on the other side the block bends but does not break, leaving a gentler slope from the base of the range to the crest. An example of a dropped block is California's Death Valley, which is below sea level and could not have been carved by erosion.

(3) ***Fold Mountains.*** Tectonic forces, in which continental plates collide or ride over each other, have given rise to the most common mountain form—fold mountains. Geologists call folds geosynclines. Upward folded strata are anticlines and downward folds are synclines. When erosion strips down the overburden of rock from folded mountain ranges, the oldest, central core is all that remains. The Alps and the Appalachians are examples of fold mountains. When the squeezing of a range is intense the rocks of the mountain mass first fold but then may break, and parts of the rocks are pushed sideways and override neighboring formations. This explains why older rocks are often found perched on top of younger ones. Isolated blocks of the over thrust mass may form when erosion strips away links connecting them with their place of origin. Almost every range of folded mountains in the world exhibits an over thrust of one sort or another.

(4) ***Volcanic Mountains.*** Along convergent plate boundaries volcanic activity increases. As it is forced underneath an overriding neighbor, continental crust melts and turns to magma within the mantle. Since it is less dense than the surrounding material it rises and erupts to form volcanoes.

(a) These volcanoes are found in belts, which correspond to continental margins around the world. The best known is the "Ring of Fire" encircling the Pacific Ocean from Katmai in Alaska through the Cascades (Mount Rainier and Mount Saint Helens) down through Mexico's Popocatepetl to the smokes of Tierra del Fuego. This belt then runs west down the Aleutian chain to Kamchatka, south to the volcanoes of Japan and the Philippines, and then east through New Guinea into the Pacific. Smaller volcanic belts are found along the Indonesian-SE Asian arc, the Caucasus region, and the Mediterranean.

(b) Volcanic activity also arises at boundaries where two plates are moving away from each other, creating deep rifts and long ridges where the crust has cracked apart and magma wells up to create new surface material. Examples of this are the Mid-Atlantic Ridge, which has created Iceland and the Azores, and the Rift Valley of East Africa with Kilimanjaro's cone.

(5) ***Complex Mountains.*** Most ranges are complex mountains with portions that have been subject to several processes. A block may have been simply pushed upward without tilting with other portions folded, domed, and faulted, often with a sprinkling of volcanoes. In addition, these processes occur both at the macro and the micro level. One massive fold can make an entire mountain peak; however, there are folds measured by a rope length, and tiny folds found within a handhold. A mountain front may be formed from a single fault, but smaller faults that form ledges and gullies may also be present.

1-6. ROUTE CLASSIFICATION

Military mountaineers must be able to assess a vertical obstacle, develop a course of action to overcome the obstacle, and have the skills to accomplish the plan. Assessment of a vertical obstacle requires experience in the classifications of routes and understanding the levels of difficulty they represent. Without a solid understanding of the difficulty of a chosen route, the mountain leader can place his life and the life of other soldiers in extreme danger. Ignorance is the most dangerous hazard in the mountain environment.

a. In North America the Yosemite Decimal System (YDS) is used to rate the difficulty of routes in mountainous terrain. The YDS classes are:

- Class 1—Hiking trail.
- Class 2—Off-trail scramble.
- Class 3—Climbing, use of ropes for beginners (moderate scrambling).
- Class 4—Belayed climbing. (This is moderate to difficult scrambling, which may have some exposure.)
- Class 5—Free climbing. (This class requires climbers to be roped up, belay and emplace intermediate protection.)

b. Class 5 is further subdivided into the following classifications:

(1) Class 5.0-5.4—Little difficulty. This is the simplest form of free climbing. Hands are necessary to support balance. This is sometimes referred to as advanced rock scrambling.

(2) Class 5.5—Moderate difficulty. Three points of contact are necessary.

(3) Class 5.6—Medium difficulty. The climber can experience vertical position or overhangs where good grips can require moderate levels of energy expenditure.

(4) Class 5.7—Great difficulty. Considerable climbing experience is necessary. Longer stretches of climbing requiring several points of intermediate protection. Higher levels of energy expenditure will be experienced.

(5) Class 5.8—Very great difficulty. Increasing amount of intermediate protection is the rule. High physical conditioning, climbing technique, and experience required.

(6) Class 5.9—Extremely great difficulty. Requires well above average ability and excellent condition. Exposed positions, often combined with small belay points. Passages of the difficult sections can often be accomplished under good conditions. Often combined with aid climbing (A0-A4).

(7) Class 5.10—Extraordinary difficulty. Climb only with improved equipment and intense training. Besides acrobatic climbing technique, mastery of refined security technique is indispensable. Often combined with aid climbing (A0-A4).

(8) Class 5.11-5.14—Greater increases of difficulty, requiring more climbing ability, experience, and energy expenditure. Only talented and dedicated climbers reach this level.

c. Additional classifications include the following.

(1) Classes are further divided into a, b, c, and d categories starting from 5.10 to 5.14 (for example, 5.10d).

(2) Classes are also further divided from 5.9 and below with +/- categories (for example, 5.8+).

(3) All class 5 climbs can also be designated with "R" or "X," which indicates a run-out on a climb. This means that placement of intermediate protection is not possible on portions of the route. (For example, in a classification of 5.8R, the "R" indicates periods of run-out where, if a fall was experienced, ground fall would occur.) Always check the local guidebook to find specific designation for your area.

(4) All class 5 climbs can also be designated with "stars." These refer to the popularity of the climb to the local area. Climbs are represented by a single "star" up to five "stars;" a five-star climb is a classic climb and is usually aesthetically pleasing.

d. Aid climb difficulty classification includes:

(1) A0—"French-free." This technique involves using a piece of gear to make progress; for example, clipping a sling into a bolt or piece of protection and then pulling up on it or stepping up in the sling. Usually only needed to get past one or two more difficult moves on advanced free climbs.

(2) A1—Easy aid. The placement of protection is straight forward and reliable. There is usually no high risk of any piece of protection pulling out. This technique requires etriers and is fast and simple.

(3) A2—Moderate aid. The placement of protection is generally straight forward, but placement can be awkward and strenuous. Usually A2 involves one or two moves that are difficult with good protection placement below and above the difficult moves. No serious fall danger.

(4) A3—Hard aid. This technique requires testing your protection. It involves several awkward and strenuous moves in a row. Generally solid placements which will hold a fall and are found within a full rope length. However, long fall potential does exist, with falls of 40 to 60 feet and intermediate protection on the awkward placements failing. These falls, however, are usually clean and with no serious bodily harm.

(5) A4—Serious aid. This technique requires lots of training and practice. More like walking on eggs so none of them break. Leads will usually take extended amounts of time which cause the lead climber to doubt and worry about each placement. Protection placed will usually only hold a climber's weight and falls can be as long as two-thirds the rope length.

(6) A5—Extreme aid. All protection is sketchy at best. Usually no protection placed on the entire route can be trusted to stop a fall.

(7) A6—Extremely severe aid. Continuous A5 climbing with A5 belay stations. If the leader falls, the whole rope team will probably experience ground fall.

(8) Aid climbing classes are also further divided into +/- categories, such as A3+ or A3-, which would simply refer to easy or hard.

e. Grade ratings (commitment grades) inform the climber of the approximate time a climber trained to the level of the climb will take to complete the route.

- I—Several hours.
- II—Half of a day.
- III—About three-fourths of a day.
- IV—Long hard day (usually not less than 5.7).
- V—1 1/2 to 2 1/2 days (usually not less than 5.8).
- VI—Greater than 2 days.

f. Climbing difficulties are rated by different systems. Table 1-1 shows a comparison of these systems.

- YDS (Yosemite Decimal System)—Used in the United States.
- UIAA (Union des International Alpine Association)—Used in Europe.
- British—The British use adjectives and numbers to designate the difficulty of climbs. This system can be confusing if the climber is not familiar with it.
- French—The French use numbers and letters to designate the difficulty of climbs.
- Brazilian—Brazil uses Roman Numerals and adjectives to designate difficulty.
- Australian—Australia uses only numbers to designate difficulty.

YDS	UIAA	BRITISH	FRENCH	BRAZIL	AUSTRALIA
Class 1	I	easy (E)			
Class 2	II	easy (E)			
Class 3	III	easy (E)	1a, b, c		
Class 4	III-	moderate (MOD)	1a, b, c		
5.0	III	moderate (MOD)	2a, b		4
5.1	III+	difficult (DIFF)	2a, b		5
5.2	IV-	hard difficult	2c, 3a		6
5.3	IV	very difficult	3b, c, 4a		7
5.4	IV+	hard very difficult	3b, c, 4a	II	8, 9
5.5	V-	mild severe	3b, c, 4a	IIsup	10, 11
5.6	V	severe, hard severe, 4a	4a, b, c	III	12, 13
5.7	V+	severe, hard severe, 4b	4a, b, c	IIIsup	14
5.8	VI-	hard severe, hard very severe, 4c	5a, b	IV	15
5.9	VI	5a	5b, c	IVsup	16, 17
5.10a	VII-	E1, 5b	5b, c	V	18
5.10b	VII	E1, 5b	5b, c	Vsup	19
5.10c	VII	E1, 5b	5b, c	VI	20
5.10d	VII+	E1/E2, 5b-5c	5b, c	VIsup	21
5.11a	VIII-	E3, 6a	6a, b, c	VII	22
5.11b	VIII	E3/E4, 6a	6a, b, c	VII	23
5.11c	VIII	E4, 6b	6a, b, c	VIIsup	24
5.12a	IX-	E5, E6/7, 6c	7a	VIII	26
5.12b	IX	E5, E6/7, 6c	7a	VIIIsup	27
5.12c	IX	E5, E6/7, 6c	7a		28
5.12d	IX+	E6/7, 7a	7a		29

Table 1-1. Rating systems.

g. Ice climbing ratings can have commitment ratings and technical ratings. The numerical ratings are often prefaced with WI (waterfall ice), AI (alpine ice), or M (mixed rock and ice).

(1) ***Commitment Ratings.*** Commitment ratings are expressed in Roman numerals.

- I—A short, easy climb near the road, with no avalanche hazard and a straightforward descent.
- II—A route of one or two pitches within a short distance of rescue assistance, with little objective hazard.
- III—A multipitch route at low elevation, or a one-pitch climb with an approach that takes about an hour. The route requires anywhere from a few hours to a long day to complete. The descent may require building rappel anchors, and the route might be prone to avalanche.
- IV—A multipitch route at higher elevations; may require several hours of approach on skis or foot. This route is subject to objective hazards, possibly with a hazardous descent.
- V—A long climb in a remote setting, requiring all day to complete the climb itself. Requires many rappels off anchors for the descent. This route has sustained exposure to avalanche or other objective hazards.

- VI—A long ice climb in an alpine setting, with sustained technical climbing. Only elite climbers will complete it in a day. A difficult and involved approach and descent, with objective hazards ever-present, all in a remote area.
- VII—Everything a grade VI has, and more of it. Possibly days to approach the climb, and objective hazards rendering survival as questionable. Difficult physically and mentally.

(2) ***Technical Ratings.*** Technical ratings are expressed as Arabic numerals.

- 1—A frozen lake or stream bed.
- 2—A pitch with short sections of ice up to 80 degrees; lots of opportunity for protection and good anchors.
- 3—Sustained ice up to 80 degrees; the ice is usually good, with places to rest, but it requires skill at placing protection and setting anchors.
- 4—A sustained pitch that is vertical or slightly less than vertical; may have special features such as chandeliers and run-outs between protection.
- 5—A long, strenuous pitch, possibly 50 meters of 85- to 90-degree ice with few if any rests between anchors. The pitch may be shorter, but on featureless ice. Good skills at placing protection are required.
- 6—A full 50-meter pitch of dead vertical ice, possibly of poor quality; requires efficiency of movement and ability to place protection while in awkward stances.
- 7—A full rope length of thin vertical or overhanging ice of dubious adhesion. An extremely tough pitch, physically and mentally, requiring agility and creativity.
- 8—Simply the hardest ice climbing ever done; extremely bold and gymnastic.

1-7. CROSS-COUNTRY MOVEMENT

Soldiers must know the terrain to determine the feasible routes for cross-country movement when no roads or trails are available.

a. A pre-operations intelligence effort should include topographic and photographic map coverage as well as detailed weather data for the area of operations. When planning mountain operations, additional information may be needed about size, location, and characteristics of landforms; drainage; types of rock and soil; and the density and distribution of vegetation. Control must be decentralized to lower levels because of varied terrain, erratic weather, and communication problems inherent to mountainous regions.

b. Movement is often restricted due to terrain and weather. The erratic weather requires that soldiers be prepared for wide variations in temperature, types, and amounts of precipitation.

(1) Movement above the timberline reduces the amount of protective cover available at lower elevations. The logistical problem is important; therefore, each man must be self-sufficient to cope with normal weather changes using materials from his rucksack.

(2) Movement during a storm is difficult due to poor visibility and bad footing on steep terrain. Although the temperature is often higher during a storm than during clear weather, the dampness of rain and snow and the penetration of wind cause soldiers to chill quickly. Although climbers should get off the high ground and seek shelter and

warmth, if possible, during severe mountain storms, capable commanders may use reduced visibility to achieve tactical surprise.

c. When the tactical situation requires continued movement during a storm, the following precautions should be observed:

- Maintain visual contact.
- Keep warm. Maintain energy and body heat by eating and drinking often; carry food that can be eaten quickly and while on the move.
- Keep dry. Wear wet-weather clothing when appropriate, but do not overdress, which can cause excessive perspiration and dampen clothing. As soon as the objective is reached and shelter secured, put on dry clothing.
- Do not rush. Hasty movement during storms leads to breaks in contact and accidents.
- If lost, stay warm, dry, and calm.
- Do not use ravines as routes of approach during a storm as they often fill with water and are prone to flash floods.
- Avoid high pinnacles and ridgelines during electrical storms.
- Avoid areas of potential avalanche or rock-fall danger.

1-8. COVER AND CONCEALMENT

When moving in the mountains, outcroppings, boulders, heavy vegetation, and intermediate terrain can provide cover and concealment. Digging fighting positions and temporary fortifications is difficult because soil is often thin or stony. The selection of dug-in positions requires detailed planning. Some rock types, such as volcanic tuff, are easily excavated. In other areas, boulders and other loose rocks can be used for building hasty fortifications. In alpine environments, snow and ice blocks may be cut and stacked to supplement dug-in positions. As in all operations, positions and routes must be camouflaged to blend in with the surrounding terrain to prevent aerial detection.

1-9. OBSERVATION

Observation in mountains varies because of weather and ground cover. The dominating height of mountainous terrain permits excellent long-range observation. However, rapidly changing weather with frequent periods of high winds, rain, snow, sleet, hail, and fog can limit visibility. The rugged nature of the terrain often produces dead space at midranges.

a. Low cloud cover at higher elevations may neutralize the effectiveness of OPs established on peaks or mountaintops. High wind speeds and sound often mask the noises of troop movement. Several OPs may need to be established laterally, in depth, and at varying altitudes to provide visual coverage of the battle area.

b. Conversely, the nature of the terrain can be used to provide concealment from observation. This concealment can be obtained in the dead space. Mountainous regions are subject to intense shadowing effects when the sun is low in relatively clear skies. The contrast from lighted to shaded areas causes visual acuity in the shaded regions to be considerably reduced. These shadowed areas can provide increased concealment when combined with other camouflage and should be considered in maneuver plans.

1-10. FIELDS OF FIRE

Fields of fire, like observation, are excellent at long ranges. However, dead space is a problem at short ranges. When forces cannot be positioned to cover dead space with direct fires, mines and obstacles or indirect fire must be used. Range determination is deceptive in mountainous terrain. Soldiers must routinely train in range estimation in mountainous regions to maintain their proficiency.

Section II. MOUNTAIN WEATHER

Most people subconsciously "forecast" the weather. If they look outside and see dark clouds they may decide to take rain gear. If an unexpected wind strikes, people glance to the sky for other bad signs. A conscious effort to follow weather changes will ultimately lead to a more accurate forecast. An analysis of mountain weather and how it is affected by mountain terrain shows that such weather is prone to patterns and is usually severe, but patterns are less obvious in mountainous terrain than in other areas. Conditions greatly change with altitude, latitude, and exposure to atmospheric winds and air masses. Mountain weather can be extremely erratic. It varies from stormy winds to calm, and from extreme cold to warmth within a short time or with a minor shift in locality. The severity and variance of the weather causes it to have a major impact on military operations.

1-11. CONSIDERATIONS FOR PLANNING

Mountain weather can be either a dangerous obstacle to operations or a valuable aid, depending on how well it is understood and to what extent advantage is taken of its peculiar characteristics.

a. Weather often determines the success or failure of a mission since it is highly changeable. Military operations plans must be flexible, especially in planning airmobile and airborne operations. The weather must be anticipated to allow enough time for planning so that the leaders of subordinate units can use their initiative in turning an important weather factor in their favor. The clouds that often cover the tops of mountains and the fogs that cover valleys are an excellent means of concealing movements that normally are made during darkness or in smoke. Limited visibility can be used as a combat multiplier.

b. The safety or danger of almost all high mountain regions, especially in winter, depends upon a change of a few degrees of temperature above or below the freezing point. Ease and speed of travel depend mainly on the weather. Terrain that can be crossed swiftly and safely one day may become impassable or highly dangerous the next due to snowfall, rainfall, or a rise in temperature. The reverse can happen just as quickly. The prevalence of avalanches depends on terrain, snow conditions, and weather factors.

c. Some mountains, such as those found in desert regions, are dry and barren with temperatures ranging from extreme heat in the summer to extreme cold in the winter. In tropical regions, lush jungles with heavy seasonal rains and little temperature variation often cover mountains. High rocky crags with glaciated peaks can be found in mountain ranges at most latitudes along the western portion of the Americas and Asia.

d. Severe weather may decrease morale and increase basic survival problems. These problems can be minimized when men have been trained to accept the weather by being self-sufficient. Mountain soldiers properly equipped and trained can use the weather to their advantage in combat operations.

1-12. MOUNTAIN AIR

High mountain air is dry and may be drier in the winter. Cold air has a reduced capacity to hold water vapor. Because of this increased dryness, equipment does not rust as quickly and organic material decomposes slowly. The dry air also requires soldiers to increase consumption of water. The reduced water vapor in the air causes an increase in evaporation of moisture from the skin and in loss of water through transpiration in the respiratory system. Due to the cold, most soldiers do not naturally consume the quantity of fluids they would at higher temperatures and must be encouraged to consciously increase their fluid intake.

a. Pressure is low in mountainous areas due to the altitude. The barometer usually drops 2.5 centimeters for every 300 meters gained in elevation (3 percent).

b. The air at higher altitudes is thinner as atmospheric pressure drops with the increasing altitude. The altitude has a natural filtering effect on the sun's rays. Rays are absorbed or reflected in part by the molecular content of the atmosphere. This effect is greater at lower altitudes. At higher altitudes, the thinner, drier air has a reduced molecular content and, consequently, a reduced filtering effect on the sun's rays. The intensity of both visible and ultraviolet rays is greater with increased altitude. These conditions increase the chance of sunburn, especially when combined with a snow cover that reflects the rays upward.

1-13. WEATHER CHARACTERISTICS

The earth is surrounded by an atmosphere that is divided into several layers. The world's weather systems are in the lower of these layers known as the "troposphere." This layer reaches as high as 40,000 feet. Weather is a result of an atmosphere, oceans, land masses, unequal heating and cooling from the sun, and the earth's rotation. The weather found in any one place depends on many things such as the air temperature, humidity (moisture content), air pressure (barometric pressure), how it is being moved, and if it is being lifted or not.

a. Air pressure is the "weight" of the atmosphere at any given place. The higher the pressure, the better the weather will be. With lower air pressure, the weather will more than likely be worse. In order to understand this, imagine that the air in the atmosphere acts like a liquid. Areas with a high level of this "liquid" exert more pressure on an area and are called high-pressure areas. Areas with a lower level are called low-pressure areas. The average air pressure at sea level is 29.92 inches of mercury (hg) or 1,013 millibars (mb). The higher in altitude, the lower the pressure.

(1) ***High Pressure.*** The characteristics of a high-pressure area are as follows:

- The airflow is clockwise and out.
- Otherwise known as an "anticyclone".
- Associated with clear skies.
- Generally the winds will be mild.
- Depicted as a blue "H" on weather maps.

(2) ***Low Pressure.*** The characteristics of a low-pressure area are as follows:

- The airflow is counterclockwise and in.
- Otherwise known as a "cyclone".
- Associated with bad weather.
- Depicted as a red "L" on weather maps.

b. Air from a high-pressure area is basically trying to flow out and equalize its pressure with the surrounding air. Low pressure, on the other hand, is building up vertically by pulling air in from outside itself, which causes atmospheric instability resulting in bad weather.

c. On a weather map, these differences in pressure are depicted as isobars. Isobars resemble contour lines and are measured in either millibars or inches of mercury. The areas of high pressure are called "ridges" and lows are called "troughs."

1-14. WIND

In high mountains, the ridges and passes are seldom calm; however, strong winds in protected valleys are rare. Normally, wind speed increases with altitude since the earth's frictional drag is strongest near the ground. This effect is intensified by mountainous terrain. Winds are accelerated when they converge through mountain passes and canyons. Because of these funneling effects, the wind may blast with great force on an exposed mountainside or summit. Usually, the local wind direction is controlled by topography.

a. The force exerted by wind quadruples each time the wind speed doubles; that is, wind blowing at 40 knots pushes four times harder than a wind blowing at 20 knots. With increasing wind strength, gusts become more important and may be 50 percent higher than the average wind speed. When wind strength increases to a hurricane force of 64 knots or more, soldiers should lay on the ground during gusts and continue moving during lulls. If a hurricane- force wind blows where there is sand or snow, dense clouds fill the air. The rocky debris or chunks of snow crust are hurled near the surface. During the winter season, or at high altitudes, commanders must be constantly aware of the wind-chill factor and associated cold-weather injuries (see Chapter 2).

b. Winds are formed due to the uneven heating of the air by the sun and rotation of the earth. Much of the world's weather depends on a system of winds that blow in a set direction.

c. Above hot surfaces, air expands and moves to colder areas where it cools and becomes denser, and sinks to the earth's surface. The results are a circulation of air from the poles along the surface of the earth to the equator, where it rises and moves to the poles again.

d. Heating and cooling together with the rotation of the earth causes surface winds. In the Northern Hemisphere, there are three prevailing winds:

(1) ***Polar Easterlies***. These are winds from the polar region moving from the east. This is air that has cooled and settled at the poles.

(2) ***Prevailing Westerlies***. These winds originate from approximately 30 degrees north latitude from the west. This is an area where prematurely cooled air, due to the earth's rotation, has settled to the surface.

(3) ***Northeast Tradewinds***. These are winds that originate from approximately 30° north from the northeast.

e. The jet stream is a long meandering current of high-speed winds often exceeding 250 miles per hour near the transition zone between the troposphere and the stratosphere known as the tropopause. These winds blow from a generally westerly direction dipping down and picking up air masses from the tropical regions and going north and bringing down air masses from the polar regions.

f. The patterns of wind mentioned above move air. This air comes in parcels called "air masses." These air masses can vary from the size of a small town to as large as a country. These air masses are named from where they originate:

- Maritime—over water.
- Continental—over land
- Polar—north of 60° north latitude.
- Tropical—south of 60° north latitude.

Combining these parcels of air provides the names and description of the four types of air masses:

- Continental Polar—cold, dry air mass.
- Maritime Polar—cold, wet air mass.
- Maritime Tropical—warm, wet air mass.
- Continental Tropical—warm, dry air mass.

g. Two types of winds are peculiar to mountain environments, but do not necessarily affect the weather.

(1) ***Anabatic Wind (Valley Winds).*** These winds blow up mountain valleys to replace warm rising air and are usually light winds.

(2) ***Katabatic Wind (Mountain Wind).*** These winds blow down mountain valley slopes caused by the cooling of air and are occasionally strong winds.

1-15. HUMIDITY

Humidity is the amount of moisture in the air. All air holds water vapor even if it cannot be seen. Air can hold only so much water vapor; however, the warmer the air, the more moisture it can hold. When air can hold all that it can the air is "saturated" or has 100 percent relative humidity.

a. If air is cooled beyond its saturation point, the air will release its moisture in one form or another (clouds, fog, dew, rain, snow, and so on). The temperature at which this happens is called the "condensation point". The condensation point varies depending on the amount of water vapor contained in the air and the temperature of the air. If the air contains a great deal of water, condensation can occur at a temperature of 68 degrees Fahrenheit, but if the air is dry and does not hold much moisture, condensation may not form until the temperature drops to 32 degrees Fahrenheit or even below freezing.

b. The adiabatic lapse rate is the rate at which air cools as it rises or warms as it descends. This rate varies depending on the moisture content of the air. Saturated (moist) air will warm and cool approximately 3.2 degrees Fahrenheit per 1,000 feet of elevation gained or lost. Dry air will warm and cool approximately 5.5 degrees Fahrenheit per 1,000 feet of elevation gained or lost.

1-16. CLOUD FORMATION

Clouds are indicators of weather conditions. By reading cloud shapes and patterns, observers can forecast weather with little need for additional equipment such as a barometer, wind meter, and thermometer. Any time air is lifted or cooled beyond its saturation point (100 percent relative humidity), clouds are formed. The four ways air gets lifted and cooled beyond its saturation point are as follows.

a. **Convective Lifting.** This effect happens due to the sun's heat radiating off the Earth's surface causing air currents (thermals) to rise straight up and lift air to a point of saturation.

b. **Frontal Lifting.** A front is formed when two air masses of different moisture content and temperature collide. Since air masses will not mix, warmer air is forced aloft over the colder air mass. From there it is cooled and then reaches its saturation point. Frontal lifting creates the majority of precipitation.

c. **Cyclonic Lifting.** An area of low pressure pulls air into its center from all over in a counterclockwise direction. Once this air reaches the center of the low pressure, it has nowhere to go but up. Air continues to lift until it reaches the saturation point.

d. **Orographic Lifting.** This happens when an air mass is pushed up and over a mass of higher ground such as a mountain. Air is cooled due to the adiabatic lapse rate until the air's saturation point is reached.

1-17. TYPES OF CLOUDS

Clouds are one of the signposts to what is happening with the weather. Clouds can be described in many ways. They can be classified by height or appearance, or even by the amount of area covered vertically or horizontally. Clouds are classified into five categories: low-, mid-, and high-level clouds; vertically-developed clouds; and less common clouds.

a. **Low-Level Clouds.** Low-level clouds (0 to 6,500 feet) are either cumulus or stratus (Figures 1-1 and 1-2, page 1-16). Low-level clouds are mostly composed of water droplets since their bases lie below 6,500 feet. When temperatures are cold enough, these clouds may also contain ice particles and snow.

(1) The two types of precipitating low-level clouds are nimbostratus and stratocumulus (Figures 1-3 and 1-4, page 1-17).

(a) Nimbostratus clouds are dark, low-level clouds accompanied by light to moderately falling precipitation. The sun or moon is not visible through nimbostratus clouds, which distinguishes them from mid-level altostratus clouds. Because of the fog and falling precipitation commonly found beneath and around nimbostratus clouds, the cloud base is typically extremely diffuse and difficult to accurately determine.

(b) Stratocumulus clouds generally appear as a low, lumpy layer of clouds that is sometimes accompanied by weak precipitation. Stratocumulus vary in color from dark gray to light gray and may appear as rounded masses with breaks of clear sky in between. Because the individual elements of stratocumulus are larger than those of altocumulus, deciphering between the two cloud types is easier. With your arm extended toward the sky, altocumulus elements are about the size of a thumbnail while stratocumulus are about the size of a fist.

Figure 1-1. Cumulus clouds.

Figure 1-2. Stratus clouds.

Figure 1-3. Nimbostratus clouds.

Figure 1-4. Stratocumulus clouds.

(2) Low-level clouds may be identified by their height above nearby surrounding relief of known elevation. Most precipitation originates from low-level clouds because rain or snow usually evaporate before reaching the ground from higher clouds. Low-level clouds usually indicate impending precipitation, especially if the cloud is more than 3,000 feet thick. (Clouds that appear dark at their bases are more than 3,000 feet thick.)

b. **Mid-Level Clouds.** Mid-level clouds (between 6,500 to 20,000 feet) have a prefix of alto. Middle clouds appear less distinct than low clouds because of their height. Alto clouds with sharp edges are warmer because they are composed mainly of water droplets. Cold clouds, composed mainly of ice crystals and usually colder than -30 degrees F, have distinct edges that grade gradually into the surrounding sky. Middle clouds usually indicate fair weather, especially if they are rising over time. Lowering middle clouds indicate potential storms, though usually hours away. There are two types of mid-level clouds, altocumulus and altostratus clouds (Figures 1-5 and 1-6).

(1) Altocumulus clouds can appear as parallel bands or rounded masses. Typically a portion of an altocumulus cloud is shaded, a characteristic which makes them distinguishable from high-level cirrocumulus. Altocumulus clouds usually form in advance of a cold front. The presence of altocumulus clouds on a warm humid summer morning is commonly followed by thunderstorms later in the day. Altocumulus clouds that are scattered rather than even, in a blue sky, are called "fair weather" cumulus and suggest arrival of high pressure and clear skies.

(2) Altostratus clouds are often confused with cirrostratus. The one distinguishing feature is that a halo is not observed around the sun or moon. With altostratus, the sun or moon is only vaguely visible and appears as if it were shining through frosted glass.

Figure 1-5. Altocumulus.

Figure 1-6. Altostratus.

c. **High-Level Clouds.** High-level clouds (more than 20,000 feet above ground level) are usually frozen clouds, indicating air temperatures at that elevation below -30 degrees Fahrenheit, with a fibrous structure and blurred outlines. The sky is often covered with a thin veil of cirrus that partly obscures the sun or, at night, produces a ring of light around the moon. The arrival of cirrus indicates moisture aloft and the approach of a traveling storm system. Precipitation is often 24 to 36 hours away. As the storm approaches, the cirrus thickens and lowers, becoming altostratus and eventually stratus. Temperatures are warm, humidity rises, and winds become southerly or south easterly. The two types of high-level clouds are cirrus and cirrostratus (Figure 1-7 and Figure 1-8, page 1-20).

(1) Cirrus clouds are the most common of the high-level clouds. Typically found at altitudes greater than 20,000 feet, cirrus are composed of ice crystals that form when super-cooled water droplets freeze. Cirrus clouds generally occur in fair weather and point in the direction of air movement at their elevation. Cirrus can be observed in a variety of shapes and sizes. They can be nearly straight, shaped like a comma, or seemingly all tangled together. Extensive cirrus clouds are associated with an approaching warm front.

(2) Cirrostratus clouds are sheet-like, high-level clouds composed of ice crystals. They are relatively transparent and can cover the entire sky and be up to several thousand feet thick. The sun or moon can be seen through cirrostratus. Sometimes the only indication of cirrostratus clouds is a halo around the sun or moon. Cirrostratus clouds tend to thicken as a warm front approaches, signifying an increased production of ice crystals. As a result, the halo gradually disappears and the sun or moon becomes less visible.

Figure 1-7. Cirrus.

Figure 1-8. Cirrostratus.

d. **Vertical-Development Clouds.** Clouds with vertical development can grow to heights in excess of 39,000 feet, releasing incredible amounts of energy. The two types of clouds with vertical development are fair weather cumulus and cumulonimbus.

(1) Fair weather cumulus clouds have the appearance of floating cotton balls and have a lifetime of 5 to 40 minutes. Known for their flat bases and distinct outlines, fair weather cumulus exhibit only slight vertical growth, with the cloud tops designating the limit of the rising air. Given suitable conditions, however, these clouds can later develop into towering cumulonimbus clouds associated with powerful thunderstorms. Fair weather cumulus clouds are fueled by buoyant bubbles of air known as thermals that rise up from the earth's surface. As the air rises, the water vapor cools and condenses forming water droplets. Young fair weather cumulus clouds have sharply defined edges and bases while the edges of older clouds appear more ragged, an artifact of erosion. Evaporation along the cloud edges cools the surrounding air, making it heavier and producing sinking motion outside the cloud. This downward motion inhibits further convection and growth of additional thermals from down below, which is why fair weather cumulus typically have expanses of clear sky between them. Without a continued supply of rising air, the cloud begins to erode and eventually disappears.

(2) Cumulonimbus clouds are much larger and more vertically developed than fair weather cumulus (Figure 1-9). They can exist as individual towers or form a line of towers called a squall line. Fueled by vigorous convective updrafts, the tops of cumulonimbus clouds can reach 39,000 feet or higher. Lower levels of cumulonimbus clouds consist mostly of water droplets while at higher elevations, where the temperatures are well below freezing, ice crystals dominate the composition. Under favorable conditions, harmless fair weather cumulus clouds can quickly develop into large cumulonimbus associated with powerful thunderstorms known as super-cells. Super-cells are large thunderstorms with deep rotating updrafts and can have a lifetime of several hours. Super-cells produce frequent lightning, large hail, damaging winds, and

tornadoes. These storms tend to develop during the afternoon and early evening when the effects of heating from the sun are the strongest.

Figure 1-9. Cumulonimbus.

e. **Other Cloud Types.** These clouds are a collection of miscellaneous types that do not fit into the previous four groups. They are orographic clouds, lenticulars, and contrails.

(1) Orographic clouds develop in response to the forced lifting of air by the earth's topography. Air passing over a mountain oscillates up and down as it moves downstream. Initially, stable air encounters a mountain, is lifted upward, and cools. If the air cools to its saturation temperature during this process, the water vapor condenses and becomes visible as a cloud. Upon reaching the mountain top, the air is heavier than the environment and will sink down the other side, warming as it descends. Once the air returns to its original height, it has the same buoyancy as the surrounding air. However, the air does not stop immediately because it still has momentum carrying it downward. With continued descent, the air becomes warmer then the surrounding air and accelerates back upwards towards its original height. Another name for this type of cloud is the lenticular cloud.

(2) Lenticular clouds are cloud caps that often form above pinnacles and peaks, and usually indicate higher winds aloft (Figure 1-10, page 1-22). Cloud caps with a lens shape, similar to a "flying saucer," indicate extremely high winds (over 40 knots). Lenticulars should always be watched for changes. If they grow and descend, bad weather can be expected.

Figure 1-10. Lenticular.

(3) Contrails are clouds that are made by water vapor being inserted into the upper atmosphere by the exhaust of jet engines (Figure 1-11). Contrails evaporate rapidly in fair weather. If it takes longer than two hours for contrails to evaporate, then there is impending bad weather (usually about 24 hours prior to a front).

Figure 1-11. Contrails.

f. **Cloud Interpretation.** Serious errors can occur in interpreting the extent of cloud cover, especially when cloud cover must be reported to another location. Cloud cover always appears greater on or near the horizon, especially if the sky is covered with cumulus clouds, since the observer is looking more at the sides of the clouds rather than between them. Cloud cover estimates should be restricted to sky areas more than 40 degrees above the horizon—that is, to the local sky. Assess the sky by dividing the 360 degrees of sky around you into eighths. Record the coverage in eighths and the types of clouds observed.

1-18. FRONTS

Fronts occur when two air masses of different moisture and temperature contents meet. One of the indicators that a front is approaching is the progression of the clouds. The four types of fronts are warm, cold, occluded, and stationary.

a. **Warm Front.** A warm front occurs when warm air moves into and over a slower or stationary cold air mass. Because warm air is less dense, it will rise up and over the cooler air. The cloud types seen when a warm front approaches are cirrus, cirrostratus, nimbostratus (producing rain), and fog. Occasionally, cumulonimbus clouds will be seen during the summer months.

b. **Cold Front.** A cold front occurs when a cold air mass overtakes a slower or stationary warm air mass. Cold air, being more dense than warm air, will force the warm air up. Clouds observed will be cirrus, cumulus, and then cumulonimbus producing a short period of showers.

c. **Occluded Front.** Cold fronts generally move faster than warm fronts. The cold fronts eventually overtake warm fronts and the warm air becomes progressively lifted from the surface. The zone of division between cold air ahead and cold air behind is called a "cold occlusion." If the air behind the front is warmer than the air ahead, it is a warm occlusion. Most land areas experience more occlusions than other types of fronts. The cloud progression observed will be cirrus, cirrostratus, altostratus, and nimbostratus. Precipitation can be from light to heavy.

d. **Stationary Front.** A stationary front is a zone with no significant air movement. When a warm or cold front stops moving, it becomes a stationary front. Once this boundary begins forward motion, it once again becomes a warm or cold front. When crossing from one side of a stationary front to another, there is typically a noticeable temperature change and shift in wind direction. The weather is usually clear to partly cloudy along the stationary front.

1-19. TEMPERATURE

Normally, a temperature drop of 3 to 5 degrees Fahrenheit for every 1,000 feet gain in altitude is encountered in motionless air. For air moving up a mountain with condensation occurring (clouds, fog, and precipitation), the temperature of the air drops 3.2 degrees Fahrenheit with every 1,000 feet of elevation gain. For air moving up a mountain with no clouds forming, the temperature of the air drops 5.5 degrees Fahrenheit for every 1,000 feet of elevation gain.

a. An expedient to this often occurs on cold, clear, calm mornings. During a troop movement or climb started in a valley, higher temperatures may often be encountered as altitude is gained. This reversal of the normal cooling with elevation is called temperature inversion. Temperature inversions are caused when mountain air is cooled by ice, snow, and heat loss through thermal radiation. This cooler, denser air settles into the valleys and low areas. The inversion continues until the sun warms the surface of the earth or a moderate wind causes a mixing of the warm and cold layers. Temperature inversions are common in the mountainous regions of the arctic, subarctic, and mid-latitudes.

b. At high altitudes, solar heating is responsible for the greatest temperature contrasts. More sunshine and solar heat are received above the clouds than below. The important effect of altitude is that the sun's rays pass through less of the atmosphere and more direct heat is received than at lower levels, where solar radiation is absorbed and reflected by dust and water vapor. Differences of 40 to 50 degrees Fahrenheit may occur between surface temperatures in the shade and surface temperatures in the sun. This is particularly true for dark metallic objects. The difference in temperature felt on the skin between the sun and shade is normally 7 degrees Fahrenheit. Special care must be taken to avoid sunburn and snow blindness. Besides permitting rapid heating, the clear air at high altitudes also favors rapid cooling at night. Consequently, the temperature rises fast after sunrise and drops quickly after sunset. Much of the chilled air drains downward, due to convection currents, so that the differences between day and night temperatures are greater in valleys than on slopes.

c. Local weather patterns force air currents up and over mountaintops. Air is cooled on the windward side of the mountain as it gains altitude, but more slowly (3.2 degrees Fahrenheit per 1,000 feet) if clouds are forming due to heat release when water vapor becomes liquid. On the leeward side of the mountain, this heat gained from the condensation on the windward side is added to the normal heating that occurs as the air descends and air pressure increases. Therefore, air and winds on the leeward slope are considerably warmer than on the windward slope, which is referred to as Chinook winds. The heating and cooling of the air affects planning considerations primarily with regard to the clothing and equipment needed for an operation.

1-20. WEATHER FORECASTING

The use of a portable aneroid barometer, thermometer, wind meter, and hygrometer help in making local weather forecasts. Reports from other localities and from any weather service, including USAF, USN, or the National Weather Bureau, are also helpful. Weather reports should be used in conjunction with the locally observed current weather situation to forecast future weather patterns.

a. Weather at various elevations may be quite different because cloud height, temperature, and barometric pressure will all be different. There may be overcast and rain in a lower area, with mountains rising above the low overcast into warmer clear weather.

b. To be effective, a forecast must reach the small-unit leaders who are expected to utilize weather conditions for assigned missions. Several different methods can be used to create a forecast. The method a forecaster chooses depends upon the forecaster's experience, the amount of data available, the level of difficulty that the forecast situation presents, and the degree of accuracy needed to make the forecast. The five ways to forecast weather are:

(1) ***Persistence Method.*** "Today equals tomorrow" is the simplest way of producing a forecast. This method assumes that the conditions at the time of the forecast will not

change; for example, if today was hot and dry, the persistence method predicts that tomorrow will be the same.

(2) ***Trends Method.*** "Nowcasting" involves determining the speed and direction of fronts, high- and low-pressure centers, and clouds and precipitation. For example, if a cold front moves 300 miles during a 24-hour period, we can predict that it will travel 300 miles in another 24-hours.

(3) ***Climatology Method.*** This method averages weather statistics accumulated over many years. This only works well when the pattern is similar to the following years.

(4) ***Analog Method.*** This method examines a day's forecast and recalls a day in the past when the weather looked similar (an analogy). This method is difficult to use because finding a perfect analogy is difficult.

(5) ***Numerical Weather Prediction.*** This method uses computers to analyze all weather conditions and is the most accurate of the five methods.

1-21. RECORDING DATA

An accurate observation is essential in noting trends in weather patterns. Ideally, under changing conditions, trends will be noted in some weather parameters. However, this may not always be the case. A minor shift in the winds may signal an approaching storm.

a. **Wind Direction.** Assess wind direction as a magnetic direction from which the wind is blowing.

b. **Wind Speed.** Assess wind speed in knots.

(1) If an anemometer is available, assess speed to the nearest knot.

(2) If no anemometer is available, estimate the speed in knots. Judge the wind speed by the way objects, such as trees, bushes, tents, and so forth, are blowing.

c. **Visibility in Meters.** Observe the farthest visible major terrain or man-made feature and determine the distance using any available map.

d. **Present Weather.** Include any precipitation or obscuring weather. The following are examples of present weather:

- Rain—continuous and steady liquid precipitation that will last at least one hour.
- Rain showers—short-term and potentially heavy downpours that rarely last more than one hour.
- Snow—continuous and steady frozen precipitation that will last at least one hour.
- Snow showers—short-term and potentially heavy frozen downpours that rarely last more than one hour.
- Fog, haze—obstructs visibility of ground objects.
- Thunderstorms—a potentially dangerous storm. Thunderstorms will produce lightning, heavy downpours, colder temperatures, tornadoes (not too frequently), hail, and strong gusty winds at the surface and aloft. Winds commonly exceed 35 knots.

e. **Total Cloud Cover.** Assess total cloud cover in eighths. Divide the sky into eight different sections measuring from horizon to horizon. Count the sections with cloud cover, which gives the total cloud cover in eighths. (For example, if half of the sections are covered with clouds, total cloud cover is 4/8.)

f. **Ceiling Height.** Estimate where the cloud base intersects elevated terrain. Note if bases are above all terrain. If clouds are not touching terrain, then estimate to the best of your ability.

g. **Temperature.** Assess temperature with or without a thermometer.

(1) With a thermometer, assess temperature in degrees Celsius (use Fahrenheit only if Celsius conversion is not available). To convert Fahrenheit to Celsius: C = F minus 32 times .55. To convert Celsius to Fahrenheit: F = 1.8 times C plus 32.

Example: 41 degrees F – 32 x .55 = 5 degrees C.
5 degrees C x 1.8 + 32 = 41 degrees F.

(2) Without a thermometer, estimate temperature as above or below freezing ($0^{\circ}C$), as well as an estimated temperature.

h. **Pressure Trend.** With a barometer or altimeter, assess the pressure trend.

(1) A high pressure moving in will cause altimeters to indicate lower elevation.

(2) A low pressure moving in will cause altimeters to indicate higher elevation.

i. **Observed Weather.** Note changes or trends in observed weather conditions.

(1) Deteriorating trends include:

- Marked wind direction shifts. A high pressure system wind flows clockwise. A low pressure system wind flows counterclockwise. The closer the isometric lines are, the greater the differential of pressure (greater wind speeds).
- Marked wind speed increases.
- Changes in obstructions to visibility.
- Increasing cloud coverage.
- Increase in precipitation. A steady drizzle is usually a long-lasting rain.
- Lowering cloud ceilings.
- Marked cooler temperature changes, which could indicate that a cold front is passing through.
- Marked increase in humidity.
- Decreasing barometric pressure, which indicates a lower pressure system is moving through the area.

(2) Improving trends include:

- Steady wind direction, which indicates no change in weather systems in the area.
- Decreasing wind speeds.
- Clearing of obstructions to visibility.
- Decreasing or ending precipitation.
- Decreasing cloud coverage.
- Increasing height of cloud ceilings.
- Temperature changes slowly warmer.
- Humidity decreases.
- Increasing barometric pressure, which indicates that a higher pressure system is moving through the area.

j. **Update.** Continue to evaluate observed conditions and update the forecast.

Section III. MOUNTAIN HAZARDS

Hazards can be termed natural (caused by natural occurrence), man-made (caused by an individual, such as lack of preparation, carelessness, improper diet, equipment misuse), or as a combination (human trigger). There are two kinds of hazards while in the mountains—subjective and objective. Combinations of objective and subjective hazards are referred to as cumulative hazards.

1-22. SUBJECTIVE HAZARDS

Subjective hazards are created by humans; for example, choice of route, companions, overexertion, dehydration, climbing above one's ability, and poor judgment.

a. **Falling.** Falling can be caused by carelessness, over-fatigue, heavy equipment, bad weather, overestimating ability, a hold breaking away, or other reasons.

b. **Bivouac Site.** Bivouac sites must be protected from rockfall, wind, lightning, avalanche run-out zones, and flooding (especially in gullies). If the possibility of falling exists, rope in, the tent and all equipment may have to be tied down.

c. **Equipment.** Ropes are not total security; they can be cut on a sharp edge or break due to poor maintenance, age, or excessive use. You should always pack emergency and bivouac equipment even if the weather situation, tour, or a short climb is seemingly low of dangers.

1-23. OBJECTIVE HAZARDS

Objective hazards are caused by the mountain and weather and cannot be influenced by man; for example, storms, rockfalls, icefalls, lightning, and so on.

a. **Altitude.** At high altitudes (especially over 6,500 feet), endurance and concentration is reduced. Cut down on smoking and alcohol. Sleep well, acclimatize slowly, stay hydrated, and be aware of signs and symptoms of high-altitude illnesses. Storms can form quickly and lightning can be severe.

b. **Visibility.** Fog, rain, darkness, and or blowing snow can lead to disorientation. Take note of your exact position and plan your route to safety before visibility decreases. Cold combined with fog can cause a thin sheet of ice to form on rocks (verglas). Whiteout conditions can be extremely dangerous. If you must move under these conditions, it is best to rope up. Have the point man move to the end of the rope. The second man will use the first man as an aiming point with the compass. Use a route sketch and march table. If the tactical situation does not require it, plan route so as not to get caught by darkness.

c. **Gullies.** Rock, snow, and debris are channeled down gullies. If ice is in the gully, climbing at night may be better because the warming of the sun will loosen stones and cause rockfalls.

d. **Rockfall.** Blocks and scree at the base of a climb can indicate recurring rockfall. Light colored spots on the wall may indicate impact chips of falling rock. Spring melt or warming by the sun of the rock/ice/snow causes rockfall.

e. **Avalanches.** Avalanches are caused by the weight of the snow overloading the slope. (Refer to paragraph 1-25 for more detailed information on avalanches.)

f. **Hanging Glaciers and Seracs.** Avoid, if at all possible, hanging glaciers and seracs. They will fall without warning regardless of the time of day or time of year. One cubic meter of glacier ice weighs 910 kilograms (about 2,000 pounds). If you must cross these danger areas, do so quickly and keep an interval between each person.

g. **Crevasses.** Crevasses are formed when a glacier flows over a slope and makes a bend, or when a glacier separates from the rock walls that enclose it. A slope of only two to three degrees is enough to form a crevasse. As this slope increases from 25 to 30 degrees, hazardous icefalls can be formed. Likewise, as a glacier makes a bend, it is likely that crevasses will form at the outside of the bend. Therefore, the safest route on a glacier would be to the inside of bends, and away from steep slopes and icefalls. Extreme care must be taken when moving off of or onto the glacier because of the moat that is most likely to be present.

1-24. WEATHER HAZARDS

Weather conditions in the mountains may vary from one location to another as little as 10 kilometers apart. Approaching storms may be hard to spot if masked by local peaks. A clear, sunny day in July could turn into a snowstorm in less than an hour. Always pack some sort of emergency gear.

a. Winds are stronger and more variable in the mountains; as wind doubles in speed, the force quadruples.

b. Precipitation occurs more on the windward side than the leeward side of ranges. This causes more frequent and denser fog on the windward slope.

c. Above approximately 8,000 feet, snow can be expected any time of year in the temperate climates.

d. Air is dryer at higher altitudes, so equipment does not rust as quickly, but dehydration is of greater concern.

e. Lightning is frequent, violent, and normally attracted to high points and prominent features in mountain storms. Signs indicative of thunderstorms are tingling of the skin, hair standing on end, humming of metal objects, crackling, and a bluish light (St. Elmo's fire) on especially prominent metal objects (summit crosses and radio towers).

(1) Avoid peaks, ridges, rock walls, isolated trees, fixed wire installations, cracks that guide water, cracks filled with earth, shallow depressions, shallow overhangs, and rock needles. Seek shelter around dry, clean rock without cracks; in scree fields; or in deep indentations (depressions, caves). Keep at least half a body's length away from a cave wall and opening.

(2) Assume a one-point-of-contact body position. Squat on your haunches or sit on a rucksack or rope. Pull your knees to your chest and keep both feet together. If half way up the rock face, secure yourself with more than one point—lightning can burn through rope. If already rappelling, touch the wall with both feet together and hurry to the next anchor.

f. During and after rain, expect slippery rock and terrain in general and adjust movement accordingly. Expect flash floods in gullies or chimneys. A climber can be washed away or even drowned if caught in a gully during a rainstorm. Be especially alert for falling objects that the rain has loosened.

g. Dangers from impending high winds include frostbite (from increased wind-chill factor), windburn, being blown about (especially while rappelling), and debris being blown about. Wear protective clothing and plan the route to be finished before bad weather arrives.

h. For each 100-meter rise in altitude, the temperature drops approximately one degree Fahrenheit. This can cause hypothermia and frostbite even in summer, especially when combined with wind, rain, and snow. Always wear or pack appropriate clothing.

i. If it is snowing, gullies may contain avalanches or snow sloughs, which may bury the trail. Snowshoes or skis may be needed in autumn or even late spring. Unexpected snowstorms may occur in the summer with accumulations of 12 to 18 inches; however, the snow quickly melts.

j. Higher altitudes provide less filtering effects, which leads to greater ultraviolet (UV) radiation intensity. Cool winds at higher altitudes may mislead one into underestimating the sun's intensity, which can lead to sunburns and other heat injuries. Use sunscreen and wear hat and sunglasses, even if overcast. Drink plenty of fluids.

1-25. AVALANCHE HAZARDS

Avalanches occur when the weight of accumulated snow on a slope exceeds the cohesive forces that hold the snow in place. (Table 1-2, page 1-32, shows an avalanche hazard evaluation checklist.)

a. **Slope Stability.** Slope stability is the key factor in determining the avalanche danger.

(1) ***Slope Angle.*** Slopes as gentle as 15 degrees have avalanched. Most avalanches occur on slopes between 30 and 45 degrees. Slopes above 60 degrees often do not build up significant quantities of snow because they are too steep.

(2) ***Slope Profile.*** Dangerous slab avalanches are more likely to occur on convex slopes, but may occur on concave slopes.

(3) ***Slope Aspect.*** Snow on north facing slopes is more likely to slide in midwinter. South facing slopes are most dangerous in the spring and on sunny, warm days. Slopes on the windward side are generally more stable than leeward slopes.

(4) ***Ground Cover.*** Rough terrain is more stable than smooth terrain. On grassy slopes or scree, the snow pack has little to anchor to.

b. **Triggers.** Various factors trigger avalanches.

(1) ***Temperature.*** When the temperature is extremely low, settlement and adhesion occur slowly. Avalanches that occur during extreme cold weather usually occur during or immediately following a storm. At a temperature just below freezing, the snowpack stabilizes quickly. At temperatures above freezing, especially if temperatures rise quickly, the potential for avalanche is high. Storms with a rise in temperature can deposit dry snow early, which bonds poorly with the heavier snow deposited later. Most avalanches occur during the warmer midday.

(2) ***Precipitation.*** About 90 percent of avalanches occur during or within twenty-four hours after a snowstorm. The rate at which snow falls is important. High rates of snowfall (2.5 centimeters per hour or greater), especially when accompanied by wind, are usually responsible for major periods of avalanche activity. Rain falling on snow will increase its weight and weakens the snowpack.

(3) ***Wind.*** Sustained winds of 15 miles per hour and over transport snow and form wind slabs on the lee side of slopes.

(4) ***Weight.*** Most victims trigger the avalanches that kill them.

(5) ***Vibration.*** Passing helicopters, heavy equipment, explosions, and earth tremors have triggered avalanches.

c. **Snow Pits.** Snow pits can be used to determine slope stability.

(1) Dig the snow pit on the suspect slope or a slope with the same sun and wind conditions. Snow deposits may vary greatly within a few meters due to wind and sun variations. (On at least one occasion, a snow pit dug across the fall line triggered the suspect slope). Dig a 2-meter by 2-meter pit across the fall line, through all the snow, to the ground. Once the pit is complete, smooth the face with a shovel.

(2) Conduct a shovel shear test.

(a) A shovel shear test puts pressure on a representative sample of the snowpack. The core of this test is to isolate a column of the snowpack from three sides. The column should be of similar size to the blade of the shovel. Dig out the sides of the column without pressing against the column with the shovel (this affects the strength). To isolate the rear of the column, use a rope or string to saw from side to side to the base of the column.

(b) If the column remained standing while cutting the rear, place the shovel face down on the top of the column. Tap with varying degrees of strength on the shovel to see what force it takes to create movement on the bed of the column. The surface that eventually slides will be the layer to look at closer. This test provides a better understanding of the snowpack strength. For greater results you will need to do this test in many areas and formulate a scale for the varying methods of tapping the shovel.

(3) Conduct a Rutschblock test. To conduct the test, isolate a column slightly longer than the length of your snowshoes or skis (same method as for the shovel shear test). One person moves on their skis or snowshoes above the block without disturbing the block. Once above, the person carefully places one showshoe or ski onto the block with no body weight for the first stage of the test. The next stage is adding weight to the first leg. Next, place the other foot on the block. If the block is still holding up, squat once, then twice, and so on. The remaining stage is to jump up and land on the block.

d. **Types of Snow Avalanches.** There are two types of snow avalanches: loose snow (point) and slab.

(1) Loose snow avalanches start at one point on the snow cover and grow in the shape of an inverted "V." Although they happen most frequently during the winter snow season, they can occur at any time of the year in the mountains. They often fall as many small sluffs during or shortly after a storm. This process removes snow from steep upper slopes and either stabilizes lower slopes or loads them with additional snow.

(2) Wet loose snow avalanches occur in spring and summer in all mountain ranges. Large avalanches of this type, lubricated and weighed down by meltwater or rain can travel long distances and have tremendous destructive power. Coastal ranges that have high temperatures and frequent rain are the most common areas for this type of avalanche.

(3) Slab avalanches occur when cohesive snow begins to slide on a weak layer. The fracture line where the moving snow breaks away from the snowpack makes this type of avalanche easy to identify. Slab release is rapid. Although any avalanche can kill you, slab avalanches are generally considered more dangerous than loose snow avalanches.

(a) Most slab avalanches occur during or shortly after a storm when slopes are loaded with new snow at a critical rate. The old rule of never travel in avalanche terrain for a few days after a storm still holds true.

(b) As slabs become harder, their behavior becomes more unpredictable; they may allow several people to ski across before releasing. Many experts believe they are susceptible to rapid temperature changes. Packed snow expands and contracts with temperature changes. For normal density, settled snow, a drop in temperature of 10 degrees Celsius (18 degrees Fahrenheit) would cause a snow slope 300 meters wide to contract 2 centimeters. Early ski mountaineers in the Alps noticed that avalanches sometimes occurred when shadows struck a previously sun-warmed slope.

d. **Protective Measures.** Avoiding known or suspected avalanche areas is the easiest method of protection. Other measures include:

(1) ***Personal Safety.*** Remove your hands from ski pole wrist straps. Detach ski runaway cords. Prepare to discard equipment. Put your hood on. Close up your clothing to prepare for hypothermia. Deploy avalanche cord. Make avalanche probes and shovels accessible. Keep your pack on at all times—do not discard. Your pack can act as a flotation device, as well as protect your spine.

(2) ***Group Safety.*** Send one person across the suspect slope at a time with the rest of the group watching. All members of the group should move in the same track from safe zone to safe zone.

e. **Route Selection.** Selecting the correct route will help avoid avalanche prone areas, which is always the best choice. Always allow a wide margin of safety when making your decision.

(1) The safest routes are on ridge tops, slightly on the windward side; the next safest route is out in the valley, far from the bottom of slopes.

(2) Avoid cornices from above or below. Should you encounter a dangerous slope, either climb to the top of the slope or descend to the bottom—well out of the way of the run-out zone. If you must traverse, pick a line where you can traverse downhill as quickly as possible. When you must ascend a dangerous slope, climb to the side of the avalanche path, and not directly up the center.

(3) Take advantage of dense timber, ridges, or rocky outcrops as islands of safety. Use them for lunch and rest stops. Spend as little time as possible on open slopes.

(4) Since most avalanches occur within twenty-four hours of a storm and or at midday, avoid moving during these periods. Moving at night is tactically sound and may be safer.

f. **Stability Analysis.** Look for nature's billboards on slopes similar to the one you are on.

(1) ***Evidence of Avalanching.*** Look for recent avalanches and for signs of wind-loading and wind-slabs.

(2) ***Fracture Lines.*** Avoid any slopes showing cracks.

(3) ***Sounds.*** Beware of hollow sounds—a "whumping" noise. They may suggest a radical settling of the snowpack.

g. **Survival.** People trigger avalanches that bury people. If these people recognized the hazard and chose a different route, they would avoid the avalanche. The following steps should be followed if caught in an avalanche.

(1) Discard equipment. Equipment can injure or burden you; discarded equipment will indicate your position to rescuers.

(2) Swim or roll to stay on tope of the snow. FIGHT FOR YOUR LIFE. Work toward the edge of the avalanche. If you feel your feet touch the ground, give a hard push and try to "pop out" onto the surface.

(3) If your head goes under the snow, shut your mouth, hold your breath, and position your hands and arms to form an air pocket in front of your face. Many avalanche victims suffocate by having their mouths and noses plugged with snow.

(4) When you sense the slowing of the avalanche, you must try your hardest to reach the surface. Several victims have been found quickly because a hand or foot was sticking above the surface.

(5) When the snow comes to rest it sets up like cement and even if you are only partially buried, it may be impossible to dig yourself out. Don't shout unless you hear rescuers immediately above you; in snow, no one can hear you scream. Don't struggle to free yourself—you will only waste energy and oxygen.

(6) Try to relax. If you feel yourself about to pass out, do not fight it. The respiration of an unconscious person is more shallow, their pulse rate declines, and the body temperature is lowered, all of which reduce the amount of oxygen needed. (See Appendix C for information on search and rescue techniques.)

AVALANCHE HAZARD EVALUATION CHECKLIST

Critical Data PARAMETERS: KEY INFORMATION	Hazard Rating G	Y	R
TERRAIN: *Is the terrain capable of producing an avalanche?*			
-Slope angle (steep enough to slide? prime time?)	☐	☐	☐
-Slope aspect (leeward, shadowed, or extremely sunny?)	☐	☐	☐
-Slope configuration (anchoring? shape?)	☐	☐	☐
Overall Terrain Rating:	☐	☐	☐
SNOWPACK: *Could the snow fail?*			
-Slab Configuration (slab? depth and distribution?)	☐	☐	☐
-Bonding Ability (weak layer? tender spots?)	☐	☐	☐
-Sensitivity (how much force to fail? shear tests? clues?)	☐	☐	☐
Overall Snowpack Rating:	☐	☐	☐
Weather: *Is the weather contributing to instability?*			
-Precipitation (type, amount, intensity? added weight?)	☐	☐	☐
-Wind (snow transport? amount and rate of deposition?)	☐	☐	☐
-Temperature (storm trends? effects on snowpack?)	☐	☐	☐
Overall Weather Rating:	☐	☐	☐
Human: *What are your alternatives and their possible consequences?*			
-Attitude (toward life? risk? goals? assumptions?)	☐	☐	☐
-Technical Skill Level (traveling? evaluating aval. hazard?)	☐	☐	☐
-Strength/Equipment (strength? prepared for the worst?)	☐	☐	☐
Overall Human Rating:	☐	☐	☐

Decision/Action:
Overall Hazard Rating/GO or NO Go? **GO** ☐ **or NOGO** ☐

***HAZARD LEVEL SYMBOLS:**
R = Red light (stop/dangerous)
G = Green light (go/OK)
Y = Yellow light (caution/potentially dangerous).

©Alaska Mountain Safety Center, Inc.

Table 1-2. Avalanche hazard evaluation checklist.

CHAPTER 2
MOUNTAIN LIVING

Units deploying to high elevations must receive advanced training to survive in the harsh mountain environment. Normal activities (navigation, communications, and movement) require specialized techniques. Training should be conducted as realistically as possible, preferably under severe conditions so the soldier gains confidence. Extended training exercises test support facilities and expose the soldier to the isolation common to mountain operations. Training should reflect the harsh mountain environment and should consider the following:

- *Temperature and altitude extremes.*
- *Hygiene and sanitation.*
- *Limited living space (difficulty of bivouac).*
- *Clothing requirements.*

Section I. SURVIVAL

The soldier trained to fight and survive in a mountain environment will have increased confidence in himself. Training should include: psychological preparation, locating water, shelter considerations, fire building, health hazards, and techniques for obtaining food (see FM 21-76).

2-1. WATER SUPPLY

Mountain water should never be assumed safe for consumption. Training in water discipline should be emphasized to ensure soldiers drink water only from approved sources. Fluids lost through respiration, perspiration, and urination must be replaced if the soldier is to operate efficiently.

a. Maintaining fluid balance is a major problem in mountain operations. The sense of thirst may be dulled by high elevations despite the greater threat of dehydration. Hyperventilation and the cool, dry atmosphere bring about a three- to four-fold increase in water loss by evaporation through the lungs. Hard work and overheating increase the perspiration rate. The soldier must make an effort to drink liquids even when he does not feel thirsty. One quart of water, or the equivalent, should be drunk every four hours; more should be drunk if the unit is conducting rigorous physical activity.

b. Three to six quarts of water each day should be consumed. About 75 percent of the human body is liquid. All chemical activities in the body occur in water solution, which assists in removing toxic wastes and in maintaining an even body temperature. A loss of two quarts of body fluid (2.5 percent of body weight) decreases physical efficiency by 25 percent, and a loss of 12 quarts (15 percent of body weight) is usually fatal. Salt lost by sweating should be replaced in meals to avoid a deficiency and subsequent cramping. Consuming the usual military rations (three meals a day) provides sufficient sodium replacement. Salt tablets are not necessary and may contribute to dehydration.

c. Even when water is plentiful, thirst should be satisfied in increments. Quickly drinking a large volume of water may actually slow the soldier. If he is hot and the water is cold, severe cramping may result. A basic rule is to drink small amounts often. Pure water should always be kept in reserve for first aid use. Emphasis must be placed on the three rules of water discipline:

- Drink only treated water.
- Conserve water for drinking. Potable water in the mountains may be in short supply.
- Do not contaminate or pollute water sources.

d. Snow, mountain streams, springs, rain, and lakes provide good sources of water supply. Purification must be accomplished, however, no matter how clear the snow or water appears. Fruits, juices, and powdered beverages may supplement and encourage water intake (do not add these until the water has been treated since the purification tablets may not work). Soldiers cannot adjust permanently to a decreased water intake. If the water supply is insufficient, physical activity must be reduced. Any temporary deficiency should be replaced to maintain maximum performance.

e. All water that is to be consumed must be potable. Drinking water must be taken only from approved sources or purified to avoid disease or the possible use of polluted water. Melting snow into water requires an increased amount of fuel and should be planned accordingly. Nonpotable water must not be mistaken for drinking water. Water that is unfit to drink, but otherwise not dangerous, may be used for other purposes such as bathing. Soldiers must be trained to avoid wasting water. External cooling (pouring water over the head and chest) is a waste of water and an inefficient means of cooling. Drinking water often is the best way to maintain a cool and functioning body.

f. Water is scarce above the timberline. After setting up a perimeter (patrol base, assembly area, defense), a watering party should be employed. After sundown, high mountain areas freeze, and snow and ice may be available for melting to provide water. In areas where water trickles off rocks, a shallow reservoir may be dug to collect water (after the sediment settles). Water should be treated with purification tablets (iodine tablets or calcium hypochlorite), or by boiling at least one to two minutes. Filtering with commercial water purification pumps can also be conducted. Solar stills may be erected if time and sunlight conditions permit (see FM 21-76). Water should be protected from freezing by storing it next to a soldier or by placing it in a sleeping bag at night. Water should be collected at midday when the sun thaw available.

2-2. NUTRITION

Success in mountain operations depends on proper nutrition. Because higher altitudes affect eating habits, precautions must be taken. If possible, at least one hot meal each day should be eaten, which may require personnel to heat their individual rations.

a. The following elements are characteristic of nutritional acclimatization in mountain operations:

- Weight loss during the first two to three days at high elevation.
- A loss of appetite with symptoms of mountain sickness.
- Loss of weight usually stops with acclimatization.
- At progressively higher elevations (greater than 14,000 feet), the tolerance of fatty/high-protein foods rapidly decreases. A high carbohydrate diet may lessen the symptoms of acute mountain sickness and is digested better than fat at high altitudes.

b. Increased fatigue may cause soldiers to become disinterested in eating properly. Decreased consumption may result in malnutrition because of the unpleasant taste of cold rations. Leaders should ensure that fuel tablets and squad stoves are available, or that natural flammable materials are used if possible. Although there is no physiological need for hot food, it does increase morale and a sense of well being. Loss of weight in the first few days occurs because of dehydration, metabolic changes, and loss of appetite. Carbohydrate-containing beverages, such as fruit juices and sports drinks, are an effective means of increasing carbohydrates, energy, and liquid intake when the normal appetite response is blunted at altitude.

c. Three major food components are required to maintain a well-functioning body: proteins, fats, and carbohydrates. These food components provide energy, amino acids, vitamins, fiber, and minerals. All three components must be provided in the correct proportions to maintain a healthy body.

(1) ***Protein.*** Proteins consist of a large number of amino acid units that are linked together to form the protein. The amino acids, resulting from digestion of protein, are absorbed through the intestine into the blood, and are used to make or replace body proteins (muscle and body tissue). Sources of readily useable animal proteins include eggs, milk, cheese, poultry, fish, and meats. Other foods such as cereals, vegetables, and legumes also provide amino acids. These proteins are not as balanced in essential amino acid composition as meat, eggs, or milk proteins. The minimum daily protein requirement, regardless of physical activity, is 8 ounces for a 154-pound man. Since amino acids are either oxidized for energy or stored as fats, consuming excess protein is inefficient and may increase the water intake needed for urea nitrogen excretion. Protein requires water for digestion and may facilitate dehydration. Proteins provide the body about four kilocalories of energy per gram and require the most energy for the body to digest.

(2) ***Fats.*** Fats are the most concentrated form of food energy. Of the total daily caloric intake, 25 to 30 percent may be supplied as fats. Main sources of fats are meats, nuts, butter, eggs, milk, and cheese. Fats require more water and oxygen, and are harder to digest at higher altitudes. Fats are the body's natural stored source of energy. Fats provide the body around 9 kilocalories of energy per gram and require less energy for the body to digest than protein but more than carbohydrates.

(3) ***Carbohydrates.*** Carbohydrates are an important source of calories. In the form of glucose, carbohydrates are found in the most important energy-producing cycles in the body's cells. If carbohydrate intake exceeds energy needs, moderate amounts are stored in the muscles and liver. Larger amounts are converted into fats and stored in that form. Carbohydrates should compose up to 50 percent of the total daily caloric intake. Nutritionally, the most useful sources of carbohydrates are foods such as unrefined grains, vegetables, and fruit. Carbohydrates provide the body around four kilocalories of energy per gram and are the easiest to digest.

(4) ***Vitamins.*** Vitamins are classified into two groups on the basis of their ability to dissolve in fat or water. The fat-soluble vitamins include vitamins A, D, E, and K. The water-soluble vitamins include the B vitamins and vitamin C, which are found in cereals, vegetables, fruits, and meats. A well-balanced diet provides all of the required vitamins. Since most water-soluble vitamins are not stored, a proper diet is necessary to ensure adequate levels of these vitamins. If an improper and unbalanced diet is likely to occur

during a deployment, vitamin supplements should be considered, especially if this period is to exceed 10 days.

(5) ***Minerals.*** Mineral elements can be divided into two groups: those needed in the diet in amounts of 100 milligrams or more a day such as calcium, phosphorous, and magnesium; and trace elements needed in amounts of only a few milligrams a day such as iodine, iron, and zinc. Required minerals are contained in a balanced diet (meats, vegetables, fruits).

d. Eating a balanced diet provides the energy needed to conduct daily activities and to maintain the internal body processes. A balanced diet containing adequate amounts of vitamins and minerals ensures an efficient metabolism. Since climbing is a strenuous activity and demands high-energy use, a balanced diet is a necessity.

(1) The efficiency of the body to work above the basal metabolism varies from 20 to 40 percent, depending on the soldier. Over 50 percent of caloric intake is released as heat and is not available when the soldier works. (About 4,500 calories are expended for strenuous work and 3,500 calories for garrison activity.) Heat is a by-product of exertion. Exertion causes excessive bodily heat loss through perspiration and increased radiation. During inactivity in cold weather, the metabolism may not provide enough heat. The "internal thermostat" initiates and causes the muscles to shiver, thus releasing heat. Shivering also requires energy and burns up to 220 calories per hour (estimated for a 100-pound man).

(2) With an abrupt ascent to high altitudes, the soldier experiences physiological acclimatization. The circulatory system labors to provide the needed oxygen to the body. Large meals require the digestive system to work harder than usual to assimilate food. Large meals may be accompanied by indigestion, shortness of breath, cramps, and illness. Therefore, relatively light meals that are high in carbohydrates are best while acclimatizing at higher elevations. Personnel should eat moderately and rest before strenuous physical activity. Since fats and protein are harder to digest, less digestive disturbances may occur if meals are eaten before resting. A diet high in carbohydrates is not as dense in energy and may require eating more often. Carbohydrates, beginning in the morning and continuing through mid-afternoon, are important in maintaining energy levels.

(3) Extra food should be carried in case resupply operations fail. Food should be lightweight and easy to digest, and be eaten hot or cold. Meals-ready-to-eat (MREs) meet these criteria and provide all of the basic food groups. Commanders may consider supplementing MREs with breakfast bars, fruits, juices, candies, cereal bars, and chocolate. Bouillon cubes can replace water and salt as well as warming cold bodies and stimulating the appetite. Hot beverages of soup, juices, powdered milk, and cider should also be considered. Since coffee, tea, and hot chocolate are diuretics, the consumption of these beverages should not be relied upon for hydration.

(4) Warm meals should be provided when possible. When cooking, the heat source must be kept away from equipment and ammunition. At higher elevations, the cooking time may be doubled. To conserve fuel, stoves, fires, and fuel tablets should be protected from the wind. Extra fuel should be stored in tightly sealed, marked, metal containers. Use stoves and heat tabs for warming food and boiling water. Canteen cups and utensils should be cleaned after use. All food items and garbage are carried with the unit. If possible, garbage should be burned or deep buried. Caution must be taken to prevent animals from foraging through rucksacks, ahkios, and burial sites. As all missions are tactical, no trace of a unit should be detected.

(5) Certain drugs, medications, alcohol, and smoking have adverse effects on the circulation, perspiration, hydration, and judgment of soldiers. Therefore, they should be avoided when operating in extremely cold conditions or at high altitudes.

2-3. PERSONAL HYGIENE AND SANITATION

The principles of personal hygiene and sanitation that govern operations on low terrain also apply in the mountains. Commanders must conduct frequent inspections to ensure that personal habits of hygiene are not neglected. Standards must be maintained as a deterrent to disease, and as reinforcement to discipline and morale.

a. **Personal Hygiene.** This is especially important in the high mountains, mainly during periods of cold weather. In freezing weather, the soldier may neglect washing due to the cold temperatures and scarcity of water. This can result in skin infections and vermin infestation. If bathing is difficult for any extended period, the soldier should examine his skin and clean it often. Snow baths in lieu of a water bath are recommended. This helps reduce skin infections and aids the comfort of the soldier.

(1) Snow may be used instead of toilet paper. Soldiers should shave at rest periods in the shelter so that oils stripped in shaving will be replenished. A beard may mask the presence of frostbite or lice. Water-based creams and lotions should be avoided in cold environments since this will further dehydrate tissues and induce frostbite by freezing. The nonwater-based creams can be used for shaving in lieu of soap. Sunscreens and chap sticks should be used on lips, nose, and eyelids. Topical steroid ointments should be carried for rashes. The teeth must also be cleaned to avoid diseases of the teeth and gums. Underwear should be changed when possible, but this should not be considered a substitute for bathing. When operating in areas where resupply is not possible, each soldier should carry a complete change of clothing. If laundering of clothing is difficult, clothes should be shaken and air-dried. Sleeping bags must be regularly cleaned and aired.

(2) The principles of foot hygiene must be followed to protect the feet from cold injuries. The causes of such injuries are present throughout the year in high mountains. Boots should be laced tightly when climbing to provide needed support but not so tight as to constrict circulation. Socks should be worn with no wrinkles since this causes blisters on the feet. Feet should be washed daily, and kept as dry and clean as possible. If regular foot washing is impossible, socks should be changed often (at halts and rest periods or at least once a day) and feet massaged, dried, and sprinkled with foot powder. Talc or antifungal powder should be used when massaging; excess powder is brushed off to avoid clumping, which may cause blisters. Feet can be cleaned with snow, but must be quickly dried. Whenever changing socks, soldiers should closely examine their feet for wrinkles, cracks, blisters, and discoloration. Nails should be trimmed but not too short. Long nails wear out socks; short nails do not provide proper support for the ends of the toes. Medical attention should be sought for any possible problems.

(3) Feet should be sprayed two or three times a day with an aluminum chlorohydrate antiperspirant for a week and then once a day for the rest of the winter. If fissures or cracks occur in the feet, it is best to discontinue spraying until they are healed or to spray less often to control sweating. This process stops about 70 percent of the sweating in the feet.

(4) During periods of extreme cold, there is a tendency for the soldier to become constipated. This condition is brought about by the desire to avoid the inconvenience and

discomfort of defecating. Adequate water intake plus a low protein, high roughage diet can be helpful in preventing constipation.

b. **Sanitation.** In rocky or frozen ground, digging latrines is usually difficult. If latrines are constructed, they should be located downwind from the position and buried after use. In tactical situations, the soldier in a designated, downwind location away from water sources may dig "cat holes." Since waste freezes, it can be covered with snow and ice or pushed down a crevasse. In rocky areas above the timberline, waste may be covered with stones.

Section II. ACCLIMATIZATION AND CONDITIONING

Terrestrial altitude can be classified into five categories. Low altitude is sea level to 5,000 feet. Here, arterial blood is 96 percent saturated with oxygen in most people. Moderate altitude is from 5,000 to 8,000 feet. At these altitudes, arterial blood is greater than 92 percent saturated with oxygen, and effects of altitude are mild and temporary. High altitude extends from 8,000 to 14,000 feet, where arterial blood oxygen saturation ranges from 92 percent down to 80 percent. Altitude illness is common here. Very high altitude is the region from 14,000 to 18,000 feet, where altitude illness is the rule. Areas above 18,000 feet are considered extreme altitudes.

Soldiers deployed to high mountainous elevations require a period of acclimatization before undertaking extensive military operations. The expectation that freshly deployed, unacclimatized troops can go immediately into action is unrealistic, and could be disastrous if the opposing force is acclimatized. Even the physically fit soldier experiences physiological and psychological degradation when thrust into high elevations. Time must be allocated for acclimatization, conditioning, and training of soldiers. Training in mountains of low or medium elevation (5,000 to 8,000 feet) does not require special conditioning and acclimatization procedures. However, some soldiers will have some impairment of operating efficiency at these low altitudes. Above 8,000 feet (high elevation), most unacclimatized soldiers may display some altitude effects. Training should be conducted at progressively higher altitudes, starting at about 8,000 feet and ending at 14,000 feet. Attempts to acclimatize beyond 17,000 feet results in a degradation of the body greater than the benefits gained. The indigenous populations can out-perform even the most acclimatized and physically fit soldier who is brought to this altitude; therefore, employment of the local population may be advantageous.

2-4. SYMPTOMS AND ADJUSTMENTS

A person is said to be acclimatized to high elevations when he can effectively perform physically and mentally. The acclimatization process begins immediately upon arrival at the higher elevation. If the change in elevation is large and abrupt, some soldiers can suffer from acute mountain sickness (AMS), high-altitude pulmonary edema (HAPE), or high-altitude cerebral edema (HACE). Disappearance of the symptoms of acute mountain sickness (from four to seven days) does not indicate complete acclimatization. The process of adjustment continues for weeks or months. The altitude at which complete acclimatization is possible is not a set point but for most soldiers with proper ascent, nutrition and physical activity it is about 14,000 feet.

a. Immediately upon arrival at high elevations, only minimal physical work can be performed because of physiological changes. The incidence and severity of AMS symptoms vary with initial altitude, the rate of ascent, and the level of exertion and individual susceptibility. Ten to twenty percent of soldiers who ascend rapidly (in less than 24 hours) to altitudes up to 6,000 feet experience some mild symptoms. Rapid ascent to 10,000 feet causes mild symptoms in 75 percent of personnel. Rapid ascent to elevations of 12,000 to 14,000 feet will result in moderate symptoms in over 50 percent of the soldiers and 12 to 18 percent may have severe symptoms. Rapid ascent to 17,500 feet causes severe, incapacitating symptoms in almost all individuals. Vigorous activity during ascent or within the first 24 hours after ascent will increase both the incidence and severity of symptoms. Some of the behavioral effects that will be encountered in unacclimatized personnel include:

- Increased errors in performing simple mental tasks.
- Decreased ability for sustained concentration.
- Deterioration of memory.
- Decreased vigilance or lethargy.
- Increased irritability in some individuals.
- Impairment of night vision and some constriction in peripheral vision (up to 30 percent at 6,000 feet).
- Loss of appetite.
- Sleep disturbances.
- Irregular breathing.
- Slurred speech.
- Headache.

b. Judgment and self-evaluation are impaired the same as a person who is intoxicated. During the first few days at a high altitude, leaders have extreme difficulty in maintaining a coordinated, operational unit. The roughness of the terrain and the harshness and variability of the weather add to the problems of unacclimatized personnel. Although strong motivation may succeed in overcoming some of the physical handicaps imposed by the environment, the total impact still results in errors of judgment. When a soldier cannot walk a straight line and has a loss of balance, or he suffers from an incapacitating headache, he should be evacuated to a lower altitude (a descent of at least 1,000 feet for at least 24 hours).

2-5. PHYSICAL AND PSYCHOLOGICAL CONDITIONING

The commander must develop a conditioning/training program to bring his unit to a level where it can operate successfully in mountain conditions. Priorities of training must be established. As with all military operations, training is a major influence on the success of mountain operations.

a. U.S. forces do not routinely train in mountainous terrain. Therefore, extensive preparations are needed to ensure individual and unit effectiveness. Units must be physically and psychologically conditioned and adjusted before undertaking rigorous mountain operations. Units must be conditioned and trained as a team to cope with the terrain, environment, and enemy situation. Certain factors must be considered:

- What are the climatic and terrain conditions of the area of operations?
- How much time is available for conditioning and training?
- Will the unit conduct operations with other U.S. or Allied forces? Are there language barriers? What assistance will be required? Will training and conditioning be required for attached personnel?

- What additional personnel will accompany the unit? Will they be available for training and conditioning?
- What is the current level of physical fitness of the unit?
- What is the current level of individual expertise in mountaineering?
- What type of operations can be expected?
- What is the composition of the advance party? Will they be available to assist in training and acclimatization?
- What areas in the U.S. most closely resemble the area of operations?
- Are predeployment areas and ranges available?
- Does the unit have instructors qualified in mountain warfare?
- What type equipment will be required (to fit the season, mission, terrain)?
- Does the unit have enough of the required equipment? Do personnel know how to use the equipment? Will the equipment go with the advance party, with the unit, or follow after the unit's arrival?
- Does equipment require modification?
- Do weapons and equipment require special maintenance?

b. When the unit arrives in the area of operations, all personnel require a period of conditioning and acclimatization. The time schedule should allow for longer and more frequent periods of rest. The rigors of establishing an assembly area exhaust most unacclimatized personnel. Water, food, and rest must be considered as priorities, ensuring sufficient amounts while individual metabolisms and bodies become accustomed to functioning at higher elevations.

c. Since the acclimatization process cannot be shortened, and the absence of acclimatization hampers the successful execution of operations, deployment to higher elevations must consider the following:

(1) Above 8,000 feet, a unit should ascend at a rate of 1,000 to 2,000 feet per day. Units can leapfrog, taking an extended rest period.

(2) Units should not resort to the use of pharmaceutical pretreatment with carbonic anhydrase inhibitors such as acetazolamide (Diamox). These drugs have side effects that mimic the signs and symptoms of AMS. Inexperienced medics may have difficulty recognizing the differences between the side effects of the drug and a condition that could possibly be life threatening. Additionally, these drugs are diuretics, which results in higher hydration levels (at least 25 percent increase per man per day). These higher hydration levels create a larger logistical demand on the unit by requiring more water, time to acquire water, water purification supplies, and, if in a winter environment, fuels for melting snow and ice for water.

(3) Carbonic anhydrase inhibitors such as acetazolamide are effective in the treatment of mild and severe AMS. These drugs should accompany attached medical personnel because they can treat the soldier suffering the symptoms of AMS and, although rest may be required evacuation may not be needed.

(4) Do not move troops directly to high altitudes even if allowances can be made for inactivity for the first three to five days before mission commitment. Moving troops directly to high altitude can increase the probability of altitude sickness. Even if inactivity follows deployment, the incidence of altitude sickness is more likely than with a gradual ascent.

d. Training on high-altitude effects can prevent psychological preconceptions. Soldiers who have lived on flat terrain may have difficulty when learning to negotiate steep slopes or cliffs, developing a sense of insecurity and fear. They must be slowly introduced to the new terrain and encouraged to develop the confidence required to negotiate obstacles with assurance and ease. They must be taught the many climbing techniques and principles of mountain movement. They overcome their fear of heights by becoming familiar with the problem. The soldier cannot be forced to disregard this fear.

e. Regardless of previous training and the amount of flat cross-country movement practice, the untrained soldier finds mountain movement hard and tiring. A different group of muscles are used, which must be developed and hardened. A new technique of rhythmic movement must be learned. Such conditioning is attained through frequent marches and climbs, while carrying TOE and special equipment loads. This conditions the back and legs, which results in increased ability and endurance. At the same time, the men acquire confidence and ability to safely negotiate the terrain. The better the physical condition of the soldier, the better the chance of avoiding exhaustion. Proper physical conditioning ensures the soldier is an asset and not a liability. The body improves its capacity for exercise, the metabolism becomes more efficient, and blood and oxygen flow quickly and effectively.

f. A physical fitness training program that gradually increases in difficulty should include marches, climbing, and calisthenics. This increases the soldier's endurance. Through a sustained high level of muscular exertion, the soldier's capacity for exertion is increased. Physical conditioning should include long-distance running for aerobic conditioning; calisthenics and weight training to strengthen the heart, lungs, abdomen, legs, back, arms, and hands; a swimming program to increase lung efficiency; and road marches over mountainous terrain with all combat equipment. Upon deploying to high elevations, caution must be exercised by units that are in superior physical condition. The heart rate, metabolism, and lungs must become accustomed to the elevation and thinner air. A conditioning program must be set up on site and integrated in gradual stages where acclimatization, conditioning, and mountaineering skills are realized.

g. Conditioning should begin with basic climbing. It is equally important to instill the will to climb. Confidence goes hand in hand with physical conditioning and skill development. Repetitive practice, to the point of instinctive reaction, is key to learning and maintaining climbing proficiency and technical skills. There are no quick and easy methods to becoming acclimatized and conditioned. Training should gradually challenge the soldier over an extended period and reinforce learning skills.

Section III. MEDICAL CONSIDERATIONS

Improper acclimatization poses many problems for medical personnel. Facilities and supplies may be inadequate to treat all victims. After acclimatization, personnel can still become injured (sprains, strains, fractures, frostbite, hypothermia, and trench foot). Mountain sickness and other illnesses may also occur. Evacuation of the sick and wounded is compounded by the terrain and weather.

2-6. ILLNESS AND INJURY

Units operating in mountainous regions are exposed to varied types of injuries and illnesses not associated with other areas. Medical considerations are like those for other environments; however, there are some unique aspects of mountain operations to be considered if effective support is to be provided. Most injuries in the mountain environment are soft tissue injuries. These include sprains, strains, abrasions, contusions and fractures. As with any other injuries, the most life threatening are treated first with the emphasis on airway control, breathing management, and circulatory support. Skills in basic first aid are essential to the mountain leader and should be reinforced with regular sustainment training.

2-7. TREATMENT AND EVACUATION

In harsh mountain weather, the most important course of action is to provide injured soldiers with medical aid as soon as possible. Immediate first aid is given on site. Due to rough terrain, medical units can seldom reach unit aid stations by vehicle to evacuate casualties. Litter bearers are required to move casualties to the rear where they can be evacuated by ground or air to clearing stations. The victim is protected from the weather and shock during transportation. Rendezvous points are coordinated with medical units as far forward as possible. Training must be accomplished with all litter bearers on evacuation techniques and first aid. Lightly wounded personnel may need assistance to move over rough terrain.

2-8. SOLAR INJURIES

Solar injuries can happen in warm weather or in cold weather. These types of injuries can be just as incapacitating as most other injuries but usually are not fatal. The peak hours of ultraviolet (UV) radiation are between the hours of 1100 and 1500. Due to the long wavelengths of ultraviolet light, cloudy days can be more dangerous than sunny days. On sunny days the soldier takes more care due to the bright conditions. On cloudy days the soldier tends not to wear sunglasses or sunscreen.

a. **Sunburn.** Sunburn is the burning of exposed skin surfaces by ultraviolet radiation.

(1) Contributing factors include fair skin, improper use of para-amino benzoic acid (PABA)-based sunscreens, and exposure to intense ultraviolet rays for extended periods.

(2) Symptoms of sunburn are painful, burning, red or blistered skin with a slight swelling. The skin may be warm to the touch. In severe cases chills, fever, and headaches may occur.

(3) To treat sunburn, apply cool saline dressings to alleviate pain and swelling. Do not pop blisters. If blisters do break, wash thoroughly, bandage, and seek medical attention. A solution of vinegar (acetic) and water can be lightly applied with sterile gauze to alleviate burning. The tannic acid in used tea bags can also be applied to alleviate burning. Administer pain medication if needed.

(4) To prevent sunburn, skin should be covered with clothing or PABA-based sunscreens (at least sun protection factor [SPF] 15) should be applied liberally to exposed skin during the peak hours of UV exposure. The SPF means that you can stay exposed to the suns UV rays that many times longer than without it. (For example, an SPF of 15 means that skin can be exposed to UV rays 15 times longer than without sunscreen.) During sustained activity, the sunscreen should be regularly reapplied to maintain the SPF.

b. **Snowblindness.** Snowblindness is sunburn of the cornea of the eye caused by exposure to ultraviolet radiation.

(1) A contributing factor is the reflection of sunlight from all directions off the snow, ice, and water. Ultraviolet rays can cause vision problems even on cloudy days. They are less filtered at high altitudes than at low altitudes.

(2) Symptoms of snowblindness are painful, red, watery eyes; a gritty feeling; blurred vision; and a headache.

(3) To treat snowblindness, patch both eyes with cold compresses for 24 hours. Topical anesthetics such as Tetracaine Ophthalmic can be used to relieve pain. Avoid rubbing the eyes. If still painful, keep the victim's eyes patched and administer oral pain medication. Snowblindness will usually resolve in about 24 hours for mild to moderate cases. Victims are rarely in need of evacuation unless the case is unusually severe.

(4) To prevent snow blindness, use quality sunglasses even on cloudy days in snow-covered terrain. Proper sunglasses should provide 100 percent UVA and UVB protection and have hoods on the sides to prevent reflected light from entering the eye. (Currently, the U.S. Army does not have these types of "glacier" sunglasses in their inventory and they must be acquired from nonmilitary sources.) In an emergency, improvise slit glasses from materials such as cardboard or birch bark.

2-9. COLD-WEATHER INJURIES

Cold-weather injuries can occur during any season of the year. Death has resulted in temperatures as high as 10 degrees Celsius (50 degrees Fahrenheit). A loss of body heat combined with shock produces devastating results. However, most of these accidents can be prevented by proper planning to include: timely requisition and receipt of supplies and proper clothing; thorough training of personnel with respect to the hazards of cold weather; effective methods for the receipt, dissemination, and use of cold-weather data; periodic inspections of clothing, personnel, and equipment; and personnel receiving a balance of water, rest, and nutrition.

a. Soldiers must be prepared to survive, move, and fight in winter conditions. Intense cold affects the mind as well as the body. Simple tasks take longer to perform, and they take more effort than in a temperate climate. When weather conditions become extreme the problems of survival become more significant. Warmth and comfort become the top priorities. The effects of extreme cold and the probability of injury are magnified due to the lack of proper diet and sleep. The most important measure in the prevention of cold-weather injuries is the education of personnel and their leaders.

b. Cold injuries may be divided into two types: freezing and nonfreezing. The freezing type is known as frostbite. The nonfreezing type includes hypothermia, dehydration, and immersion foot. Cold injuries result from impaired circulation and the action of ice formation and cold upon the tissues of the body. Temperature alone is not a reliable guide as to whether a cold injury can occur. Low temperatures are needed for cold injuries to occur, but freezing temperatures are not. Wind speed can accelerate body heat loss under both wet and cold conditions. All commanders and subordinate leaders/instructors must be familiar with and carry GTA 5-8-12, which includes a wind chill equivalent temperature chart (Figure 2-1, page 2-12).

WIND CHILL FACTOR CHART												
COOLING POWER OF WIND EXPRESSED AS AN EQUIVALENT CHILL TEMPERATURE (UNDER CALM CONDITIONS)												
ESTIMATED WIND SPEED (IN MPH)	ACTUAL THERMOMETER READING (F)											
	50	40	30	20	10	0	-10	-20	-30	-40	-50	-60
	EQUIVALENT TEMPERATURES (F)											
Calm	50	40	30	20	10	0	-10	-20	-30	-40	-50	-60
5	48	37	27	16	6	-5	-15	-26	-36	-47	-57	-68
10	40	28	16	4	-9	-24	-33	-46	-58	-70	-83	-95
15	36	22	9	-5	-18	-32	-45	-58	-72	-85	-99	-112
20	32	18	4	-10	-25	-39	-53	-67	-82	-96	-110	-124
25	30	16	0	-15	-29	-44	-59	-74	-88	-104	-118	-133
30	28	13	-2	-18	-33	-48	-63	-79	-94	-109	-125	-140
35	27	11	-4	-21	-35	-51	-67	-82	-98	-113	-129	-145
40	26	10	-6	-21	-37	-53	-69	-85	-100	-116	-132	-148
Winds greater than 40 MPH have little additional effect.	LITTLE DANGER (For properly clothed person) Maximum danger of false sense of security.				INCREASING DANGER Danger from freezing of exposed flesh.			GREAT DANGER				
	Trench foot and immersion foot may occur at any point on this chart.											

Figure 2-1. Wind chill chart.

c. Many other factors in various combinations determine if cold injuries will occur.

(1) ***Previous Cold Injuries.*** If a soldier has had a cold injury before, he is at higher risk for subsequent cold injuries.

(2) ***Race.*** Blacks are more susceptible to cold-weather injuries than Caucasians.

(3) ***Geographic Origin.*** Personnel from warmer climates are more susceptible to cold injury than those from colder climates.

(4) ***Ambient Temperature.*** The temperature of the air (or water) surrounding the body is critical to heat regulation. For example, the body uses more heat to maintain the temperature of the skin when the temperature of the surrounding air is 37 degrees Fahrenheit than when it is 50 degrees Fahrenheit.

(5) ***Wind Chill Factor.*** The commander should know the wind chill factor. When the forecast gives a figure that falls within the increased danger zone or beyond, caution must be taken to minimize cold injury. The equivalent wind chill temperature is especially important when the ambient temperature is 0 degrees Celsius (32 degrees Fahrenheit) or less. Tissue can freeze if exposed for a prolonged period and if frequent warming is not practiced. The lower the wind chill, the faster tissue freezing can occur. Wind chill is the rate of cooling. Wind does not lower the ambient temperature. The ambient temperature alone determines freezing or nonfreezing injuries. Frostbite Wind chill may cause faster cooling due to increased convection, but not below the ambient temperature.

(6) ***Type of Mission.*** Combat action requiring prolonged immobility and long hours of exposure to low temperatures, or not having an opportunity to warm up increases the possibility of cold injuries.

(7) ***Terrain.*** Minimal cover and wet conditions increase the potential for cold injury.

(8) ***Clothing.*** Clothing for cold weather should be worn with the acronym **C.O.L.D.** in mind.

- **C**—Clothing should be clean since prolonged wear reduces its air-trapping abilities and clogs air spaces with dirt and body oils.
- **O**—Overheating. Avoid overheating. Appropriate measures should be taken when a change in weather or activity alters the amount of clothing needed to prevent overheating and, therefore, accumulation of perspiration.
- **L**—Loose and in layers (to trap air and to conserve body heat). The uniform should be worn completely and correctly to avoid injury to exposed body surfaces. The cold-weather uniform is complete when worn with gloves and inserts.
- **D**—Dry. Keep dry. Wet clothing loses insulation value.

(9) ***Moisture.*** Water conducts heat more rapidly than air (25 percent). When the skin or clothing becomes damp or wet, the risk of cold injury is greatly increased.

(10) ***Dehydration.*** The most overlooked factor causing cold injuries is dehydration. Individuals must retain their body fluids. In cold weather the human body needs special care, and the consumption of water is important to retain proper hydration.

(11) ***Age.*** Within the usual age range of combat personnel, age is not a significant factor.

(12) ***Fatigue.*** Mental weariness may cause apathy leading to neglect of duties vital to survival.

(13) ***Concomitant Injury.*** Injuries resulting in shock or blood loss reduce blood flow to extremities and may cause the injured individual to be susceptible to cold injury, which in turn can accelerate shock.

(14) ***Discipline, Training, and Experience.*** Well-trained and disciplined soldiers suffer less than others from the cold.

(15) ***Nutrition.*** Good nutrition is essential for providing the body with fuel to produce heat in cold weather. The number of calories consumed normally increases as the temperature becomes colder.

(16) ***Excess Activity.*** Excess activity (overheating) results in loss of large amounts of body heat by perspiration. This loss of body heat combined with the loss of insulation value provided by the clothing (due to perspiration dampening the clothing) can subject a soldier to cold injuries.

(17) ***Radical Changes in the Weather.*** Weather conditions in mountainous terrain are known to change considerably throughout the day. Weather can quickly change to extremely cold and wet conditions, especially in higher elevations.

d. Commanders should ensure that the following measures are taken.

(1) Soldiers' uniforms are kept as dry as possible and are protected from the elements.

(2) Soldiers are educated on proper use of clothing systems to avoid the effects of overheating and perspiration (layer dressing and ventilate).

(3) The buddy system is used to watch for early signs of cold-weather injuries.

(4) All soldiers waterproof their equipment.

(5) The rate of movement should be slow, deliberate, and careful. Soldiers should not move out at a force march pace and then be stationary after they have perspired heavily. Soldiers should not wear excessive cold-weather clothing while moving.

e. Medical procedures are needed when sickness and injuries occur. Leaders should—

- Assess the situation (tactical and environmental).
- Approach the victim safely (avoid rock or snow slide).
- Perform emergency first aid.
- Treat for shock (always assume that shock is present).
- Check for other injuries/cold injuries.
- Develop a course of action (decide on a means of evacuation).
- Execute the plan and monitor the victim's condition.

f. Body heat may be lost through radiation, conduction, convection, or evaporation.

(1) ***Radiation.*** The direct heat loss from the body to its surrounding atmosphere is called radiation heat loss. The head can radiate up to 80 percent of the total body heat output. On cold days, personnel must keep all extremities covered to retain heat. This accounts for the largest amount of heat lost from the body.

(2) ***Conduction.*** Conduction is the direct transfer of heat from one object in contact with another (being rained on or sitting in snow).

(3) ***Convection.*** Convection is the loss of heat due to moving air or water in contact with the skin. Wind chill is convection cooling. Clothing that ventilates, insulates, and protects must control the layer of warm air next to the skin.

(4) ***Evaporation.*** The evaporation of perspiration causes heat loss. Wet clothing can cause heat loss by conduction and evaporation. Dressing in layers allows soldiers to remove or add clothing as needed.

g. Some of the most common cold-weather injuries are described in the following paragraphs.

(1) ***Shock.*** Shock is the depressed state of vital organs due to the cardiovascular (heart) system not providing enough blood. Although shock is not a cold-weather injury, it is a symptom or a result of other injuries. Any illness or injury can produce shock, which increases the instance and severity of a cold-weather injury. Shock should be assumed in all injuries and treated accordingly. Even minor injuries can produce shock due to cold, pain, fear, and loss of blood.

(a) *Symptoms.* Initial symptoms of shock include apprehension, shortness of breath, sweating, cold skin, rapid and faint pulse, and excessive thirst. If the victim is not given adequate first aid immediately, his condition may digress into incoherence, slower heart beat, unconsciousness, and possibly death.

(b) *Treatment.* To treat shock, restore breathing and heart rate through artificial respiration or cardiopulmonary resuscitation. Treat the injury and control hemorrhaging. Make the victim as comfortable as possible and try to relieve the pain. Keep the victim warm but do not overheat him. Elevate the back and head, or feet. If the victim is conscious and has no abdominal injuries, administer water. The victim should receive proper medical attention as soon as possible.

(2) ***Dehydration.*** Dehydration is the loss of body fluids to the point that normal body functions are prevented or slowed. This is usually caused by overexertion and improper water intake. Dehydration precedes all cold-weather injuries and is a major symptom in acute mountain sickness. It contributes to poor performance in all physical activities—even more so than lack of food. Cold weather requirements for water are no different than in the desert. They may, in fact, exceed desert requirements because of the increased difficulty in moving with extra clothing and through the snow. At high altitudes, the air is dry. Combined

with a rapid rate of breathing, as much as two liters of liquid may be lost each day through respiration. A soldier needs about three to six quarts of water each day to prevent dehydration when living and performing physical labor in a cold or mountainous environment. Coffee and tea are diuretics and cause excessive urination and should be avoided. The adequacy of liquid intake can best be judged by the urine color and volume. Dark amber colored urine instead of light yellow or the absence of a need to urinate upon awakening from a night's sleep are indicators of dehydration. Thirst is not a good indicator of hydration.

(a) *Contributing Factors.* Factors that contribute to dehydration in cold weather are:

- The thirst mechanism does not function properly in cold weather.
- Water is often inconvenient to obtain and purify.
- The air in cold climates and at high altitudes lacks moisture.
- Cold causes frequent urination.

(b) *Symptoms.* Symptoms of dehydration include darkening urine, decreased amounts of urine being produced, dry mouth, tiredness, mental sluggishness, lack of appetite, headache, fainting, rapid heartbeat, dizziness, higher temperature, upset stomach, and unconsciousness. The symptoms of dehydration are similar to those of hypothermia. To distinguish between them, open the victim's clothes and feel the stomach. If the stomach is cold, the victim is probably hypothermic; if it is warm, he is probably dehydrated. However, this test is not conclusive since cold-weather dehydrating can also lead to total body cooling. The cold environment may act as a diuretic and impair the body's ability to conserve fluid (cold-induced diuresis and increased rate of urination).

(c) *Treatment.* Prevent dehydration by consuming three to six quarts of fluids each day (forced drinking in the absence of thirst is mandatory) and avoid caffeine and alcohol, which may chemically contribute to dehydration. Keep the victim warm and treat for shock. In advanced cases, administer fluids by mouth if the victim is conscious. Do not let him eat snow; eating snow uses body heat. Allow the victim to rest. If he fails to improve within one hour or is unconscious, evacuate him to a medical facility immediately.

(3) ***Hypothermia.*** Hypothermia is the lowering of the body core temperature at a rate faster than the body can produce heat. Hypothermia may be caused by exposure or by sudden wetting of the body such as falling into a lake or being sprayed with fuel or other liquid. Hypothermia can occur even on moderate days with temperatures of 40 to 50 degrees Fahrenheit with little precipitation if heat loss exceeds heat gain and the condition of the soldier is allowed to deteriorate. Hypothermia is classified as mild (core temperature above 90 degrees Fahrenheit or 32 degrees Celsius) or severe (core temperature below 90 degrees Fahrenheit or 32 degrees Celsius). An individual is considered to be "clinically hypothermic" when the core temperature is less than or equal to 95 degrees Fahrenheit.

(a) *Contributing Factors.* Factors that contribute to hypothermia are:

- Dehydration.
- Poor nutrition.
- Diarrhea.
- Decreased physical activity.
- Accidental immersion in water.
- Change in weather.
- High winds.
- Inadequate types or amounts of clothing.

(b) *Symptoms* The first symptom of hypothermia is when the body core (rectal) temperature falls to about 96 degrees Fahrenheit. Other symptoms include:

- Shivering, which may progress to an uncontrollable point making it hard for an individual to care for himself. Shivering begins after a drop in body temperature of one to two degrees. This is followed by clumsiness (stumbling or falling), slow reactions, mental confusion, and difficulty in speaking.
- Body temperature drop from 95 degrees Fahrenheit to 90 degrees Fahrenheit, which can cause sluggish thinking, irrational thought, apathy, and a false sense of warmth. The victim becomes cold and pale; cannot perform simple tasks; experiences amnesia and hallucinations; develops blueness of skin and decreased heart and respiratory rate with a weak pulse; pupils of the eyes dilate; speech becomes slurred; and visual disturbance occurs.
- Body temperature drop from 90 degrees Fahrenheit to 85 degrees Fahrenheit, which causes irrationality, incoherence, loss of contact with the environment, muscular rigidity, disorientation, and exhaustion. The soldier might stop shivering after his core temperature drops below 90 degrees Fahrenheit.
- Body temperature drop from 85 degrees Fahrenheit and below, which causes muscle rigidity, unconsciousness, comatose state, and faint vital signs. The pulse may be faint or impalpable, and breathing is too shallow to observe.

(c) *Prevention.* Prevent hypothermia by using the buddy system to watch each other for symptoms; consume adequate amounts of liquids daily; rest; and eat properly.

(d) *Avoidance.* Hypothermia can be avoided by dressing in layers, which permits easy additions or deletions to prevent overheating, becoming too cold, or getting wet or windblown. If the soldier is in a situation that precludes staying warm and dry, he should seek shelter. Sweets and physical activity help to produce body heat.

(e) *Treatment.* Treatment methods vary based on the severity of the hypothermia.

- Mild cases: If a soldier shows symptoms of hypothermia, prevent additional heat loss by getting the victim into a shelter; removing wet clothing and replacing it with dry, insulated clothing; insulating the victim from the ground; and sharing a sleeping bag (cover head) to transfer body heat. Make a diagnosis (rectal temperature). Rehydrate the victim with warm liquids, sweets, and food. If the tactical situation allows, build a fire. Above all else, keep the victim conscious until his vital signs are normal, and seek medical assistance. If possible, keep the victim physically active to produce body heat.
- Severe cases: If the victim is unconscious or appears dead without any obvious injury, prevent further heat loss. Rapid rewarming of an unconscious victim may create problems and should not be attempted. It is best to evacuate as soon as possible. At all times, the victim should be handled gently so as not to cause the cold blood from the extremities to rush to the heart. Do not allow the victim to perform ANY physical activity. Immediately transport the victim to the nearest medical facility. Field reheating is not effective and may be hazardous. Provide artificial respiration if breathing stops. If no pulse is detectable, be aware that in hypothermia there is often effective circulation for the victim's hypothermic state. In such a case, cardiac compression (such as CPR) may be fatal. The exception is acute hypothermia with near drowning.

- Breathing warm, moist air is the fastest way to warm the inside of the body. If breathing steam is not possible, place tubing under the rescuer's shirt so the victim will still breathe warm, moist air. This process can be done while on the move. In addition to breathing moist, warm air the victim must be gradually rewarmed using external heat sources. Padded hot water bottles or heated stones should be placed in the armpits.
- If conscious, the victim can be given warm, sweet drinks.
- The Hibler Pack is an improvised method of rewarming hypothermic victims in the field. This is used to heat the body core first so the vital organs are warmed and not the extremities. As the body warms up the warm blood will eventually warm all parts of the body. First lay out a blanket or sleeping bag and place a poncho or space blanket inside of it. The poncho or space blanket should go from the base of the skull to the base of the butt. This keeps the sleeping bag/blanket dry and acts like a vapor barrier. Lay the hypothermic patient inside the sleeping bag/blanket. Using a stove, warm water until it is hot to the touch (but not hot enough to burn the patient) and completely dampen any absorbable materials (such as T-shirt, towel, BDU top, and so on). Place the warm, wet items inside a plastic bag or directly in the armpits and chest of the patient. After the warm, wet item has been placed on the patient, wrap the patient tightly inside the poncho/space blanket and the blanket/sleeping bag. Continually check the temperature of the wet material and keep it warm.
- All bodily systems in hypothermia are brittle so treat the victim gently. As these attempts are being made, try to evacuate the victim. Severe complications may arise as the body temperature rises, which may result in cardiac arrest even though the victim seems to be doing well.

(4) ***Immersion or Trench Foot.*** This is damage to the circulatory and nervous systems of the feet that occurs from prolonged exposure to cold and wet at above freezing temperatures. This can happen wearing boots or not. A soldier may not feel uncomfortable until the injury has already begun.

(a) *Contributing Factors.* Factors that contribute to immersion or trench foot are:

- Stepping into water over the boot tops.
- Not changing socks often enough.
- Improper hygiene.
- Prolonged exposure (three to five days).

(b) *Symptoms.* Symptoms of immersion or trench foot include the sensation of tingling, numbness, and then pain. The toes are pale, and feel cold and stiff. The skin is wet and soggy with the color turning from red to bright red, progressing to pale and mottled, and then grayish blue. As symptoms progress and damage appears, the skin becomes red and then bluish or black. Swelling may occur. Because the early stages of trench foot are not painful, soldiers must be constantly aware to prevent it.

(c) *Treatment.* To prevent this condition, keep the feet dry and clean. Change socks often, drying the insides of boots, massaging the feet, and using foot powder. Drying the feet for 24 hours usually heals mild cases. Moderate cases usually heal within three to five days. The feet should be handled gently—NOT rubbed or massaged. They should be cleaned with soap and water, dried, elevated, and exposed to room temperature. The victim must stay off his feet and seek medical attention. Severe cases, when feet are not allowed to dry, are evacuated as a litter casualty.

(5) ***Blisters.*** When first noticed and before the formation of a blister, cover a hotspot with moleskin (over the area and beyond it). Use tincture benzoin to help the moleskin adhere to and toughen the skin. Once a blister has formed, cover it with a dressing large enough to fit over the blister, and then tape it. Never drain blisters unless they are surrounded by redness, or draining pus indicates infection. If this occurs, drain the blister from the side with a clean sterile needle. After cleaning with soap and water, gently press out the fluid leaving the skin intact. Make a doughnut of moleskin to go around the blister and apply to the skin. For toe blisters, wrap the entire toe with adhesive tape over the moleskin. (Toenails should be trimmed straight across the top, leaving a 90-degree angle on the sides. This provides an arch so that the corners do not irritate the skin.)

(6) ***Frostbite.*** Frostbite is the freezing or crystallization of living tissues due to heat being lost faster than it can be replaced by blood circulation, or from direct exposure to extreme cold or high winds. Exposure time can be minutes or instantaneous. The extremities are usually the first to be affected. Damp hands and feet may freeze quickly since moisture conducts heat away from the body and destroys the insulating value of clothing. Heat loss is compounded with intense cold and inactivity. With proper clothing and equipment, properly maintained and used, frostbite can be prevented. The extent of frostbite depends on temperature and duration of exposure. Frostbite is one of the major nonfatal cold-weather injuries encountered in military operations, but does not occur above an ambient temperature of 32 degrees Fahrenheit.

(a) *Categories of Frostbite.* Superficial (mild) frostbite involves only the skin (Figure 2-2). The layer immediately below usually appears white to grayish with the surface feeling hard, but the underlying tissue is soft. Deep (severe) frostbite extends beyond the first layer of skin and may include the bone (Figure 2-3). Discoloration continues from gray to black, and the texture becomes hard as the tissue freezes deeper. This condition requires immediate evacuation to a medical facility.

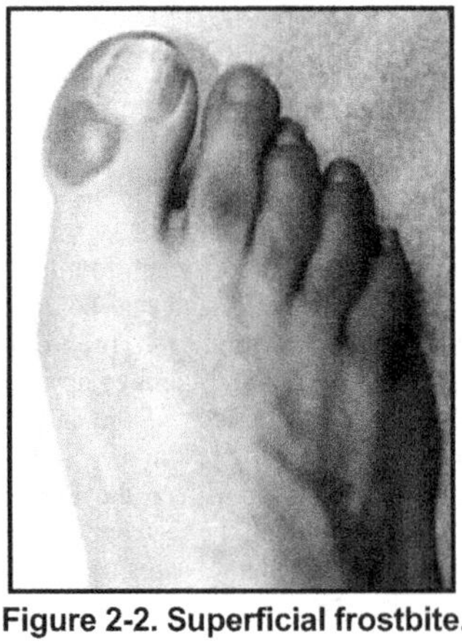

Figure 2-2. Superficial frostbite.

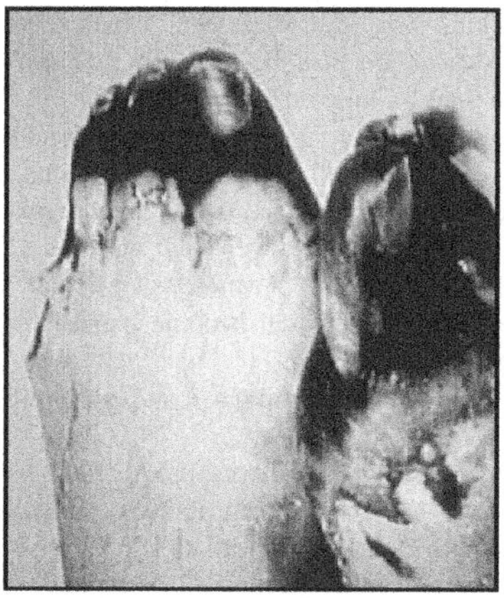

Figure 2-3. Deep frostbite.

(b) *Contributing Factors.* Factors that contribute to frostbite are:

- Dehydration.
- Below-freezing temperatures.
- Skin contact with super cooled metals or liquids.
- Use of caffeine, tobacco, or alcohol.
- Neglect.

(c) *Symptoms.* Symptoms of frostbite vary and may include a cold feeling, pain, burning, numbness, and, in the final stages, a false sense of warmth. The skin first turns red, then pale. It may be bluish in color and then may appear frosty or waxy white. The skin may feel hard, may not be movable over the joints and bony prominences, or may be frozen. Identification of deep versus superficial frostbite is difficult to determine and often requires three to seven days after rewarming for medical personnel to diagnose. Blisters, swelling, and pain may occur after thawing.

(d) *Treatment.* Using the buddy system is one of the primary ways to prevent frostbite. Buddies must watch each other for symptoms of frostbite and provide mutual aid if frostbite occurs. Frostbite should be identified early with prompt first-aid care applied to prevent further damage.

- Treat early signs of frostbite by rewarming with skin-to-skin contact or by sheltering the body part under the clothing next to the body. *Do this immediately.* If tissues have frozen, evacuate the victim before they thaw. If the feet are involved, evacuate the victim as a litter patient.
- Thawing of a frostbitten victim is a hospital procedure. If the victim has frostbite with frozen extremities, protect the frozen parts and evacuate as a litter patient.
- If frostbite is not recognized before it thaws, do not let the area refreeze since this causes more damage. The most often-affected body parts are the hands, fingers, toes, feet, ears, chin, and nose. If evacuation of the victim as a litter case

is not possible and the body part has not yet thawed, have the victim walk out on his own. Walking out on frozen feet is better than having them thaw and refreeze. Self-evacuation may be tactically necessary. Walking on frozen feet does less harm than walking on thawed feet.

- If reheating is inevitable, do not overheat the affected body parts near flame; the warming temperature should not be greater than normal body temperature. Do not rub the parts—the crystallized tissues may break internally and cause more damage. Do not pop blisters; cover them with a dry, sterile dressing. Keep the victim warm (apply loose, bulky bandages to separate toes and fingers.)
- Once a part is rewarmed it will become painful. Pain may be managed with narcotic analgesics.
- Once the foot is rewarmed it will swell and putting the boot back on will not be possible.

(7) ***Constipation.*** Constipation is the infrequent or difficult passage of stools.

(a) *Contributing Factors.* Factors that contribute to constipation are a lack of fluids, improper nutrition, and not defecating when needed.

(b) *Symptoms.* Symptoms include headache, cramping, lack of bowel movement, painful bowel movement, and loss of appetite.

(c) *Treatment.* Constipation is prevented by consuming adequate amounts and varieties of food, drinking from four to six liters of liquid each day, and defecating regularly. If allowed to progress beyond self-care stages, victims will need medical aid.

(8) ***Carbon Monoxide Poisoning.*** This is the replacement of oxygen in the blood with carbon monoxide.

(a) *Contributing Factor.* A contributing factor is inhaling fumes from burning fuel, such as fires, stoves, heaters, and running engines, without proper ventilation.

(b) *Symptoms.* Symptoms are similar to other common illnesses and include headaches, fatigue, excessive yawning, nausea, dizziness, drowsiness, confusion, and unconsciousness. Death may occur. The one visible symptom is bright red lips, mouth, and inside of the eyelids.

(c) *Treatment.* Remove the victim from the source of contamination; administer oxygen, if available; and evacuate to a medical facility. Severe complications may develop even in casualties who appear to have recovered. If the victim is unconscious, administer rescue breathing and CPR as needed.

2-10. HEAT INJURIES

Heat injuries, although associated with hot weather, can occur in cold-weather environments. Most heat injuries can be avoided by planning, periodic inspections of personnel clothing (ventilation) and equipment, a balance of water and food intake, and rest.

a. **Heat Cramps.** Heat cramps are caused by an accumulation of lactic acid in the muscles and a loss of salt through perspiration.

(1) ***Contributing Factor.*** Strenuous exertion causes the body to heat up and to produce heavy perspiration.

(2) ***Symptoms.*** Symptoms of heat cramps include pain and cramping in the arms, legs, back, and stomach. The victim sweats profusely and cannot quench his thirst.

(3) ***Treatment.*** Have the victim rest in a cool, shady area, breath deeply, and stretch the cramped muscle as soon as possible to obtain relief. Loosen the victim's clothing and have him drink cool water. Monitor his condition and seek medical attention if pain and cramps continue.

b. **Heat Exhaustion.** Heat exhaustion may occur when a soldier exerts himself in any environment and he overheats. The blood vessels in the skin become so dilated that the blood flow to the brain and other organs is reduced.

(1) ***Contributing Factors.*** Factors that contribute to heat exhaustion are strenuous activity in hot areas, unacclimatized troops, inappropriate diet, and not enough water or rest.

(2) ***Symptoms.*** Symptoms of heat exhaustion may be similar to fainting but may also include weakness; dizziness; confusion; headache; cold, clammy skin; and nausea. The victim may also have a rapid but weak pulse.

(3) ***Treatment.*** Move the victim to a cool, shady area and loosen his clothes and boots. Have the victim drink water and, if possible, immerse him in water to aid in cooling. Elevate the victim's legs to help restore proper circulation. Monitor his condition and seek medical attention if the symptoms persist.

c. **Heat Stroke.** Heat stroke is a life-threatening situation caused by overexposure to the sun. The body is so depleted of liquids that its internal cooling mechanisms fail to function.

(1) ***Contributing Factors.*** Factors that contribute to heat stroke are prolonged exposure to direct sunlight, overexertion, dehydration, and depletion of electrolytes.

(2) ***Symptoms.*** Symptoms of heat stroke include hot, dry skin; dizziness; confusion and incoherency; headache; nausea; seizures; breathing difficulty; a slow pulse; and loss of consciousness.

(3) ***Treatment.*** Cool the victim at once, and restore breathing and circulation. If the victim is conscious, administer water. If possible, submerge the victim in water to reduce his temperature, treat for shock, and prepare for immediate evacuation.

2-11. ACUTE MOUNTAIN SICKNESS

Acute mountain sickness is a temporary illness that may affect both the beginner and experienced climber. Soldiers are subject to this sickness in altitudes as low as 5,000 feet. Incidence and severity increases with altitude, and when quickly transported to high altitudes. Disability and ineffectiveness can occur in 50 to 80 percent of the troops who are rapidly brought to altitudes above 10,000 feet. At lower altitudes, or where ascent to altitudes is gradual, most personnel can complete assignments with moderate effectiveness and little discomfort.

a. Personnel arriving at moderate elevations (5,000 to 8,000 feet) usually feel well for the first few hours; a feeling of exhilaration or well-being is not unusual. There may be an initial awareness of breathlessness upon exertion and a need for frequent pauses to rest. Irregular breathing can occur, mainly during sleep; these changes may cause apprehension. Severe symptoms may begin 4 to 12 hours after arrival at higher altitudes with symptoms of nausea, sluggishness, fatigue, headache, dizziness, insomnia, depression, uncaring attitude, rapid and labored breathing, weakness, and loss of appetite.

b. A headache is the most noticeable symptom and may be severe. Even when a headache is not present, some loss of appetite and a decrease in tolerance for food occurs. Nausea, even without food intake, occurs and leads to less food intake. Vomiting may occur and contribute to dehydration. Despite fatigue, personnel are unable to sleep. The symptoms usually develop and increase to a peak by the second day. They gradually subside over the next several days so that the total course of AMS may extend from five to seven days. In some instances, the headache may become incapacitating and the soldier should be evacuated to a lower elevation.

c. Treatment for AMS includes the following:

- Oral pain medications such as ibuprofen or aspirin.
- Rest.
- Frequent consumption of liquids and light foods in small amounts.
- Movement to lower altitudes (at least 1,000 feet) to alleviate symptoms, which provides for a more gradual acclimatization.
- Realization of physical limitations and slow progression.
- Practice of deep-breathing exercises.
- Use of acetazolamide in the first 24 hours for mild to moderate cases.

d. AMS is nonfatal, although if left untreated or further ascent is attempted, development of high-altitude pulmonary edema (HAPE) and or high-altitude cerebral edema (HACE) can be seen. A severe persistence of symptoms may identify soldiers who acclimatize poorly and, thus, are more prone to other types of mountain sickness.

2-12. CHRONIC MOUNTAIN SICKNESS

Although not commonly seen in mountaineers, chronic mountain sickness (CMS) (or Monge's disease) can been seen in people who live at sufficiently high altitudes (usually at or above 10,000 feet) over a period of several years. CMS is a right-sided heart failure characterized by chronic pulmonary edema that is caused by years of strain on the right ventricle.

2-13. UNDERSTANDING HIGH-ALTITUDE ILLNESSES

As altitude increases, the overall atmospheric pressure decreases. Decreased pressure is the underlying source of altitude illnesses. Whether at sea level or 20,000 feet the surrounding atmosphere has the same percentage of oxygen. As pressure decreases the body has a much more difficult time passing oxygen from the lungs to the red blood cells and thus to the tissues of the body. This lower pressure means lower oxygen levels in the blood and increased carbon dioxide levels. Increased carbon dioxide levels in the blood cause a systemic vasodilatation, or expansion of blood vessels. This increased vascular size stretches the vessel walls causing leakage of the fluid portions of the blood into the interstitial spaces, which leads to cerebral edema or HACE. Unless treated, HACE will continue to progress due to the decreased atmospheric pressure of oxygen. Further ascent will hasten the progression of HACE and could possibly cause death.

While the body has an overall systemic vasodilatation, the lungs initially experience pulmonary vasoconstriction. This constricting of the vessels in the lungs causes increased workload on the right ventricle, the chamber of the heart that receives de-oxygenated blood from the right atrium and pushes it to the lungs to be re-oxygenated. As the right ventricle works harder to force blood to the lungs, its overall output is decreased thus decreasing the overall pulmonary perfusion. Decreased pulmonary perfusion causes decreased cellular respiration—the transfer of oxygen from the alveoli to the red blood cells. The body is now experiencing increased carbon dioxide levels due to the decreased oxygen levels, which now causes pulmonary vasodilatation. Just as in HACE, this expanding of the vascular structure causes leakage into interstitial space resulting in pulmonary edema or HAPE. As the edema or fluid in the lungs increases, the capability to pass oxygen to the red blood cells decreases thus creating a vicious cycle, which can quickly become fatal if left untreated.

2-14. HIGH-ALTITUDE PULMONARY EDEMA

HAPE is a swelling and filling of the lungs with fluid, caused by rapid ascent. It occurs at high altitudes and limits the oxygen supply to the body.

a. HAPE occurs under conditions of low oxygen pressure, is encountered at high elevations (over 8,000 feet), and can occur in healthy soldiers. HAPE may be considered a form of, or manifestation of, AMS since it occurs during the period of susceptibility to this disorder.

b. HAPE can cause death. Incidence and severity increase with altitude. Except for acclimatization to altitude, no known factors indicate resistance or immunity. Few cases have been reported after 10 days at high altitudes. When remaining at the same altitude, the incidence of HAPE is less frequent than that of AMS. No common indicator dictates how a soldier will react from one exposure to another. Contributing factors are:

- A history of HAPE.
- A rapid or abrupt transition to high altitudes.
- Strenuous physical exertion.
- Exposure to cold.
- Anxiety.

c. Symptoms of AMS can mask early pulmonary difficulties. Symptoms of HAPE include:

- Progressive dry coughing with frothy white or pink sputum (this is usually a later sign) and then coughing up of blood.
- Cyanosis—a blue color to the face, hands, and feet.
- An increased ill feeling, labored breathing, dizziness, fainting, repeated clearing of the throat, and development of a cough.
- Respiratory difficulty, which may be sudden, accompanied by choking and rapid deterioration.
- Progressive shortness of breath, rapid heartbeat (pulse 120 to 160), and coughing (out of contrast to others who arrived at the same time to that altitude).
- Crackling, cellophane-like noises (rales) in the lungs caused by fluid buildup (a stethoscope is usually needed to hear them).
- Unconsciousness, if left untreated. Bubbles form in the nose and mouth, and death results.

d. HAPE is prevented by good nutrition, hydration, and gradual ascent to altitude (no more than 1,000 to 2,000 feet per day to an area of sleep). A rest day, with no gain in altitude or heavy physical exertion, is planned for every 3,000 feet of altitude gained. If a soldier develops symptoms despite precautions, immediate descent is mandatory where he receives prompt treatment, rest, warmth, and oxygen. He is quickly evacuated to lower altitudes as a litter patient. A descent of 300 meters may help; manual descent is not delayed to await air evacuation. If untreated, HAPE may become irreversible and cause death. Cases that are recognized early and treated promptly may expect to recover with no aftereffects. Soldiers who have had previous attacks of HAPE are prone to second attacks.

e. Treatment of HAPE includes:

- Immediate descent (2,000 to 3,000 feet minimum) if possible; if not, then treatment in a monoplace hyperbaric chamber.
- Rest (litter evacuation)
- Supplemental oxygen if available.
- Morphine for the systemic vasodilatation and reduction of preload. This should be carefully considered due to the respiratory depressive properties of the drug.
- Furosemide (Lasix), which is a diuretic, given orally can also be effective.
- The use of mannitol should not be considered due to the fact that it crystallizes at low temperatures. Since almost all high-altitude environments are cold, using mannitol could be fatal.
- Nifidipine (Procardia), which inhibits calcium ion flux across cardiac and smooth muscle cells, decreasing contractility and oxygen demand. It may also dilate coronary arteries and arterioles.
- Diphenhydramine (Benadryl), which can help alleviate the histamine response that increases mucosal secretions.

2-15. HIGH-ALTITUDE CEREBRAL EDEMA

HACE is the accumulation of fluid in the brain, which results in swelling and a depression of brain function that may result in death. It is caused by a rapid ascent to altitude without progressive acclimatization. Prevention of HACE is the same as for HAPE. HAPE and HACE may occur in experienced, well-acclimated mountaineers without warning or obvious predisposing conditions. They can be fatal; when the first symptoms occur, immediate descent is mandatory.

a. Contributing factors include rapid ascent to heights over 8,000 feet and aggravation by overexertion.

b. Symptoms of HACE include mild personality changes, paralysis, stupor, convulsions, coma, inability to concentrate, headaches, vomiting, decrease in urination, and lack of coordination. The main symptom of HACE is a severe headache. A headache combined with any other physical or psychological disturbances should be assumed to be manifestations of HACE. Headaches may be accompanied by a loss of coordination, confusion, hallucinations, and unconsciousness. These may be combined with symptoms of HAPE. The victim is often mistakenly left alone since others may think he is only irritable or temperamental; no one should ever be ignored. The symptoms may rapidly progress to death. Prompt descent to a lower altitude is vital.

c. Preventive measures include good eating habits, maintaining hydration, and using a gradual ascent to altitude. Rest, warmth, and oxygen at lower elevations enhance recovery. Left untreated, HACE can cause death.

d. Treatment for HACE includes:

- Dexamethasone injection immediately followed by oral dexamethasone.
- Supplemental oxygen.
- Rapid descent and medical attention.
- Use of a hyberbaric chamber if descent is delayed.

2-16. HYDRATION IN HAPE AND HACE

HAPE and HACE cause increased proteins in the plasma, or the fluid portion of the blood, which in turn increases blood viscosity. Increased viscosity increases vascular pressure. Vascular leakage caused by stretching of the vessel walls is made worse because of this increased vascular pressure. From this, edema, both cerebral and pulmonary, occurs. Hydration simply decreases viscosity.

CHAPTER 3
MOUNTAINEERING EQUIPMENT

Commanders at every level must understand the complexity of operations in a mountainous environment where every aspect of combat operations becomes more difficult. Leaders must understand that each individual has a different metabolism and, therefore, cools down and heats up differently, which requires soldiers to dress-up and dress-down at different intervals. Provided all tactical concerns are met, the concept of uniformity is outdated and only reduces the unit's ability to fight and function at an optimum level. The extreme cold weather clothing system (ECWCS) is specifically designed to allow for rapid moisture transfer and optimum heat retention while protecting the individual from the elements. Every leader is responsible for ensuring that the ECWCS is worn in accordance with the manufacturers' recommendations. Commanders at all levels must also understand that skills learned at an Army mountaineering school are perishable and soldiers need constant practice to remain proficient. The properly trained mountain soldier of today can live better, move faster, and fight harder in an environment that is every bit as hostile as the enemy.

Section I. EQUIPMENT DESCRIPTION AND MAINTENANCE

With mountainous terrain encompassing a large portion of the world's land mass, the proper use of mountaineering equipment will enhance a unit's combat capability and provide a combat multiplier. The equipment described in this chapter is produced by many different manufacturers; however, each item is produced and tested to extremely high standards to ensure safety when being used correctly. The weak link in the safety chain is the user. Great care in performing preventative maintenance checks and services and proper training in the use of the equipment is paramount to ensuring safe operations. The manufacturers of each and every piece of equipment provide recommendations on how to use and care for its product. It is imperative to follow these instructions explicitly.

3-1. FOOTWEAR

Currently, CTA 50-900 provides adequate footwear for most operations in mountainous terrain. In temperate climates a combination of footwear is most appropriate to accomplish all tasks.

a. The hot weather boot provides an excellent all-round platform for movement and climbing techniques and should be the boot of choice when the weather permits. The intermediate cold weather boot provides an acceptable platform for operations when the weather is less than ideal. These two types of boots issued together will provide the unit with the footwear necessary to accomplish the majority of basic mountain missions.

b. Mountain operations are encumbered by extreme cold, and the extreme cold weather boot (with vapor barrier) provides an adequate platform for many basic mountain missions. However, plastic mountaineering boots should be incorporated into training as soon as possible. These boots provide a more versatile platform for any condition that would be encountered in the mountains, while keeping the foot dryer and warmer.

c. Level 2 and level 3 mountaineers will need mission-specific footwear that is not currently available in the military supply system. The two types of footwear they will need are climbing shoes and plastic mountaineering boots.

(1) Climbing shoes are made specifically for climbing vertical or near vertical rock faces. These shoes are made with a soft leather upper, a lace-up configuration, and a smooth "sticky rubber" sole (Figure 3-1). The smooth "sticky rubber" sole is the key to the climbing shoe, providing greater friction on the surface of the rock, allowing the climber access to more difficult terrain.

(2) The plastic mountaineering boot is a double boot system (Figure 3-1). The inner boot provides support, as well as insulation against the cold. The inner boot may or may not come with a breathable membrane. The outer boot is a molded plastic (usually with a lace-up configuration) with a lug sole. The welt of the boot is molded in such a way that crampons, ski bindings, and snowshoes are easily attached and detached.

Note: Maintenance of all types of footwear must closely follow the manufacturers' recommendations.

Figure 3-1. Climbing shoes and plastic mountaineering boots.

3-2. CLOTHING

Clothing is perhaps the most underestimated and misunderstood equipment in the military inventory. The clothing system refers to every piece of clothing placed against the skin, the insulation layers, and the outer most garments, which protect the soldier from the elements. When clothing is worn properly, the soldier is better able to accomplish his tasks. When worn improperly, he is, at best, uncomfortable and, at worst, develops hypothermia or frostbite.

a. **Socks.** Socks are one of the most under-appreciated part of the entire clothing system. Socks are extremely valuable in many respects, if worn correctly. As a system, socks provide cushioning for the foot, remove excess moisture, and provide insulation from cold temperatures. Improper wear and excess moisture are the biggest causes of hot spots and blisters. Regardless of climatic conditions, socks should always be worn in layers.

(1) The first layer should be a hydrophobic material that moves moisture from the foot surface to the outer sock.

(2) The outer sock should also be made of hydrophobic materials, but should be complimented with materials that provide cushioning and abrasion resistance.

(3) A third layer can be added depending upon the climatic conditions.

(a) In severe wet conditions, a waterproof type sock can be added to reduce the amount of water that would saturate the foot. This layer would be worn over the first two layers if conditions were extremely wet.

(b) In extremely cold conditions a vapor barrier sock can be worn either over both of the original pairs of socks or between the hydrophobic layer and the insulating layer. If the user is wearing VB boots, the vapor barrier sock is not recommended.

b. **Underwear.** Underwear should also be made of materials that move moisture from the body. Many civilian companies manufacture this type of underwear. The primary material in this product is polyester, which moves moisture from the body to the outer layers keeping the user drier and more comfortable in all climatic conditions. In colder environments, several pairs of long underwear of different thickness should be made available. A lightweight set coupled with a heavyweight set will provide a multitude of layering combinations.

c. **Insulating Layers.** Insulating layers are those layers that are worn over the underwear and under the outer layers of clothing. Insulating layers provide additional warmth when the weather turns bad. For the most part, today's insulating layers will provide for easy moisture movement as well as trap air to increase the insulating factor. The insulating layers that are presently available are referred to as pile or fleece. The ECWCS (Figure 3-2, page 3-4) also incorporates the field jacket and field pants liner as additional insulating layers. However, these two components do not move moisture as effectively as the pile or fleece.

d. **Outer Layers.** The ECWCS provides a jacket and pants made of a durable waterproof fabric. Both are constructed with a nylon shell with a laminated breathable membrane attached. This membrane allows the garment to release moisture to the environment while the nylon shell provides a degree of water resistance during rain and snow. The nylon also acts as a barrier to wind, which helps the garment retain the warm air trapped by the insulating layers. Leaders at all levels must understand the importance of wearing the ECWCS correctly.

Note: Cotton layers must not be included in any layer during operations in a cold environment.

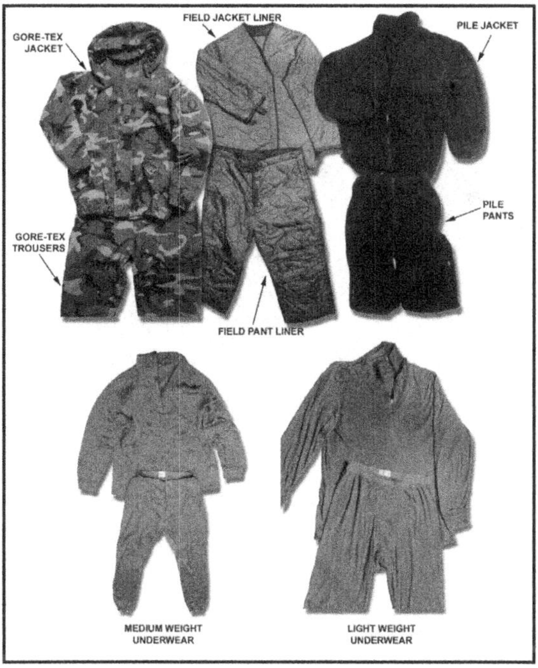

Figure 3-2. Extreme cold weather clothing system.

e. **Gaiters.** Gaiters are used to protect the lower leg from snow and ice, as well as mud, twigs, and stones. The use of waterproof fabrics or other breathable materials laminated to the nylon makes the gaiter an integral component of the cold weather clothing system. Gaiters are not presently fielded in the standard ECWCS and, in most cases, will need to be locally purchased. Gaiters are available in three styles (Figure 3-3).

(1) The most common style of gaiter is the open-toed variety, which is a nylon shell that may or may not have a breathable material laminated to it. The open front allows the boot to slip easily into it and is closed with a combination of zipper, hook-pile tape, and snaps. It will have an adjustable neoprene strap that goes under the boot to keep it snug to the boot. The length should reach to just below the knee and will be kept snug with a drawstring and cord lock.

(2) The second type of gaiter is referred to as a full or randed gaiter. This gaiter completely covers the boot down to the welt. It can be laminated with a breathable material and can also be insulated if necessary. This gaiter is used with plastic mountaineering boots and should be glued in place and not removed.

(3) The third type of gaiter is specific to high-altitude mountaineering or extremely cold temperatures and is referred to as an overboot. It is worn completely over the boot and must be worn with crampons because it has no traction sole.

Figure 3-3. Three types of gaiters.

f. **Hand Wear.** During operations in mountainous terrain the use of hand wear is extremely important. Even during the best climatic conditions, temperatures in the mountains will dip below the freezing point. While mittens are always warmer than gloves, the finger dexterity needed to do most tasks makes gloves the primary cold weather hand wear (Figure 3-4, page 3-6).

(1) The principals that apply to clothing also apply to gloves and mittens. They should provide moisture transfer from the skin to the outer layers—the insulating layer must insulate the hand from the cold and move moisture to the outer layer. The outer layer must be weather resistant and breathable. Both gloves and mittens should be required for all soldiers during mountain operations, as well as replacement liners for both. This will provide enough flexibility to accomplish all tasks and keep the users' hands warm and dry.

(2) Just as the clothing system is worn in layers, gloves and mittens work best using the same principle. Retention cords that loop over the wrist work extremely well when the wearer needs to remove the outer layer to accomplish a task that requires fine finger dexterity. Leaving the glove or mitten dangling from the wrist ensures the wearer knows where it is at all times.

Figure 3-4. Hand wear.

g. **Headwear.** A large majority of heat loss (25 percent) occurs through the head and neck area. The most effective way to counter heat loss is to wear a hat. The best hat available to the individual soldier through the military supply system is the black watch cap. Natural fibers, predominately wool, are acceptable but can be bulky and difficult to fit under a helmet. As with clothes and hand wear, man-made fibers are preferred. For colder climates a neck gaiter can be added. The neck gaiter is a tube of man-made material that fits around the neck and can reach up over the ears and nose (Figure 3-5). For extreme cold, a balaclava can be added. This covers the head, neck, and face leaving only a slot for the eyes (Figure 3-5). Worn together the combination is warm and provides for moisture movement, keeping the wearer drier and warmer.

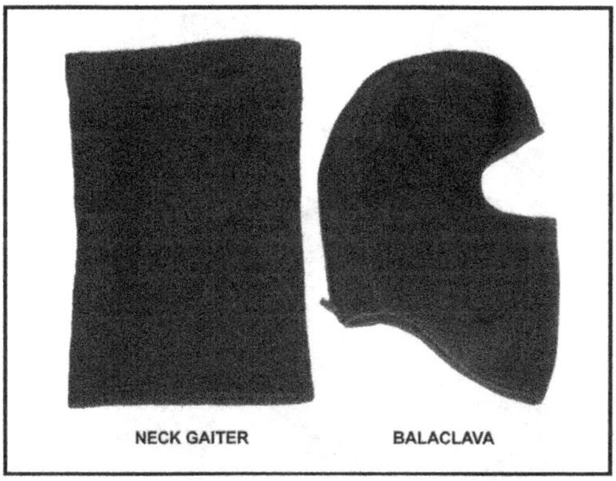

Figure 3-5. Neck gaiter and balaclava.

h. **Helmets**. The Kevlar ballistic helmet can be used for most basic mountaineering tasks. It must be fitted with parachute retention straps and the foam impact pad (Figure 3-6). The level 2 and 3 mountaineer will need a lighter weight helmet for specific climbing scenerios. Several civilian manufacturers produce an effective helmet. Whichever helmet is selected, it should be designed specifically for mountaineering and adjustable so the user can add a hat under it when needed.

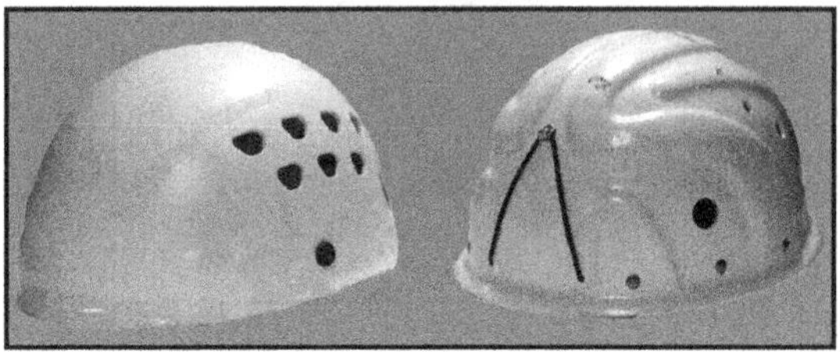

Figure 3-6. Helmets.

i. **Eyewear.** The military supply system does not currently provide adequate eyewear for mountaineering. Eyewear is divided into two catagories: glacier glasses and goggles (Figure 3-7). Glacier glasses are sunglasses that cover the entire eye socket. Many operations in the mountains occur above the tree line or on ice and snow surfaces where the harmful UV rays of the sun can bombard the eyes from every angle increasing the likelihood of snowblindness. Goggles for mountain operations should be antifogging. Double or triple lenses work best. UV rays penetrate clouds so the goggles should be UV protected. Both glacier glasses and goggles are required equipment in the mountains. The lack of either one can lead to severe eye injury or blindness.

Figure 3-7. Glacier glasses and goggles.

j. **Maintenance of Clothing.** Clothing and equipment manufacturers provide specific instructions for proper care. Following these instructions is necessary to ensure the equipment works as intended.

3-3. CLIMBING SOFTWARE

Climbing software refers to rope, cord, webbing, and harnesses. All mountaineering specific equipment, to include hardware (see paragraph 3-4), should only be used if it has the UIAA certificate of safety. UIAA is the organization that oversees the testing of mountaineering equipment. It is based in Paris, France, and comprises several commissions. The safety commission has established standards for mountaineering and climbing equipment that have become well recognized throughout the world. Their work continues as new equipment develops and is brought into common use. Community Europe (CE) recognizes UIAA testing standards and, as the broader-based testing facility for the combined European economy, meets or exceeds the UIAA standards for all climbing and mountaineering equipment produced in Europe. European norm (EN) and CE have been combined to make combined European norm (CEN). While the United States has no specific standards, American manufacturers have their equipment tested by UIAA to ensure safe operating tolerances.

a. **Ropes and Cord.** Ropes and cords are the most important pieces of mountaineering equipment and proper selection deserves careful thought. These items are your lifeline in the mountains, so selecting the right type and size is of the utmost importance. All ropes and cord used in mountaineering and climbing today are constructed with the same basic configuration. The construction technique is referred to as Kernmantle, which is, essentially, a core of nylon fibers protected by a woven sheath, similar to parachute or 550 cord (Figure 3-8).

Figure 3-8. Kernmantle construction.

(1) Ropes come in two types: static and dynamic. This refers to their ability to stretch under tension. A static rope has very little stretch, perhaps as little as one to two percent, and is best used in rope installations. A dynamic rope is most useful for climbing and general mountaineering. Its ability to stretch up to 1/3 of its overall length makes it the right choice any time the user might take a fall. Dynamic and static ropes come in various

diameters and lengths. For most military applications, a standard 10.5- or 11-millimeter by 50-meter dynamic rope and 11-millimeter by 45-meter static rope will be sufficient.

(2) When choosing dynamic rope, factors affecting rope selection include intended use, impact force, abrasion resistance, and elongation. Regardless of the rope chosen, it should be UIAA certified.

(3) Cord or small diameter rope is indispensable to the mountaineer. Its many uses make it a valuable piece of equipment. All cord is static and constructed in the same manner as larger rope. If used for Prusik knots, the cord's diameter should be 5 to 7 millimeters when used on an 11-mm rope.

b. **Webbing and Slings.** Loops of tubular webbing or cord, called slings or runners, are the simplest pieces of equipment and some of the most useful. The uses for these simple pieces are endless, and they are a critical link between the climber, the rope, carabiners, and anchors. Runners are predominately made from either 9/16-inch or 1-inch tubular webbing and are either tied or sewn by a manufacturer (Figure 3-9). Runners can also be made from a high-performance fiber known as spectra, which is stronger, more durable, and less susceptible to ultraviolet deterioration. Runners should be retired regularly following the same considerations used to retire a rope. For most military applications, a combination of different lengths of runners is adequate.

(1) Tied runners have certain advantages over sewn runners—they are inexpensive to make, can be untied and threaded around natural anchors, and can be untied and retied to other pieces of webbing to create extra long runners.

(2) Sewn runners have their own advantages—they tend to be stronger, are usually lighter, and have less bulk than the tied version. They also eliminate a major concern with the homemade knotted runner—the possibility of the knot untying. Sewn runners come in four standard lengths: 2 inches, 4 inches, 12 inches, and 24 inches. They also come in three standard widths: 9/16 inch, 11/16 inch, and 1 inch.

Figure 3-9. Tied or sewn runners.

c. **Harnesses.** Years ago climbers secured themselves to the rope by wrapping the rope around their bodies and tying a bowline-on-a-coil. While this technique is still a viable way of attaching to a rope, the practice is no longer encouraged because of the increased possibility of injury from a fall. The bowline-on-a-coil is best left for low-angle climbing or an emergency situation where harness material is unavailable. Climbers today can select from a wide range of manufactured harnesses. Fitted properly, the harness should ride high on the hips and have snug leg loops to better distribute the force of a fall to the entire pelvis. This type of harness, referred to as a seat harness, provides a comfortable seat for rappelling (Figure 3-10).

(1) Any harness selected should have one very important feature—a double-passed buckle. This is a safety standard that requires the waist belt to be passed over and back through the main buckle a second time. At least 2 inches of the strap should remain after double-passing the buckle.

(2) Another desirable feature on a harness is adjustable leg loops, which allows a snug fit regardless of the number of layers of clothing worn. Adjustable leg loops allow the soldier to make a latrine call without removing the harness or untying the rope.

(3) Equipment loops are desirable for carrying pieces of climbing equipment. For safety purposes always follow the manufacturer's directions for tying-in.

(4) A field-expedient version of the seat harness can be constructed by using 22 feet of either 1-inch or 2-inch (preferred) tubular webbing (Figure 3-10). Two double-overhand knots form the leg loops, leaving 4 to 5 feet of webbing coming from one of the leg loops. The leg loops should just fit over the clothing. Wrap the remaining webbing around the waist ensuring the first wrap is routed through the 6- to 10-inch long strap between the double-overhand knots. Finish the waist wrap with a water knot tied as tightly as possible. With the remaining webbing, tie a square knot without safeties over the water knot ensuring a minimum of 4 inches remains from each strand of webbing.

(5) The full body harness incorporates a chest harness with a seat harness (Figure 3-10). This type of harness has a higher tie-in point and greatly reduces the chance of flipping backward during a fall. This is the only type of harness that is approved by the UIAA. While these harnesses are safer, they do present several disadvantages—they are more expensive, are more restrictive, and increase the difficulty of adding or removing clothing. Most mountaineers prefer to incorporate a separate chest harness with their seat harness when warranted.

(6) A separate chest harness can be purchased from a manufacturer, or a field-expedient version can be made from either two runners or a long piece of webbing. Either chest harness is then attached to the seat harness with a carabiner and a length of webbing or cord.

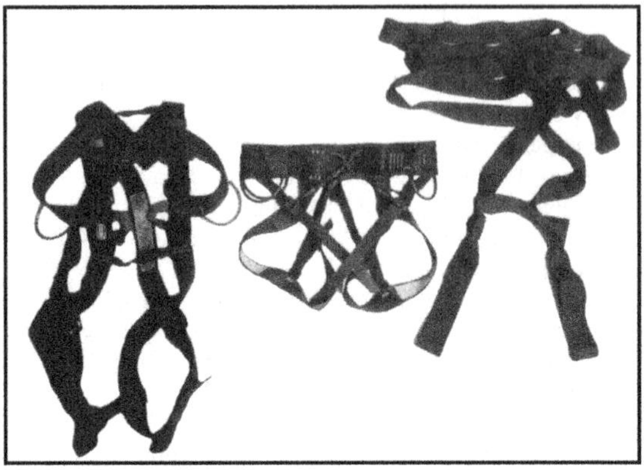

Figure 3-10. Seat harness, field-expedient harness, and full body harness.

3-4. CLIMBING HARDWARE

Climbing hardware refers to all the parts and pieces that allow the trained mountain soldier to accomplish many tasks in the mountains. The importance of this gear to the mountaineer is no less than that of the rifle to the infantryman.

a. **Carabiners.** One of the most versatile pieces of equipment available to the mountaineer is the carabiner. This simple piece of gear is the critical connection between the climber, his rope, and the protection attaching him to the mountain. Carabiners must be strong enough to hold hard falls, yet light enough for the climber to easily carry a quantity of them. Today's high tech metal alloys allow carabiners to meet both of these requirements. Steel is still widely used, but is not preferred for general mountaineering, given other options. Basic carabiner construction affords the user several different shapes. The oval, the D-shaped, and the pear-shaped carabiner are just some of the types currently available. Most models can be made with or without a locking mechanism for the gate opening (Figure 3-11, page 3-12). If the carabiner does have a locking mechanism, it is usually referred to as a locking carabiner. When using a carabiner, great care should be taken to avoid loading the carabiner on its minor axis and to avoid three-way loading (Figure 3-12, page 3-12).

Note: Great care should be used to ensure all carabiner gates are closed and locked during use.

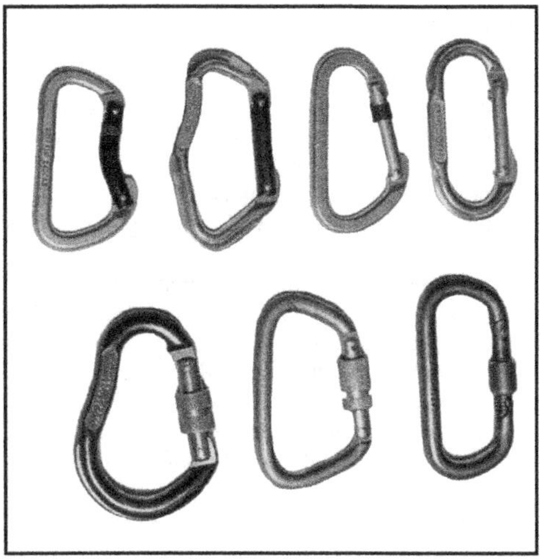

Figure 3-11. Nonlocking and locking carabiners.

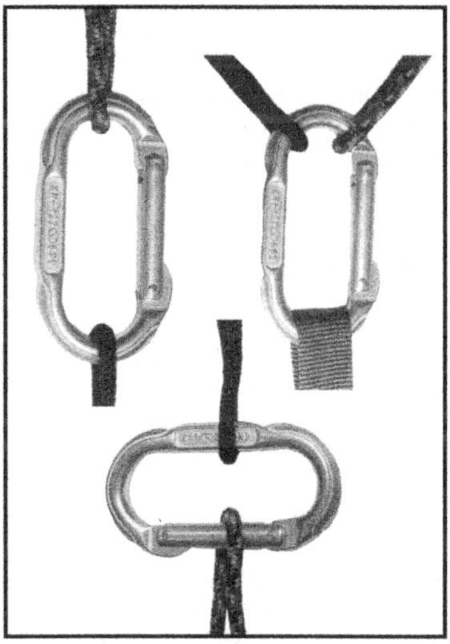

Figure 3-12. Major and minor axes and three-way loading.

(1) The major difference between the oval and the D-shaped carabiner is strength. Because of the design of the D-shaped carabiner, the load is angled onto the spine of the

carabiner thus keeping it off the gate. The down side is that racking any gear or protection on the D-shaped carabiner is difficult because the angle of the carabiner forces all the gear together making it impossible to separate quickly.

(2) The pear-shaped carabiner, specifically the locking version, is excellent for clipping a descender or belay device to the harness. They work well with the munter hitch belaying knot.

(3) Regardless of the type chosen, all carabiners should be UIAA tested. This testing is extensive and tests the carabiner in three ways—along its major axis, along its minor axis, and with the gate open.

b. **Pitons.** A piton is a metal pin that is hammered into a crack in the rock. They are described by their thickness, design, and length (Figure 3-13, page 3-14). Pitons provide a secure anchor for a rope attached by a carabiner. The many different kinds of pitons include: vertical, horizontal, wafer, and angle. They are made of malleable steel, hardened steel, or other alloys. The strength of the piton is determined by its placement rather than its rated tensile strength. The two most common types of pitons are: blades, which hold when wedged into tight-fitting cracks, and angles, which hold blade compression when wedged into a crack.

(1) ***Vertical Pitons***. On vertical pitons, the blade and eye are aligned. These pitons are used in flush, vertical cracks.

(2) ***Horizontal Pitons***. On horizontal pitons, the eye of the piton is at right angles to the blade. These pitons are used in flush, horizontal cracks and in offset or open-book type vertical or horizontal cracks. They are recommended for use in vertical cracks instead of vertical pitons because the torque on the eye tends to wedge the piton into place. This provides more holding power than the vertical piton under the same circumstances.

(3) ***Wafer Pitons***. These pitons are used in shallow, flush cracks. They have little holding power and their weakest points are in the rings provided for the carabiner.

(4) ***Knife Blade Pitons***. These are used in direct-aid climbing. They are small and fit into thin, shallow cracks. They have a tapered blade that is optimum for both strength and holding power.

(5) ***Realized Ultimate Reality Pitons***. Realized ultimate reality pitons (RURPs) are hatchet-shaped pitons about 1-inch square. They are designed to bite into thin, shallow cracks.

(6) ***Angle Pitons***. These are used in wide cracks that are flush or offset. Maximum strength is attained only when the legs of the piton are in contact with the opposite sides of the crack.

(7) ***Bong Pitons***. These are angle pitons that are more than 3.8 centimeters wide. Bongs are commonly made of steel or aluminum alloy and usually contain holes to reduce weight and accommodate carabiners. They have a high holding power and require less hammering than other pitons.

(8) ***Skyhook (Cliffhangers).*** These are small hooks that cling to tiny rock protrusions, ledges, or flakes. Skyhooks require constant tension and are used in a downward pull direction. The curved end will not straighten under body weight. The base is designed to prevent rotation and aid stability.

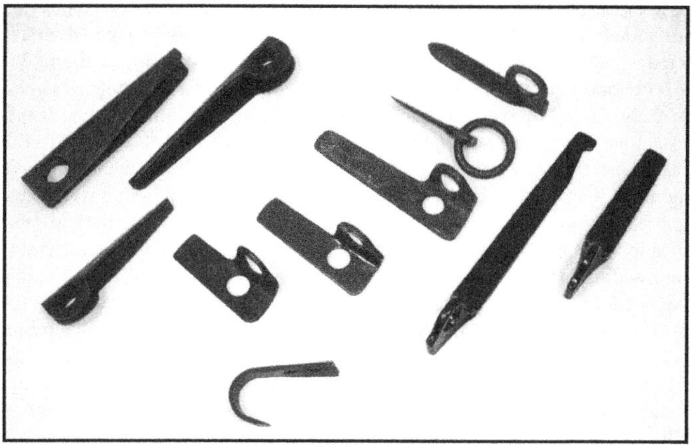

Figure 3-13. Various pitons.

c. **Piton Hammers.** A piton hammer has a flat metal head; a handle made of wood, metal, or fiberglass; and a blunt pick on the opposite side of the hammer (Figure 3-14). A safety lanyard of nylon cord, webbing, or leather is used to attach it to the climber The lanyard should be long enough to allow for full range of motion. Most hammers are approximately 25.5 centimeters long and weigh 12 to 25 ounces. The primary use for a piton hammer is to drive pitons, to be used as anchors, into the rock. The piton hammer can also be used to assist in removing pitons, and in cleaning cracks and rock surfaces to prepare for inserting the piton. The type selected should suit individual preference and the intended use.

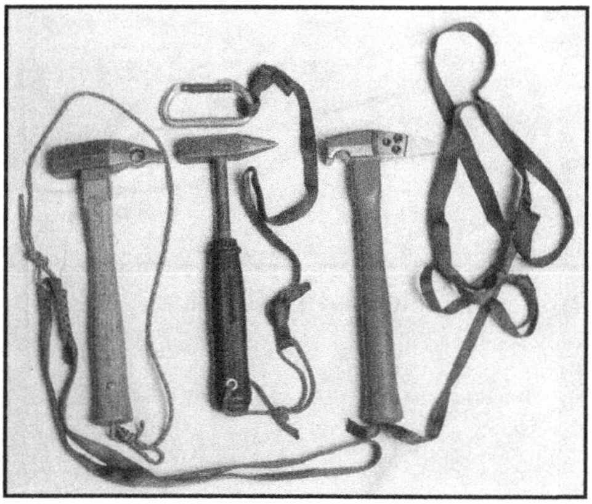

Figure 3-14. Piton hammer.

d. **Chocks.** "Chocks" is a generic term used to describe the various types of artificial protection other than bolts or pitons. Chocks are essentially a tapered metal wedge constructed in various sizes to fit different sized openings in the rock (Figure 3-15). The design of a chock will determine whether it fits into one of two categories—wedges or cams. A wedge holds by wedging into a constricting crack in the rock. A cam holds by slightly rotating in a crack, creating a camming action that lodges the chock in the crack or pocket. Some chocks are manufactured to perform either in the wedging mode or the camming mode. One of the chocks that falls into the category of both a wedge and cam is the hexagonal-shaped or "hex" chock. This type of chock is versatile and comes with either a cable loop or is tied with cord or webbing. All chocks come in different sizes to fit varying widths of cracks. Most chocks come with a wired loop that is stronger than cord and allows for easier placement. Bigger chocks can be threaded with cord or webbing if the user ties the chock himself. Care should be taken to place tubing in the chock before threading the cord. The cord used with chocks is designed to be stiffer and stronger than regular cord and is typically made of Kevlar. The advantage of using a chock rather than a piton is that a climber can carry many different sizes and use them repeatedly.

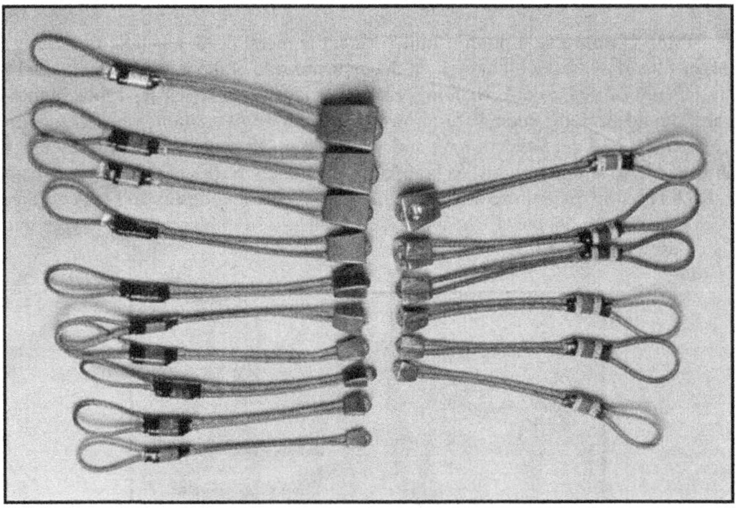

Figure 3-15. Chocks.

e. **Three-Point Camming Device.** The three-point camming device's unique design allows it to be used both as a camming piece and a wedging piece (Figure 3-16). Because of this design it is extremely versatile and, when used in the camming mode, will fit a wide range of cracks. The three-point camming device comes in several different sizes with the smaller sizes working in pockets that no other piece of gear would fit in.

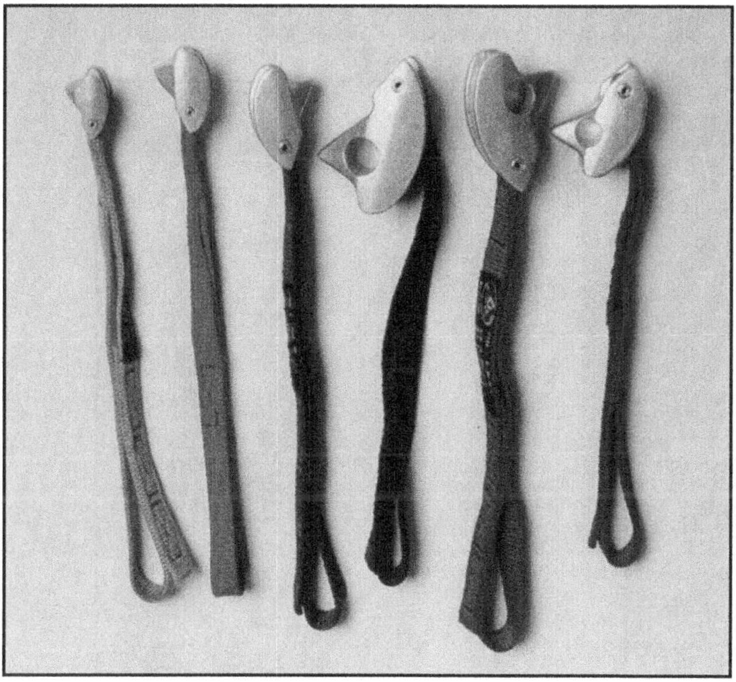

Figure 3-16. Three-point camming device.

f. **Spring-Loaded Camming Devices.** Spring-loaded camming devices (SLCDs) (Figure 3-17) provide convenient, reliable placement in cracks where standard chocks are not practical (parallel or flaring cracks or cracks under roofs). SLCDs have three or four cams rotating around a single or double axis with a rigid or semi-rigid point of attachment. These are placed quickly and easily, saving time and effort. SLCDs are available in many sizes to accommodate different size cracks. Each fits a wide range of crack widths due to the rotating cam heads. The shafts may be rigid metal or semi-rigid cable loops. The flexible cable reduces the risk of stem breakage over an edge in horizontal placements.

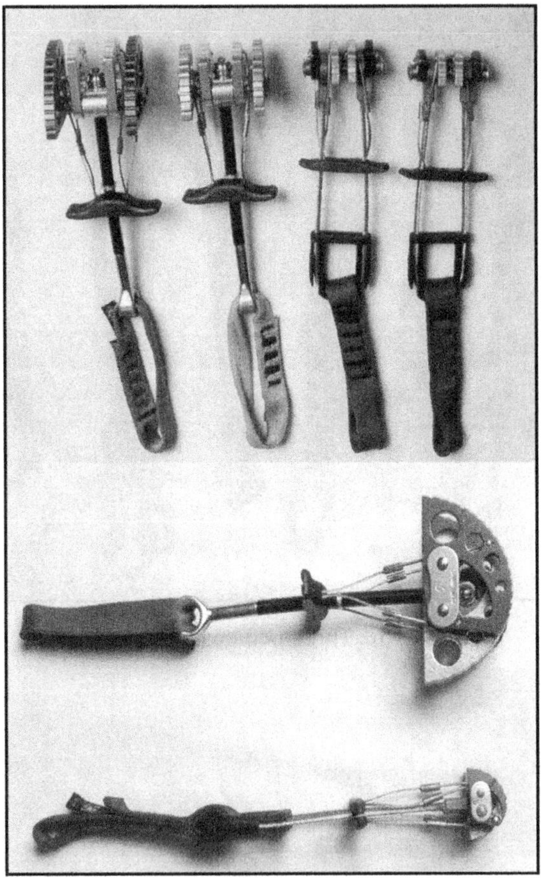

Figure 3-17. Spring-loaded camming devices.

g. **Chock Picks.** Chock picks are primarily used to extract chocks from rock when the they become severely wedged (Figure 3-18). They are also handy to clean cracks with. Made from thin metal, they can be purchased or homemade. When using a chock pick to extract a chock be sure no force is applied directly to the cable juncture. One end of the chock pick should have a hook to use on jammed SLCDs.

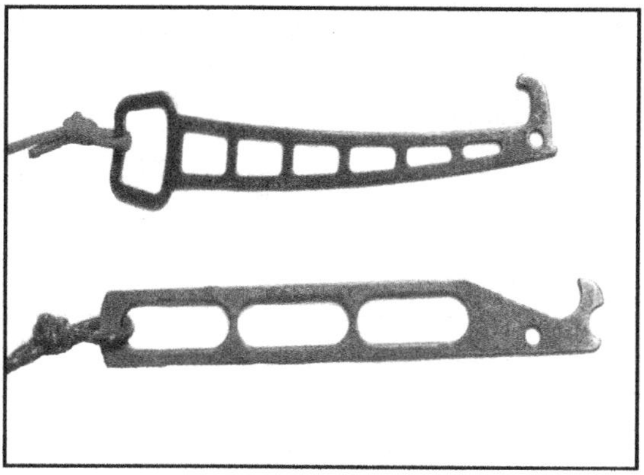

Figure18. Chock picks.

h. **Bolts.** Bolts are screw-like shafts made from metal that are drilled into rock to provide protection (Figure 3-19). The two types are contraction bolts and expansion bolts. Contraction bolts are squeezed together when driven into a rock. Expansion bolts press around a surrounding sleeve to form a snug fit into a rock. Bolts require drilling a hole into a rock, which is time-consuming, exhausting, and extremely noisy. Once emplaced, bolts are the most secure protection for a multidirectional pull. Bolts should be used only when chocks and pitons cannot be emplaced. A bolt is hammered only when it is the nail or self-driving type.

(1) A hanger (for carabiner attachment) and nut are placed on the bolt. The bolt is then inserted and driven into the hole. Because of this requirement, a hand drill must be carried in addition to a piton hammer. Hand drills (also called star drills) are available in different sizes, brands, and weights. A hand drill should have a lanyard to prevent loss.

(2) Self-driving bolts are quicker and easier to emplace. These require a hammer, bolt driver, and drilling anchor, which is driven into the rock. A bolt and carrier are then secured to the emplaced drilling anchor. All metal surfaces should be smooth and free of rust, corrosion, dirt, and moisture. Burrs, chips, and rough spots should be filed smooth and wire-brushed or rubbed clean with steel wool. Items that are cracked or warped indicate excessive wear and should be discarded.

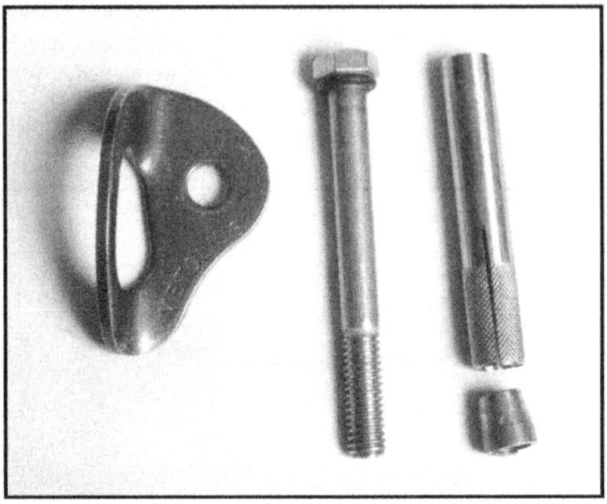

Figure 3-19. Bolts and hangers.

i. **Belay Devices.** Belay devices range from the least equipment intensive (the body belay) to high-tech metal alloy pieces of equipment. Regardless of the belay device choosen, the basic principal remains the same—friction around or through the belay device controls the ropes' movement. Belay devices are divided into three categories: the slot, the tuber, and the mechanical camming device (Figure 3-20).

(1) The slot is a piece of equipment that attaches to a locking carabiner in the harness; a bight of rope slides through the slot and into the carabiner for the belay. The most common slot type belay device is the Sticht plate.

(2) The tuber is used exactly like the slot but its shape is more like a cone or tube.

(3) The mechanical camming device is a manufactured piece of equipment that attaches to the harness with a locking carabiner. The rope is routed through this device so that when force is applied the rope is cammed into a highly frictioned position.

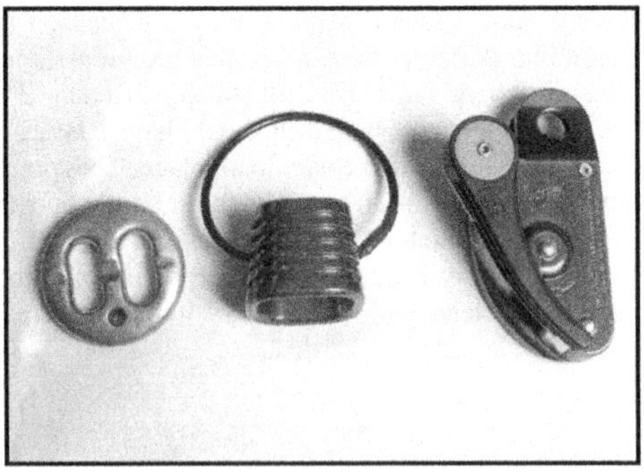

Figure 3-20. Slot, tuber, mechanical camming device.

j. **Descenders.** One piece of equipment used for generations as a descender is the carabiner. A figure-eight is another useful piece of equipment and can be used in conjunction with the carabiner for descending (Figure 3-21).

Note: All belay devices can also be used as descending devices.

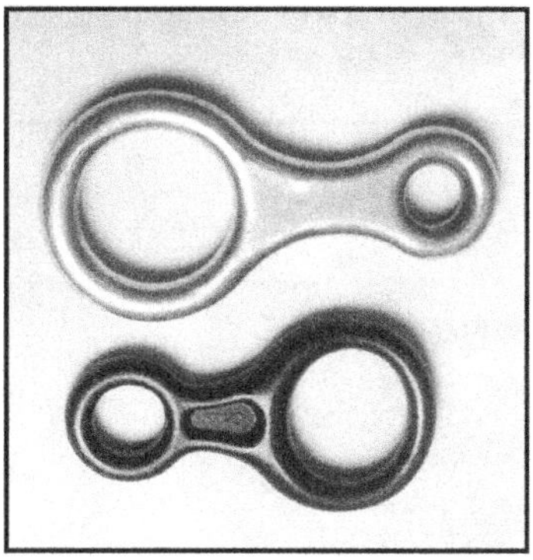

Figure 3-21. Figure-eights.

k. **Ascenders.** Ascenders may be used in other applications such as a personal safety or hauling line cam. All modern ascenders work on the principle of using a cam-like device to allow movement in one direction. Ascenders are primarily made of metal alloys and come in a variety of sizes (Figure 3-22). For difficult vertical terrain, two ascenders work best. For lower angle movement, one ascender is sufficient. Most manufacturers make ascenders as a right and left-handed pair.

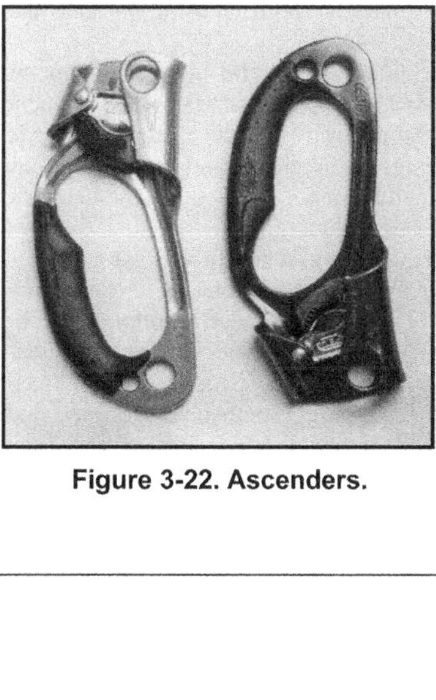

Figure 3-22. Ascenders.

l. **Pulleys.** Pulleys are used to change direction in rope systems and to create mechanical advantage in hauling systems. A pulley should be small, lightweight, and strong. They should accommodate the largest diameter of rope being used. Pulleys are made with several bearings, different-sized sheaves (wheel), and metal alloy sideplates (Figure 3-23). Plastic pulleys should always be avoided. The sideplate should rotate on the pulley axle to allow the pulley to be attached at any point along the rope. For best results, the sheave diameter must be at least four times larger than the rope's diameter to maintain high rope strength.

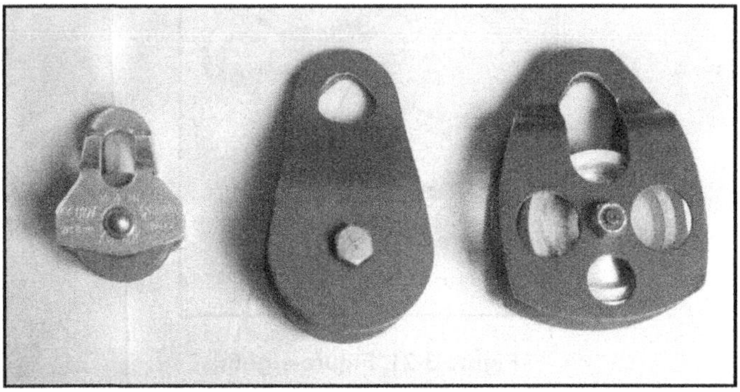

Figure 3-23. Pulley.

3-5. SNOW AND ICE CLIMBING HARDWARE

Snow and ice climbing hardware is the equipment that is particular to operations in some mountainous terrain. Specific training on this type of equipment is essential for safe use. Terrain that would otherwise be inaccessible—snowfields, glaciers, frozen waterfalls—can now be considered avenues of approach using the snow and ice climbing gear listed in this paragraph.

a. **Ice Ax.** The ice ax is one of the most important tools for the mountaineer operating on snow or ice. The climber must become proficient in its use and handling. The versatility of the ax lends itself to balance, step cutting, probing, self-arrest, belays, anchors, direct-aid climbing, and ascending and descending snow and ice covered routes.

(1) Several specific parts comprise an ice ax: the shaft, head (pick and adze), and spike (Figure 3-24, page 3-22).

(a) The shaft (handle) of the ax comes in varying lengths (the primary length of the standard mountaineering ax is 70 centimeters). It can be made of fiberglass, hollow aluminum, or wood; the first two are stronger, therefore safer for mountaineering.

(b) The head of the ax, which combines the pick and the adze, can have different configurations. The pick should be curved slightly and have teeth at least one-fourth of its length. The adze, used for chopping, is perpendicular to the shaft. It can be flat or curved along its length and straight or rounded from side to side. The head can be of one-piece construction or have replaceable picks and adzes. The head should have a hole directly above the shaft to allow for a leash to be attached.

(c) The spike at the bottom of the ax is made of the same material as the head and comes in a variety of shapes.

(2) As climbing becomes more technical, a shorter ax is much more appropriate, and adding a second tool is a must when the terrain becomes vertical. The shorter ax has all the attributes of the longer ax, but it is anywhere from 40 to 55 centimeters long and can have a straight or bent shaft depending on the preference of the user.

b. **Ice Hammer.** The ice hammer is as short or shorter than the technical ax (Figure 3-24). It is used for pounding protection into the ice or pitons into the rock. The only difference between the ice ax and the ice hammer is the ice hammer has a hammerhead instead of an adze. Most of the shorter ice tools have a hole in the shaft to which a leash is secured, which provides a more secure purchase in the ice.

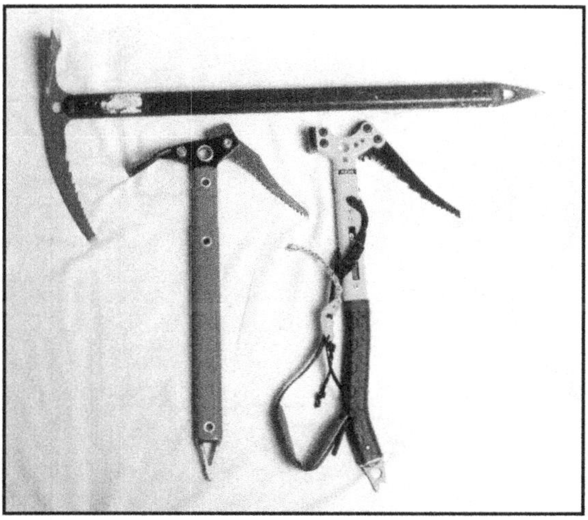

Figure 3-24. Ice ax and ice hammers.

c. **Crampons.** Crampons are used when the footing becomes treacherous. They have multiple spikes on the bottom and spikes protruding from the front (Figure 3-25). Two types of crampons are available: flexible and rigid. Regardless of the type of crampon chosen, fit is the most important factor associated with crampon wear. The crampon should fit snugly on the boot with a minimum of 1 inch of front point protruding. Straps should fit snugly around the foot and any long, loose ends should be trimmed. Both flexible and rigid crampons come in pairs, and any tools needed for adjustment will be provided by the manufacturer.

(1) The hinged or flexible crampon is best used when no technical ice climbing will be done. It is designed to be used with soft, flexible boots, but can be attached to plastic

mountaineering boots. The flexible crampon gets its name from the flexible hinge on the crampon itself. All flexible crampons are adjustable for length while some allow for width adjustment. Most flexible crampons will attach to the boot by means of a strap system. The flexible crampon can be worn with a variety of boot types.

(2) The rigid crampon, as its name implies, is rigid and does not flex. This type of crampon is designed for technical ice climbing, but can be used on less vertical terrain. The rigid crampon can only be worn with plastic mountaineering boots. Rigid crampons will have a toe and heel bail attachment with a strap that wraps around the ankle.

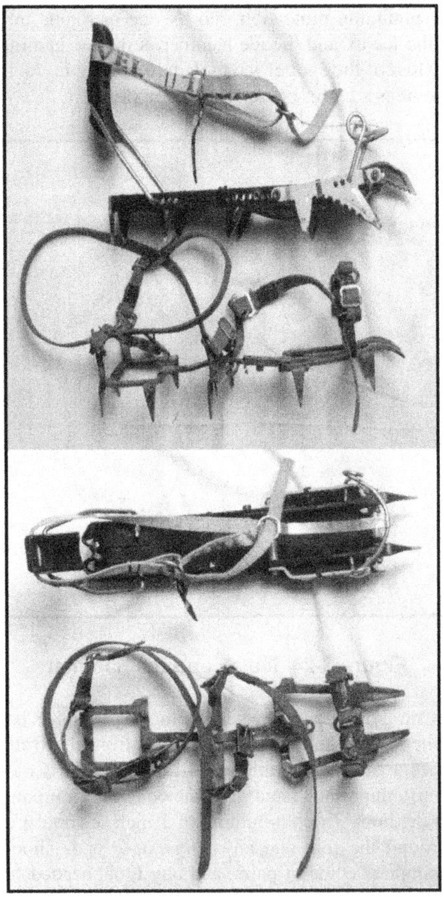

Figure 3-25. Crampons.

d. **Ice Screws.** Ice screws provide artificial protection for climbers and equipment for operations in icy terrain. They are screwed into ice formations. Ice screws are made of chrome-molybdenum steel and vary in lengths from 11 centimeters to 40 centimeters (Figure 3-26). The eye is permanently affixed to the top of the ice screw. The tip consists of milled or hand-ground teeth, which create sharp points to grab the ice when being emplaced. The ice screw has right-hand threads to penetrate the ice when turned clockwise.

(1) When selecting ice screws, choose a screw with a large thread count and large hollow opening. The close threads will allow for ease in turning and better strength. The large hollow opening will allow snow and ice to slide through when turning.

- Type I is 17 centimeters in length with a hollow inner tube.
- Type II is 22 centimeters in length with a hollow inner tube.
- Other variations are hollow alloy screws that have a tapered shank with external threads, which are driven into ice and removed by rotation.

(2) Ice screws should be inspected for cracks, bends, and other deformities that may impair strength or function. If any cracks or bends are noticed, the screw should be turned in. A file may be used to sharpen the ice screw points. Steel wool should be rubbed on rusted surfaces and a thin coat of oil applied when storing steel ice screws.

Note: Ice screws should always be kept clean and dry. The threads and teeth should be protected and kept sharp for ease of application.

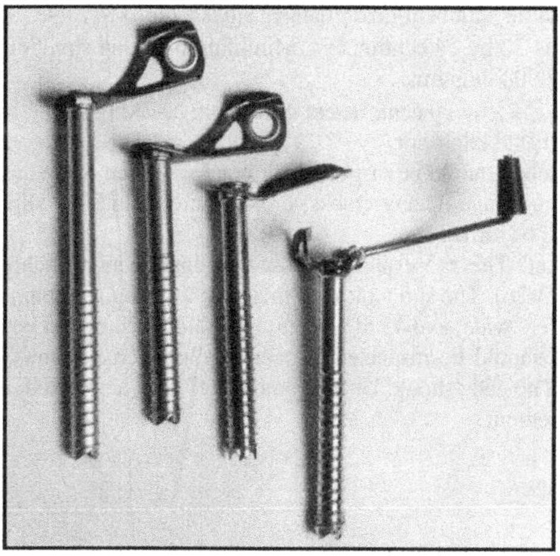

Figure 3-26. Ice screws.

e. **Ice Pitons.** Ice pitons are used to establish anchor points for climbers and equipment when conducting operations on ice. They are made of steel or steel alloys (chrome-molybdenum), and are available in various lengths and diameters (Figure 3-27). They are tubular with a hollow core and are hammered into ice with an ice hammer. The eye is permanently fixed to the top of the ice piton. The tip may be beveled to help grab the ice to facilitate insertion. Ice pitons are extremely strong when placed properly in hard ice. They can, however, pull out easily on warm days and require a considerable amount of effort to extract in cold temperatures.

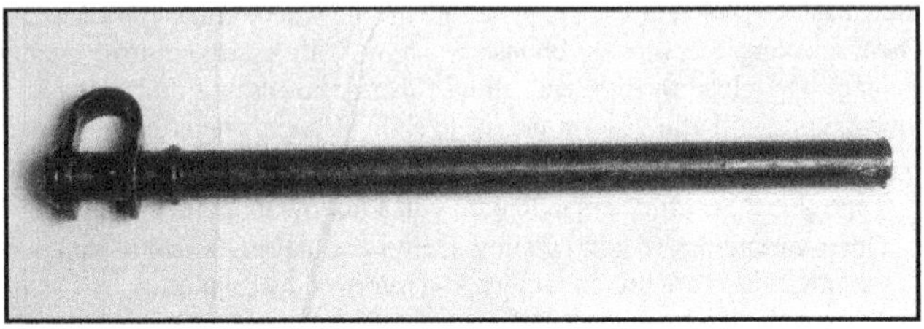

Figure 3-27. Ice piton.

f. **Wired Snow Anchors.** The wired snow anchor (or fluke) provides security for climbers and equipment in operations involving steep ascents by burying the snow anchor into deep snow (Figure 3-28, page 3-26). The fluted anchor portion of the snow anchor is made of aluminum. The wired portion is made of either galvanized steel or stainless steel. Fluke anchors are available in various sizes—their holding ability generally increases with size. They are available with bent faces, flanged sides, and fixed cables. Common types are:

- Type I is 22 by 14 centimeters. Minimum breaking strength of the swaged wire loop is 600 kilograms.
- Type II is 25 by 20 centimeters. Minimum breaking strength of the swaged wire loop is 1,000 kilograms.

The wired snow anchor should be inspected for cracks, broken wire strands, and slippage of the wire through the swage. If any cracks, broken wire strands, or slippage is noticed, the snow anchor should be turned in.

g. **Snow Picket.** The snow picket is used in constructing anchors in snow and ice (Figure 3-28, page 3-26). The snow picket is made of a strong aluminum alloy 3 millimeters thick by 4 centimeters wide, and 45 to 90 centimeters long. They can be angled or T–section stakes. The picket should be inspected for bends, chips, cracks, mushrooming ends, and other deformities. The ends should be filed smooth. If bent or cracked, the picket should be turned in for replacement.

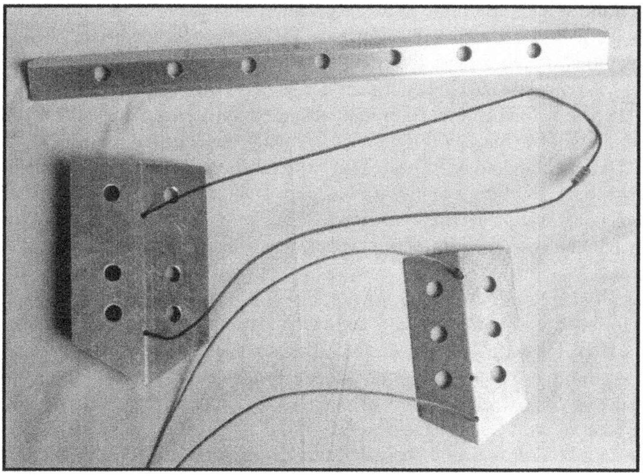

Figure 3-28. Snow anchors, flukes, and pickets.

3-6. SUSTAINABILITY EQUIPMENT

This paragraph describes all additional equipment not directly involved with climbing. This equipment is used for safety (avalanche equipment, wands), bivouacs, movement, and carrying gear. While not all of it will need to be carried on all missions, having the equipment available and knowing how to use it correctly will enhance the unit's capability in mountainous terrain.

a. **Snow Saw.** The snow saw is used to cut into ice and snow. It can be used in step cutting, in shelter construction, for removing frozen obstacles, and for cutting snow stability test pits. The special tooth design of the snow saw easily cuts into frozen snow and ice. The blade is a rigid aluminum alloy of high strength about 3 millimeters thick and 38 centimeters long with a pointed end to facilitate entry on the forward stroke. The handle is either wooden or plastic and is riveted to the blade for a length of about 50 centimeters. The blade should be inspected for rust, cracks, warping, burrs, and missing or dull teeth. A file can repair most defects, and steel wool can be rubbed on rusted areas. The handle should be inspected for cracks, bends, and stability. On folding models, the hinge and nuts should be secure. If the saw is beyond repair, it should not be used.

b. **Snow Shovel.** The snow shovel is used to cut and remove ice and snow. It can be used for avalanche rescue, shelter construction, step cutting, and removing obstacles. The snow shovel is made of a special, lightweight aluminum alloy. The handle should be telescopic, folding, or removable to be compact when not in use. The shovel should have a flat or rounded bottom and be of strong construction. The shovel should be inspected for cracks, bends, rust, and burrs. A file and steel wool can remove rust and put an edge on the blade of the shovel. The handle should be inspected for cracks, bends, and stability. If the shovel is beyond repair, it should be turned in.

c. **Wands.** Wands are used to identify routes, crevasses, snow-bridges, caches, and turns on snow and glaciers. Spacing of wands depends on the number of turns, number of hazards identified, weather conditions (and visibility), and number of teams in the climbing party. Carry too many wands is better than not having enough if they become lost. Wands are 1 to 1.25 meters long and made of lightweight bamboo or plastic shafts pointed on one end with a plastic or nylon flag (bright enough in color to see at a distance) attached to the other end. The shafts should be inspected for cracks, bends, and deformities. The flag should be inspected for tears, frays, security to the shaft, fading, and discoloration. If any defects are discovered, the wands should be replaced.

d. **Avalanche rescue equipment.** Avalanche rescue equipment (Figure 3-29) includes the following:

(1) ***Avalanche Probe.*** Although ski poles may be used as an emergency probe when searching for a victim in an avalanche, commercially manufactured probes are better for a thorough search. They are 9-millimeter thick shafts made of an aluminum alloy, which can be joined to probe up to 360 centimeters. The shafts must be strong enough to probe through avalanche debris. Some manufacturers of ski poles design poles that are telescopic and mate with other poles to create an avalanche probe.

(2) ***Avalanche Transceivers.*** These are small, compact radios used to identify avalanche burial sites. They transmit electromagnetic signals that are picked up by another transceiver on the receive mode.

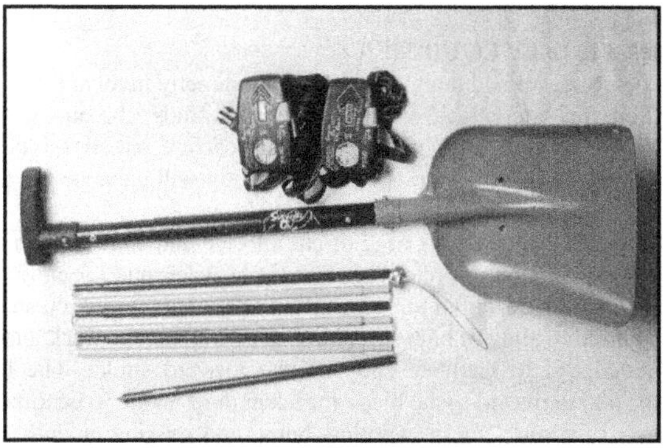

Figure 3-29. Avalanche rescue equipment.

e. **Packs.** Many types and brands of packs are used for mountaineering. The two most common types are internal and external framed packs.

(1) Internal framed packs have a rigid frame within the pack that help it maintain its shape and hug the back. This assists the climber in keeping their balance as they climb or ski. The weight in an internal framed pack is carried low on the body assisting with balance. The body-hugging nature of this type pack also makes it uncomfortable in warm weather.

(2) External framed packs suspend the load away from the back with a ladder-like frame. The frame helps transfer the weight to the hips and shoulders easier, but can be cumbersome when balance is needed for climbing and skiing.

(3) Packs come in many sizes and should be sized appropriately for the individual according to manufacturer's specifications. Packs often come with many unneeded features. A good rule of thumb is: The simpler the pack, the better it will be.

f. **Stoves.** When selecting a stove one must define its purpose—will the stove be used for heating, cooking or both? Stoves or heaters for large elements can be large and cumbersome. Stoves for smaller elements might just be used for cooking and making water, and are simple and lightweight. Stoves are a necessity in mountaineering for cooking and making water from snow and ice. When choosing a stove, factors that should be considered are weight, altitude and temperature where it will be used, fuel availability, and its reliability.

(1) There are many choices in stove design and in fuel types. White gas, kerosene, and butane are the common fuels used. All stoves require a means of pressurization to force the fuel to the burner. Stoves that burn white gas or kerosene have a hand pump to generate the pressurization and butane stoves have pressurized cartridges. All stoves need to vaporize the liquid fuel before it is burned. This can be accomplished by burning a small amount of fuel in the burner cup assembly, which will vaporize the fuel in the fuel line.

(2) Stoves should be tested and maintained prior to a mountaineering mission. They should be easy to clean and repair during an operation. The reliability of the stove has a huge impact on the success of the mission and the morale of personnel.

g. **Tents.** When selecting a tent, the mission must be defined to determine the number of people the tent will accommodate. The climate the tents will be used in is also of concern. A tent used for warmer temperatures will greatly differ from tents used in a colder, more harsh environment. Manufacturers of tents offer many designs of different sizes, weights, and materials.

(1) Mountaineering tents are made out of a breathable or weatherproof material. A single-wall tent allows for moisture inside the tent to escape through the tent's material. A double-wall tent has a second layer of material (referred to as a fly) that covers the tent. The fly protects against rain and snow and the space between the fly and tent helps moisture to escape from inside. Before using a new tent, the seams should be treated with seam sealer to prevent moisture from entering through the stitching.

(2) The frame of a tent is usually made of an aluminum or carbon fiber pole. The poles are connected with an elastic cord that allows them to extend, connect, and become long and rigid. When the tent poles are secured into the tent body, they create the shape of the tent.

(3) Tents are rated by a "relative strength factor," the speed of wind a tent can withstand before the frame deforms. Temperature and expected weather for the mission should be determined before choosing the tent.

h. **Skis.** Mountaineering skis are wide and short. They have a binding that pivots at the toe and allows for the heel to be free for uphill travel or locked for downhill. Synthetic skins with fibers on the bottom can be attached to the bottom of the ski and allow the ski to travel forward and prevent slipping backward. The skins aid in traveling uphill and slow down the rate of descents. Wax can be applied to the ski to aid in ascents instead of skins. Skis can decrease the time needed to reach an objective depending on the ability of the user. Skis can make crossing crevasses easier because of the load distribution, and they can become a makeshift stretcher for casualties. Ski techniques can be complicated and require thorough training for adequate proficiency.

i. **Snowshoes.** Snowshoes are the traditional aid to snow travel that attach to most footwear and have been updated into small, lightweight designs that are more efficient than older models. Snowshoes offer a large displacement area on top of soft snow preventing tiresome post-holing. Some snowshoes come equipped with a crampon like binding that helps in ascending steep snow and ice. Snowshoes are slower than skis, but are better suited for mixed terrain, especially if personnel are not experienced with the art of skiing. When carrying heavy packs, snowshoes can be easier to use than skis.

j. **Ski poles**. Ski poles were traditionally designed to assist in balance during skiing. They have become an important tool in mountaineering for aid in balance while hiking, snowshoeing, and carrying heavy packs. They can take some of the weight off of the lower body when carrying a heavy pack. Some ski poles are collapsible for ease of packing when not needed (Figure 3-30). The basket at the bottom prevents the pole from plunging deep into the snow and, on some models, can be detached so the pole becomes an avalanche or crevasse probe. Some ski poles come with a self-arrest grip, but should not be the only means of protection on technical terrain.

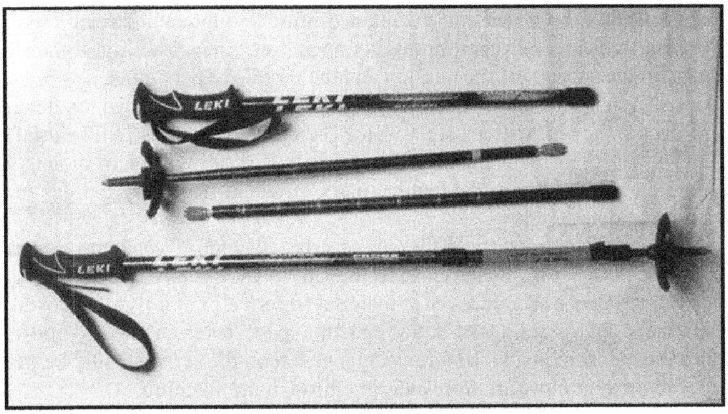

Figure 3-30. Collapsible ski poles.

k. **Sleds**. Sleds vary greatly in size, from the squad-size Ahkio, a component of the 10-man arctic tent system, to the one-person skow. Regardless of the size, sleds are an invaluable asset during mountainous operations when snow and ice is the primary surface on which to travel. Whichever sled is chosen, it must be attachable to the person or people that will be pulling it. Most sleds are constructed using fiberglass bottoms with or without exterior runners. Runners will aid the sleds ability to maintain a true track in the snow. The sled should also come with a cover of some sort—whether nylon or canvas, a cover is essential for keeping the components in the sled dry. Great care should be taken when packing the sled, especially when hauling fuel. Heavier items should be carried towards the rear of the sled and lighter items towards the front.

l. **Headlamps.** A headlamp is a small item that is not appreciated until it is needed. It is common to need a light source and the use of both hands during limited light conditions in mountaineering operations. A flashlight can provide light, but can be cumbersome when both hands are needed. Most headlamps attach to helmets by means of elastic bands.

(1) When choosing a headlamp, ensure it is waterproof and the battery apparatus is small. All components should be reliable in extreme weather conditions. When the light is being packed, care should be taken that the switch doesn't accidentally activate and use precious battery life.

(2) The battery source should compliment the resupply available. Most lights will accept alkaline, nickel-cadmium, or lithium batteries. Alkaline battery life diminishes quickly in cold temperatures, nickel-cadmium batteries last longer in cold but require a recharging unit, and lithium batteries have twice the voltage so modifications are required.

Section II. EQUIPMENT PACKING

Equipment brought on a mission is carried in the pack, worn on the body, or hauled in a sled (in winter). Obviously, the rucksack and sled (or Ahkio) can hold much more than a climber can carry. They would be used for major bivouac gear, food, water, first aid kits, climbing equipment, foul weather shells, stoves, fuel, ropes, and extra ammunition and demolition materials, if needed.

3-7. CHOICE OF EQUIPMENT

Mission requirements and unit SOP will influence the choice of gear carried but the following lists provide a sample of what should be considered during mission planning.

a. **Personal Gear.** Personal gear includes emergency survival kit containing signaling material, fire starting material, food procurement material, and water procurement material. Pocket items should include a knife, whistle, pressure bandage, notebook with pen or pencil, sunglasses, sunblock and lip protection, map, compass and or altimeter.

b. **Standard Gear.** Standard gear that can be individually worn or carried includes cushion sole socks; combat boots or mountain boots, if available; BDU and cap; LCE with canteens, magazine pouches, and first aid kit; individual weapon; a large rucksack containing waterproof coat and trousers, polypropylene top, sweater, or fleece top; helmet; poncho; and sleeping bag.

CAUTION
Cotton clothing, due to its poor insulating and moisture-wicking characteristics, is virtually useless in most mountain climates, the exception being hot, desert, or jungle mountain environments. Cotton clothing should be replaced with synthetic fabric clothing.

c. **Mountaineering Equipment and Specialized Gear.** This gear includes:

- Sling rope or climbing harness.
- Utility cord(s).
- Nonlocking carabiners.
- Locking carabiner(s).
- Rappelling gloves.
- Rappel/belay device.
- Ice ax.
- Crampons.
- Climbing rope, one per climbing team.
- Climbing rack, one per climbing team.

d. **Day Pack.** When the soldier plans to be away from the bivouac site for the day on a patrol or mountaineering mission, he carries a light day pack. This pack should contain the following items:

- Extra insulating layer: polypropylene, pile top, or sweater.
- Protective layer: waterproof jacket and pants, rain suit, or poncho.
- First aid kit.
- Flashlight or headlamp.
- Canteen.
- Cold weather hat or scarf.
- Rations for the time period away from the base camp.
- Survival kit.
- Sling rope or climbing harness.
- Carabiners.
- Gloves.
- Climbing rope, one per climbing team.
- Climbing rack, one per climbing team.

e. **Squad or Team Safety Pack.** When a squad-sized element leaves the bivouac site, squad safety gear should be carried in addition to individual day packs. This can either be loaded into one rucksack or cross-loaded among the squad members. In the event of an injury, casualty evacuation, or unplanned bivouac, these items may make the difference between success and failure of the mission.

- Sleeping bag.
- Sleeping mat.

- Squad stove.
- Fuel bottle.

f. **The Ten Essentials.** Regardless of what equipment is carried, the individual military mountaineer should always carry the "ten essentials" when moving through the mountains.

(1) ***Map.***

(2) ***Compass, Altimeter, and or GPS.***

(3) ***Sunglasses and Sunscreen.***

(a) In alpine or snow-covered sub-alpine terrain, sunglasses are a vital piece of equipment for preventing snow blindness. They should filter 95 to 100 percent of ultraviolet light. Side shields, which minimize the light entering from the side, should permit ventilation to help prevent lens fogging. At least one extra pair of sunglasses should be carried by each independent climbing team.

(b) Sunscreens should have an SPF factor of 15 or higher. For lip protection, a total UV blocking lip balm that resists sweating, washing, and licking is best. This lip protection should be carried in the chest pocket or around the neck to allow frequent reapplication.

(4) ***Extra Food.*** One day's worth extra of food should be carried in case of delay caused by bad weather, injury, or navigational error.

(5) ***Extra Clothing.*** The clothing used during the active part of a climb, and considered to be the basic climbing outfit, includes socks, boots, underwear, pants, blouse, sweater or fleece jacket, hat, gloves or mittens, and foul weather gear (waterproof, breathable outerwear or waterproof rain suit).

(a) Extra clothing includes additional layers needed to make it through the long, inactive hours of an unplanned bivouac. Keep in mind the season when selecting this gear.

- Extra underwear to switch out with sweat-soaked underwear.
- Extra hats or balaclavas.
- Extra pair of heavy socks.
- Extra pair of insulated mittens or gloves.
- In winter or severe mountain conditions, extra insulation for the upper body and the legs.

(b) To back up foul weather gear, bring a poncho or extra-large plastic trash bag. A reflective emergency space blanket can be used for hypothermia first aid and emergency shelter. Insulated foam pads prevent heat loss while sitting or lying on snow. Finally, a bivouac sack can help by protecting insulating layers from the weather, cutting the wind, and trapping essential body heat inside the sack.

(6) ***Headlamp and or Flashlight.*** Headlamps provide the climber a hands-free capability, which is important while climbing, working around the camp, and employing weapons systems. Miniature flashlights can be used, but commercially available headlamps are best. Red lens covers can be fabricated for tactical conditions. Spare batteries and spare bulbs should also be carried.

(7) ***First-aid Kit.*** Decentralized operations, the mountain environment—steep, slick terrain and loose rock combined with heavy packs, sharp tools, and fatigue—requires each climber to carry his own first-aid kit. Common mountaineering injuries that can be expected are punctures and abrasions with severe bleeding, a broken bone, serious sprain,

and blisters. Therefore, the kit should contain at least enough material to stabilize these conditions. Pressure dressings, gauze pads, elastic compression wrap, small adhesive bandages, butterfly bandages, moleskin, adhesive tape, scissors, cleanser, latex gloves and splint material (if above tree line) should all be part of the kit.

(8) ***Fire Starter.*** Fire starting material is key to igniting wet wood for emergency campfires. Candles, heat tabs, and canned heat all work. These can also be used for quick warming of water or soup in a canteen cup. In alpine zones above tree line with no available firewood, a stove works as an emergency heat source.

(9) ***Matches and Lighter.*** Lighters are handy for starting fires, but they should be backed up by matches stored in a waterproof container with a strip of sandpaper.

(10) ***Knife.*** A multipurpose pocket tool should be secured with cord to the belt, harness, or pack.

g. **Other Essential Gear.** Other essential gear may be carried depending on mission and environmental considerations.

(1) ***Water and Water Containers.*** These include wide-mouth water bottles for water collection; camel-back type water holders for hands-free hydration; and a small length of plastic tubing for water procurement at snow-melt seeps and rainwater puddles on bare rock.

(2) ***Ice Ax.*** The ice ax is essential for travel on snowfields and glaciers as well as snow-covered terrain in spring and early summer. It helps for movement on steep scree and on brush and heather covered slopes, as well as for stream crossings.

(3) ***Repair Kit.*** A repair kit should include:

- Stove tools and spare parts.
- Duct tape.
- Patches.
- Safety pins.
- Heavy-duty thread.
- Awl and or needles.
- Cord and or wire.
- Small pliers (if not carrying a multipurpose tool).
- Other repair items as needed.

(4) ***Insect Repellent.***

(5) ***Signaling Devices.***

(6) ***Snow Shovel.***

3-8. TIPS ON PACKING

When loading the internal frame pack the following points should be considered.

a. In most cases, speed and endurance are enhanced if the load is carried more by the hips (using the waist belt) and less by the shoulders and back. This is preferred for movement over trails or less difficult terrain. By packing the lighter, more compressible items (sleeping bag, clothing) in the bottom of the rucksack and the heavier gear (stove, food, water, rope, climbing hardware, extra ammunition) on top, nearer the shoulder blades, the load is held high and close to the back, thus placing the most weight on the hips.

b. In rougher terrain it pays to modify the pack plan. Heavy articles of gear are placed lower in the pack and close to the back, placing more weight on the shoulders and back. This lowers the climber's center of gravity and helps him to better keep his balance.

c. Equipment that may be needed during movement should be arranged for quick access using either external pockets or placing immediately underneath the top flap of the pack. As much as possible, this placement should be standardized across the team so that necessary items can be quickly reached without unnecessary unpacking of the pack in emergencies.

d. The pack and its contents should be soundly waterproofed. Clothing and sleeping bag are separately sealed and then placed in the larger wet weather bag that lines the rucksack. Zip-lock plastic bags can be used for small items, which are then organized into color-coded stuffsacks. A few extra-large plastic garbage bags should be carried for a variety of uses—spare waterproofing, emergency bivouac shelter, and water procurement, among others.

e. The ice ax, if not carried in hand, should be stowed on the outside of the pack with the spike up and the adze facing forward or to the outside, and be securely fastened. Mountaineering packs have ice ax loops and buckle fastening systems for this. If not, the ice ax is placed behind one of the side pockets, as stated above, and then tied in place.

f. Crampons should be secured to the outside rear of the pack with the points covered.

CHAPTER 4
ROPE MANAGEMENT AND KNOTS

The rope is a vital piece of equipment to the mountaineer. When climbing, rappelling, or building various installations, the mountaineer must know how to properly utilize and maintain this piece of equipment. If the rope is not managed or maintained properly, serious injury may occur. This chapter discusses common rope terminology, management techniques, care and maintenance procedures, and knots.

SECTION I. PREPARATION, CARE AND MAINTENANCE, INSPECTION, TERMINOLOGY

The service life of a rope depends on the frequency of use, applications (rappelling, climbing, rope installations), speed of descent, surface abrasion, terrain, climate, and quality of maintenance. Any rope may fail under extreme conditions (shock load, sharp edges, misuse).

4-1. PREPARATION

The mountaineer must select the proper rope for the task to be accomplished according to type, diameter, length, and tensile strength. It is important to prepare all ropes before departing on a mission. Avoid rope preparation in the field.

a. **Packaging.** New rope comes from the manufacturer in different configurations—boxed on a spool in various lengths, or coiled and bound in some manner. Precut ropes are usually packaged in a protective cover such as plastic or burlap. Do not remove the protective cover until the rope is ready for use.

b. **Securing the Ends of the Rope:** If still on a spool, the rope must be cut to the desired length. All ropes will fray at the ends unless they are bound or seared. Both static and dynamic rope ends are secured in the same manner. The ends must be heated to the melting point so as to attach the inner core strands to the outer sheath. By fusing the two together, the sheath cannot slide backward or forward. Ensure that this is only done to the ends of the rope. If the rope is exposed to extreme temperatures, the sheath could be weakened, along with the inner core, reducing overall tensile strength. The ends may also be dipped in enamel or lacquer for further protection.

4-2. CARE AND MAINTENANCE

The rope is a climber's lifeline. It must be cared for and used properly. These general guidelines should be used when handling ropes.

a. Do not step on or drag ropes on the ground unnecessarily. Small particles of dirt will be ground between the inner strands and will slowly cut them.

b. While in use, do not allow the rope to come into contact with sharp edges. Nylon rope is easily cut, particularly when under tension. If the rope must be used over a sharp edge, pad the edge for protection.

c. Always keep the rope as dry as possible. Should the rope become wet, hang it in large loops off the ground and allow it to dry. Never dry a rope with high heat or in direct sunlight.

d. Never leave a rope knotted or tightly stretched for longer than necessary. Over time it will reduce the strength and life of the rope.

e. Never allow one rope to continuously rub over or against another. Allowing rope-on-rope contact with nylon rope is extremely dangerous because the heat produced by the friction will cause the nylon to melt.

f. Inspect the rope before each use for frayed or cut spots, mildew or rot, or defects in construction (new rope).

g. The ends of the rope should be whipped or melted to prevent unraveling.

h. Do not splice ropes for use in mountaineering.

i. Do not mark ropes with paints or allow them to come in contact with oils or petroleum products. Some of these will weaken or deteriorate nylon.

j. Never use a mountaineering rope for any purpose except mountaineering.

k. Each rope should have a corresponding rope log (DA Form 5752-R, Rope History and Usage), which is also a safety record. It should annotate use, terrain, weather, application, number of falls, dates, and so on, and should be annotated each time the rope is used (Figure 4-1). DA Form 5752-R is authorized for local reproduction on 8 1/2- by 11-inch paper.

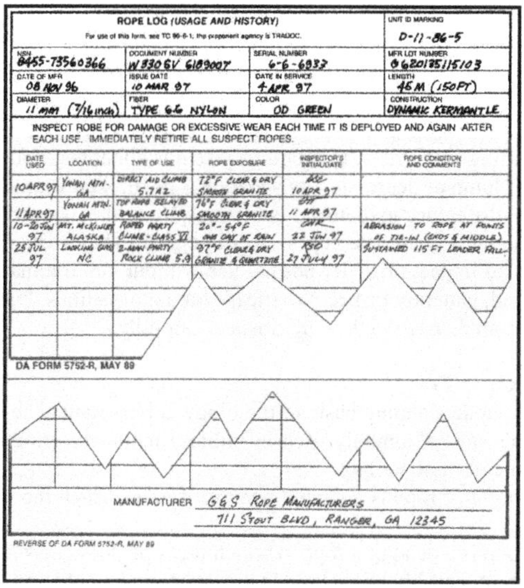

ROPE LOG (USAGE AND HISTORY)
For use of this form, see TC 90-6-1; the proponent agency is TRADOC.

UNIT ID MARKING: D-11-86-5			
NSN: 8455-73560366	DOCUMENT NUMBER: W330 6V 6189007	SERIAL NUMBER: 6-6-6937	MFR LOT NUMBER: 0620135115103
DATE OF MFR: 08 NOV 96	ISSUE DATE: 10 MAR 97	DATE IN SERVICE: 4 APR 97	LENGTH: 45 M (150 FT)
DIAMETER: 11 mm (7/16 inch)	FIBER: TYPE 6.6 NYLON	COLOR: OD GREEN	CONSTRUCTION: DYNAMIC KERNMANTLE

INSPECT ROBE FOR DAMAGE OR EXCESSIVE WEAR EACH TIME IT IS DEPLOYED AND AGAIN AFTER EACH USE. IMMEDIATELY RETIRE ALL SUSPECT ROPES.

DATE USED	LOCATION	TYPE OF USE	ROPE EXPOSURE	INSPECTOR'S INITIAL/DATE	ROPE CONDITION AND COMMENTS
10 APR 97	YONAH MTN. GA	DIRECT AID CLIMB 5.7 A2	72°F CLEAR & DRY SMOOTH GRANITE	10 APR 97	
11 APR 97	YONAH MTN. GA	TOP ROPE BELAYED BALANCE CLIMB	76°F CLEAR & DRY SMOOTH GRANITE	11 APR 97	
10-20 JUN 97	MT. MCKINLEY ALASKA	ROPED PARTY CLIMB - CLASS VI	20° - 54°F ONE DAY OF RAIN	22 JUN 97	ABRASION TO ROPE AT POINTS OF TIE-IN (ENDS & MIDDLE)
25 JUL 97	LOOKING GLASS NC	2-MAN PARTY ROCK CLIMB 5.9	97°F CLEAR & DRY GRANITE & QUARTZITE	27 JULY 97	SUSTAINED 115 FT LEADER FALL

DA FORM 5752-R, MAY 89

MANUFACTURER G & S ROPE MANUFACTURERS
711 STOUT BLVD, RANGER, GA 12345

REVERSE OF DA FORM 5752-R, MAY 89

Figure 4-1. Example of completed DA Form 5752-R.

l. Never subject the rope to high heat or flame. This will significantly weaken it.

m. All ropes should be washed periodically to remove dirt and grit, and rinsed thoroughly. Commercial rope washers are made from short pieces of modified pipe that connect to any faucet. Pinholes within the pipe force water to circulate around and scrub the rope as you slowly feed it through the washer. Another method is to machine wash, on a gentle cycle, in cold water with a nylon safe soap, never bleach or harsh cleansers. Ensure that only front loading washing machine are used to wash ropes.

n. Ultraviolet radiation (sunlight) tends to deteriorate nylon over long periods of time. This becomes important if rope installations are left in place over a number of months.

o. When not in use, ropes should be loosely coiled and hung on wooden pegs rather than nails or other metal objects. Storage areas should be relatively cool with low humidity levels to prevent mildew or rotting. Rope may also be loosely stacked and placed in a rope bag and stored on a shelf. Avoid storage in direct sunlight, as the ultraviolet radiation will deteriorate the nylon over long periods

4-3. INSPECTION

Ropes should be inspected before and after each use, especially when working around loose rock or sharp edges.

a. Although the core of the kernmantle rope cannot be seen, it is possible to damage the core without damaging the sheath. Check a kernmantle rope by carefully inspecting the sheath before and after use while the rope is being coiled. When coiling, be aware of how the rope feels as it runs through the hands. Immediately note and tie off any lumps or depressions felt.

b. Damage to the core of a kernmantle rope usually consists of filaments or yarn breakage that results in a slight retraction. If enough strands rupture, a localized reduction in the diameter of the rope results in a depression that can be felt or even seen.

c. Check any other suspected areas further by putting them under tension (the weight of one person standing on a Prusik tensioning system is about maximum). This procedure will emphasize the lump or depression by separating the broken strands and enlarging the dip. If a noticeable difference in diameter is obvious, retire the rope immediately.

d. Many dynamic kernmantle ropes are quite soft. They may retain an indention occasionally after an impact or under normal use without any trauma to the core. When damage is suspected, patiently inspect the sheath for abnormalities. Damage to the sheath does not always mean damage to the core. Inspect carefully.

4-4. TERMINOLOGY

When using ropes, understanding basic terminology is important. The terms explained in this section are the most commonly used in military mountaineering. (Figure 4-2, page 4-4, illustrates some of these terms.)

a. **Bight**. A bight of rope is a simple bend of rope in which the rope does not cross itself.

b. **Loop.** A loop is a bend of a rope in which the rope does cross itself.

c. **Half Hitch**. A half hitch is a loop that runs around an object in such a manner as to lock or secure itself.

d. **Turn.** A turn wraps around an object, providing 360-degree contact.

e. **Round Turn**. A round turn wraps around an object one and one-half times. A round turn is used to distribute the load over a small diameter anchor (3 inches or less). It may also be used around larger diameter anchors to reduce the tension on the knot, or provide added friction.

f. **Running End**. A running end is the loose or working end of the rope.

g. **Standing Part**. The standing part is the static, stationary, or nonworking end of the rope.

h. **Lay.** The lay is the direction of twist used in construction of the rope.

i. **Pigtail.** The pigtail (tail) is the portion of the running end of the rope between the safety knot and the end of the rope.

j. **Dress.** Dress is the proper arrangement of all the knot parts, removing unnecessary kinks, twists, and slack so that all rope parts of the knot make contact.

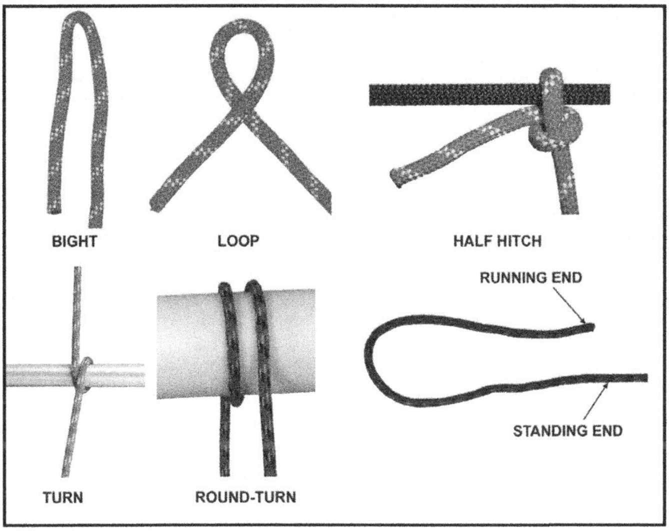

Figure 4-2. Examples of roping terminology.

Section II. COILING, CARRYING, THROWING

The ease and speed of rope deployment and recovery greatly depends upon technique and practice.

4-5. COILING AND CARRYING THE ROPE

Use the butterfly or mountain coil to coil and carry the rope. Each is easy to accomplish and results in a minimum amount of kinks, twists, and knots later during deployment.

a. **Mountain Coil.** To start a mountain coil, grasp the rope approximately 1 meter from the end with one hand. Run the other hand along the rope until both arms are outstretched. Grasping the rope firmly, bring the hands together forming a loop, which is laid in the hand closest to the end of the rope. This is repeated, forming uniform loops that run in a clockwise direction, until the rope is completely coiled. The rope may be given a 1/4 twist as each loop is formed to overcome any tendency for the rope to twist or form figure-eights.

(1) In finishing the mountain coil, form a bight approximately 30 centimeters long with the starting end of the rope and lay it along the top of the coil. Uncoil the last loop and, using this length of the rope, begin making wraps around the coil and the bight, wrapping toward the closed end of the bight and making the first wrap bind across itself so as to lock it into place. Make six to eight wraps to adequately secure the coil, and then route the end of the rope through the closed end of the bight. Pull the running end of the bight tight, securing the coil.

(2) The mountain coil may be carried either in the pack (by forming a figure eight), doubling it and placing it under the flap, or by placing it over the shoulder and under the opposite arm, slung across the chest. (Figure 4-3 shows how to coil a mountain coil.)

Figure 4-3. Mountain coil.

b. **Butterfly Coil.** The butterfly coil is the quickest and easiest technique for coiling (Figure 4-4).

Figure 4-4. Butterfly coil.

(1) ***Coiling.*** To start the double butterfly, grasp both ends of the rope and begin back feeding. Find the center of the rope forming a bight. With the bight in the left hand, grasp both ropes and slide the right hand out until there is approximately one arms length of rope. Place the doubled rope over the head, draping it around the neck and on top of the shoulders. Ensure that it hangs no lower than the waist. With the rest of the doubled rope in front of you, make doubled bights placing them over the head in the same manner as the first bight. Coil alternating from side to side (left to right, right to left) while maintaining equal-length bights. Continue coiling until approximately two arm-lengths of rope remain. Remove the coils from the neck and shoulders carefully, and hold the center in one hand. Wrap the two ends around the coils a minimum of three doubled wraps, ensuring that the first wrap locks back on itself.

(2) ***Tie-off and Carrying.*** Take a doubled bight from the loose ends of rope and pass it through the apex of the coils. Pull the loose ends through the doubled bight and dress it down. Place an overhand knot in the loose ends, dressing it down to the apex of the bight securing coils. Ensure that the loose ends do not exceed the length of the coils. In this configuration the coiled rope is secure enough for hand carrying or carrying in a rucksack, or for storage. (Figure 4-5 shows a butterfly coil tie-off.)

Figure 4-5. Butterfly coil tie-off.

c. **Coiling Smaller Diameter Rope.** Ropes of smaller diameters may be coiled using the butterfly or mountain coil depending on the length of the rope. Pieces 25 feet and shorter (also known as cordage, sling rope, utility cord) may be coiled so that they can be hung from the harness. Bring the two ends of the rope together, ensuring no kinks are in the rope. Place the ends of the rope in the left hand with the two ends facing the body. Coil the doubled rope in a clockwise direction forming 6- to 8-inch coils (coils may be larger depending on the length of rope) until an approximate 12-inch bight is left. Wrap that bight around the coil, ensuring that the first wrap locks on itself. Make three or more wraps. Feed the bight up through the bights formed at the top of the coil. Dress it down tightly. Now the piece of rope may be hung from a carabiner on the harness.

e. **Uncoiling, Back-feeding, and Stacking.** When the rope is needed for use, it must be uncoiled and stacked on the ground properly to avoid kinks and snarls.

(1) Untie the tie-off and lay the coil on the ground. Back-feed the rope to minimize kinks and snarls. (This is also useful when the rope is to be moved a short distance and

coiling is not desired.) Take one end of the rope in the left hand and run the right hand along the rope until both arms are outstretched. Next, lay the end of the rope in the left hand on the ground. With the left hand, re-grasp the rope next to the right hand and continue laying the rope on the ground.

(2) The rope should be laid or stacked in a neat pile on the ground to prevent it from becoming tangled and knotted when throwing the rope, feeding it to a lead climber, and so on. This technique can also be started using the right hand.

4-6. THROWING THE ROPE

Before throwing the rope, it must be properly managed to prevent it from tangling during deployment. The rope should first be anchored to prevent complete loss of the rope over the edge when it is thrown. Several techniques can be used when throwing a rope. Personal preference and situational and environmental conditions should be taken into consideration when determining which technique is best.

a. Back feed and neatly stack the rope into coils beginning with the anchored end of the rope working toward the running end. Once stacked, make six to eight smaller coils in the left hand. Pick up the rest of the larger coils in the right hand. The arm should be generally straight when throwing. The rope may be thrown underhanded or overhanded depending on obstacles around the edge of the site. Make a few preliminary swings to ensure a smooth throw. Throw the large coils in the right hand first. Throw up and out. A slight twist of the wrist, so that the palm of the hand faces up as the rope is thrown, allows the coils to separate easily without tangling. A smooth follow through is essential. When a slight tug on the left hand is felt, toss the six to eight smaller coils out. This will prevent the ends of the rope from becoming entangled with the rest of the coils as they deploy. As soon as the rope leaves the hand, the thrower should sound off with a warning of "ROPE" to alert anyone below the site.

b. Another technique may also be used when throwing rope. Anchor, back feed, and stack the rope properly as described above. Take the end of the rope and make six to eight helmet-size coils in the right hand (more may be needed depending on the length of the rope). Assume a "quarterback" simulated stance. Aiming just above the horizon, vigorously throw the rope overhanded, up and out toward the horizon. The rope must be stacked properly to ensure smooth deployment.

c. When windy weather conditions prevail, adjustments must be made. In a strong cross wind, the rope should be thrown angled into the wind so that it will land on the desired target. The stronger the wind, the harder the rope must be thrown to compensate.

SECTION III. KNOTS

All knots used by a mountaineer are divided into four classes: Class I—joining knots, Class II—anchor knots, Class III—middle rope knots, and Class IV—special knots. The variety of knots, bends, bights, and hitches is almost endless. These classes of knots are intended only as a general guide since some of the knots discussed may be appropriate in more than one class. The skill of knot tying can perish if not used and practiced. With experience and practice, knot tying becomes instinctive and helps the mountaineer in many situations.

4-7. SQUARE KNOT

The square knot is used to tie the ends of two ropes of equal diameter (Figure 4-6). It is a joining knot.

a. **Tying the Knot.**

STEP 1. Holding one working end in each hand, place the working end in the right hand over the one in the left hand.

STEP 2. Pull it under and back over the top of the rope in the left hand.

STEP 3. Place the working end in the left hand over the one in the right hand and repeat STEP 2.

STEP 4. Dress the knot down and secure it with an overhand knot on each side of the square knot.

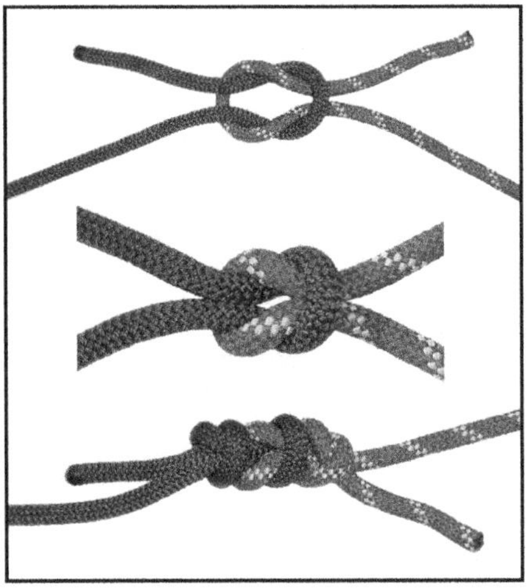

Figure 4-6. Square knot.

b. **Checkpoints.**

(1) There are two interlocking bights.

(2) The running end and standing part are on the same side of the bight formed by the other rope.

(3) The running ends are parallel to and on the same side of the standing ends with 4-inch minimum pig tails after the overhand safeties are tied.

4-8. FISHERMAN'S KNOT

The fisherman's knot is used to tie two ropes of the same or approximately the same diameter (Figure 4-7, page 4-10). It is a joining knot.

a. **Tying the Knot.**

STEP 1. Tie an overhand knot in one end of the rope.

STEP 2. Pass the working end of the other rope through the first overhand knot. Tie an overhand knot around the standing part of the first rope with the working end of the second rope.

STEP 3. Tightly dress down each overhand knot and tightly draw the knots together.

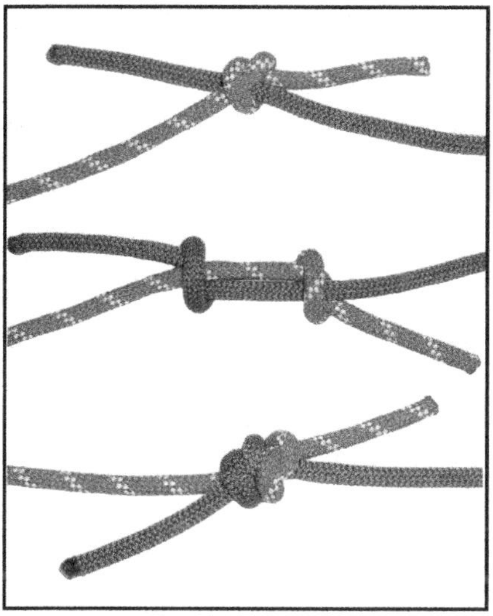

Figure 4-7. Fisherman's knot.

b. **Checkpoints.**

(1) The two separate overhand knots are tied tightly around the long, standing part of the opposing rope.

(2) The two overhand knots are drawn snug.

(3) Ends of rope exit knot opposite each other with 4-inch pigtails.

4-9. DOUBLE FISHERMAN'S KNOT

The double fisherman's knot (also called double English or grapevine) is used to tie two ropes of the same or approximately the same diameter (Figure 4-8). It is a joining knot.

a. **Tying the Knot.**

STEP 1. With the working end of one rope, tie two wraps around the standing part of another rope.

STEP 2. Insert the working end (STEP 1) back through the two wraps and draw it tight.

STEP 3. With the working end of the other rope, which contains the standing part (STEPS 1 and 2), tie two wraps around the standing part of the other rope (the working end in STEP 1). Insert the working end back through the two wraps and draw tight.

STEP 4. Pull on the opposing ends to bring the two knots together.

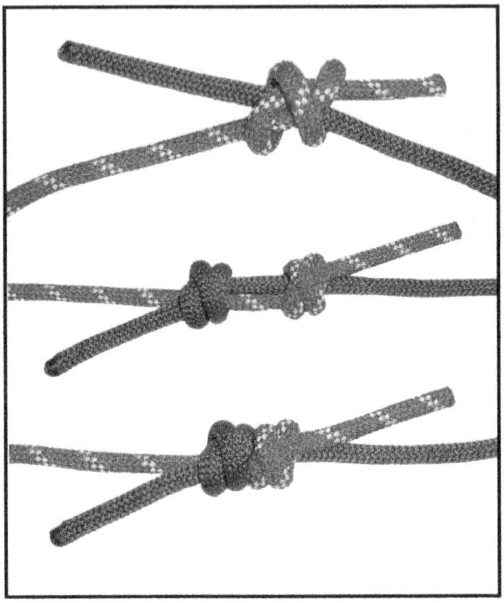

Figure 4-8. Double fisherman's knot.

b. **Checkpoints.**

(1) Two double overhand knots securing each other as the standing parts of the rope are pulled apart.

(2) Four rope parts on one side of the knot form two "x" patterns, four rope parts on the other side of the knot are parallel.

(3) Ends of rope exit knot opposite each other with 4-inch pigtails.

4-10. FIGURE-EIGHT BEND

The figure-eight bend is used to join the ends of two ropes of equal or unequal diameter within 5-mm difference (Figure 4-9, page 4-12).

a. **Tying the Knot.**

STEP 1. Grasp the top of a 2-foot bight.

STEP 2. With the other hand, grasp the running end (short end) and make a 360-degree turn around the standing end.

STEP 3. Place the running end through the loop just formed creating an in-line figure eight.

STEP 4. Route the running end of the other ripe back through the figure eight starting from the original rope's running end. Trace the original knot to the standing end.

STEP 5. Remove all unnecessary twists and crossovers. Dress the knot down.

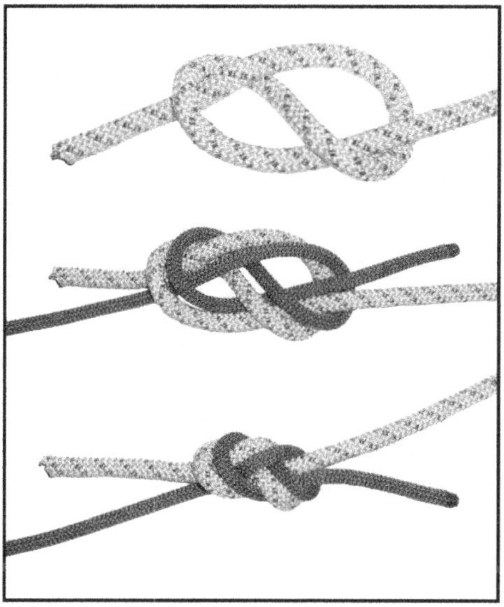

Figure 4-9. Figure-eight bend.

b. **Checkpoints.**

(1) There is a figure eight with two ropes running side by side.

(2) The running ends are on opposite sides of the knot.

(3) There is a minimum 4-inch pigtail.

4-11. WATER KNOT

The water knot is used to attach two webbing ends (Figure 4-10). It is also called a ring bend, overhand retrace, or tape knot. It is used in runners and harnesses and is a joining knot.

a. **Tying the Knot.**

STEP 1. Tie an overhand knot in one of the ends.

STEP 2. Feed the other end back through the knot, following the path of the first rope in reverse.

STEP 3. Draw tight and pull all of the slack out of the knot. The remaining tails must extend at least 4 inches beyond the knot in both directions.

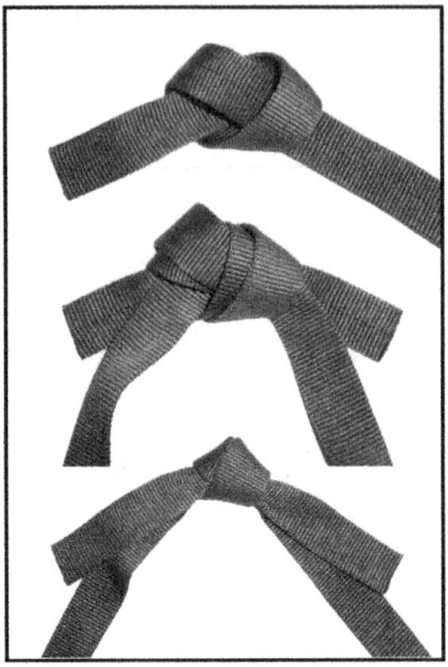

Figure 4-10. Water knot.

b. **Checkpoints.**

(1) There are two overhand knots, one retracing the other.

(2) There is no slack in the knot, and the working ends come out of the knot in opposite directions.

(3) There is a minimum 4-inch pigtail.

4-12. BOWLINE

The bowline is used to tie the end of a rope around an anchor. It may also be used to tie a single fixed loop in the end of a rope (Figure 4-11, page 4-14). It is an anchor knot.

a. **Tying the Knot.**

STEP 1. Bring the working end of the rope around the anchor, from right to left (as the climber faces the anchor).

STEP 2. Form an overhand loop in the standing part of the rope (on the climber's right) toward the anchor.

STEP 3. Reach through the loop and pull up a bight.

STEP 4. Place the working end of the rope (on the climber's left) through the bight, and bring it back onto itself. Now dress the knot down.

STEP 5. Form an overhand knot with the tail from the bight.

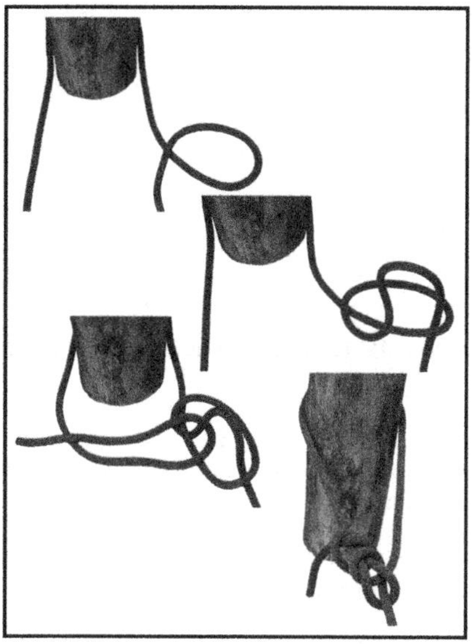

Figure 4-11. Bowline knot.

b. **Checkpoints.**

(1) The bight is locked into place by a loop.

(2) The short portion of the bight is on the inside and on the loop around the anchor (or inside the fixed loop).

(3) There is a minimum 4-inch pigtail after tying the overhand safety.

4-13. ROUND TURN AND TWO HALF HITCHES

This knot is used to tie the end of a rope to an anchor, and it must have constant tension (Figure 4-12). It is an anchor knot.

a. **Tying the Knot.**

STEP 1. Route the rope around the anchor from right to left and wrap down (must have two wraps in the rear of the anchor, and one in the front). Run the loop around the object to provide 360-degree contact, distributing the load over the anchor.

STEP 2. Bring the working end of the rope left to right and over the standing part, forming a half hitch (first half hitch).

STEP 3. Repeat STEP 2 (last half hitch has a 4 inch pigtail).

STEP 4. Dress the knot down.

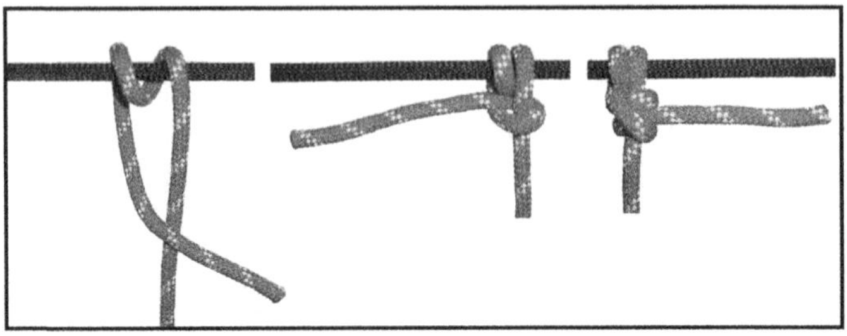

Figure 4-12. Round turn and two half hitches.

b. **Checkpoints.**

(1) A complete round turn should exist around the anchor with no crosses.

(2) Two half hitches should be held in place by a diagonal locking bar with no less than a 4-inch pigtail remaining.

4-14. FIGURE-EIGHT RETRACE (REROUTED FIGURE-EIGHT)

The figure-eight retrace knot produces the same result as a figure-eight loop. However, by tying the knot in a retrace, it can be used to fasten the rope to trees or to places where the loop cannot be used (Figure 4-13, page 4-16). It is also called a rerouted figure-eight and is an anchor knot.

a. **Tying the Knot.**

STEP 1. Use a length of rope long enough to go around the anchor, leaving enough rope to work with.

STEP 2. Tie a figure-eight knot in the standing part of the rope, leaving enough rope to go around the anchor. To tie a figure-eight knot form a loop in the rope, wrap the working end around the standing part, and route the working end through the loop. The finished knot is dressed loosely.

STEP 3. Take the working end around the anchor point.

STEP 4. With the working end, insert the rope back through the loop of the knot in reverse.

STEP 5. Keep the original figure eight as the outside rope and retrace the knot around the wrap and back to the long-standing part.

STEP 6. Remove all unnecessary twists and crossovers; dress the knot down.

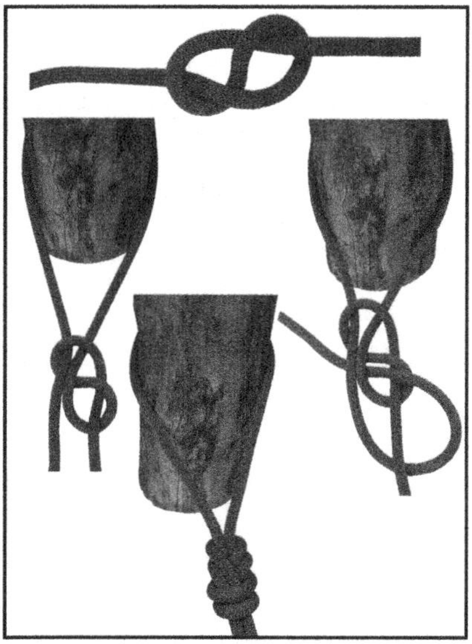

Figure 4-13. Figure-eight retrace.

b. **Checkpoints**

(1) A figure eight with a doubled rope running side by side, forming a fixed loop around a fixed object or harness.

(2) There is a minimum 4-inch pigtail.

4-15. CLOVE HITCH

The clove hitch is an anchor knot that can be used in the middle of the rope as well as at the end (Figure 4-14). The knot must have constant tension on it once tied to prevent slipping. It can be used as either an anchor or middle of the rope knot, depending on how it is tied.

a. **Tying the Knot.**

(1) ***Middle of the Rope.***

STEP 1. Hold rope in both hands, palms down with hands together. Slide the left hand to the left from 20 to 25 centimeters.

STEP 2. Form a loop away from and back toward the right.

STEP 3. Slide the right hand from 20 to 25 centimeters to the right. Form a loop inward and back to the left hand.

STEP 4. Place the left loop on top of the right loop. Place both loops over the anchor and pull both ends of the rope in opposite directions. The knot is tied.

(2) ***End of the Rope.***

Note: For instructional purposes, assume that the anchor is horizontal.

STEP 1. Place 76 centimeters of rope over the top of the anchor. Hold the standing end in the left hand. With the right hand, reach under the horizontal anchor, grasp the working end, and bring it inward.

STEP 2. Place the working end of the rope over the standing end (to form a loop). Hold the loop in the left hand. Place the working end over the anchor from 20 to 25 centimeters to the left of the loop.

STEP 3. With the right hand, reach down to the left hand side of the loop under the anchor. Grasp the working end of the rope. Bring the working end up and outward.

STEP 4. Dress down the knot.

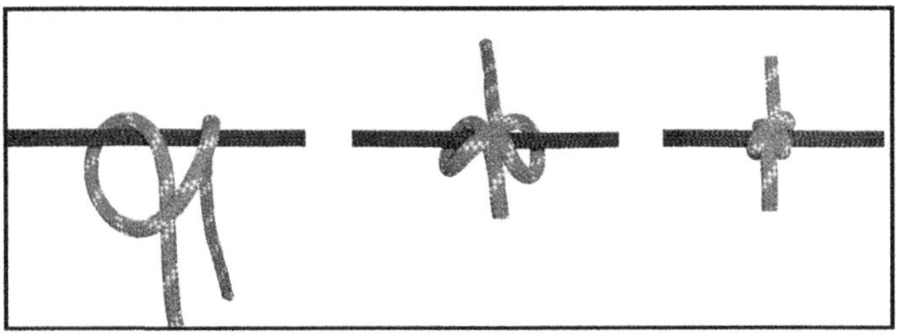

Figure 4-14. Clove hitch.

b. **Checkpoints.**

(1) The knot has two round turns around the anchor with a diagonal locking bar.

(2) The locking bar is facing 90 degrees from the direction of pull.

(3) The ends exit 180 degrees from each other.

(4) The knot has more than a 4-inch pigtail remaining.

4-16. WIREMAN'S KNOT

The wireman's knot forms a single, fixed loop in the middle of the rope (Figure 4-15, page 4-18). It is a middle rope knot.

a. **Tying the Knot.**

STEP 1. When tying this knot, face the anchor that the tie-off system will be tied to. Take up the slack from the anchor, and wrap two turns around the left hand (palm up) from left to right.

STEP 2. A loop of 30 centimeters is taken up in the second round turn to create the fixed loop of the knot.

STEP 3. Name the wraps from the palm to the fingertips: heel, palm, and fingertip.

STEP 4. Secure the palm wrap with the right thumb and forefinger, and place it over the heel wrap.

STEP 5. Secure the heel wrap and place it over the fingertip wrap.

STEP 6. Secure the fingertip wrap and place it over the palm wrap.

STEP 7. Secure the palm wrap and pull up to form a fixed loop.

STEP 8. Dress the knot down by pulling on the fixed loop and the two working ends.

STEP 9. Pull the working ends apart to finish the knot.

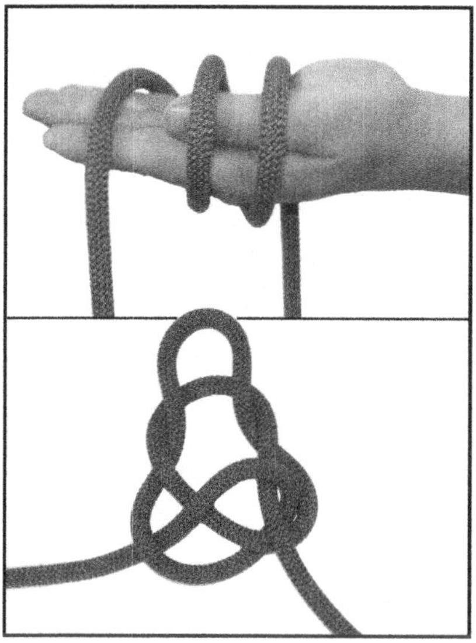

Figure 4-15. Wireman's knot.

b. **Checkpoints.**

(1) The completed knot should have four separate bights locking down on themselves with the fixed loop exiting from the top of the knot and laying toward the near side anchor point.

(2) Both ends should exit opposite each other without any bends.

4-17. DIRECTIONAL FIGURE-EIGHT

The directional figure-eight knot forms a single, fixed loop in the middle of the rope that lays back along the standing part of the rope (Figure 4-16). It is a middle rope knot.

a. **Tying the Knot.**

STEP 1. Face the far side anchor so that when the knot is tied, it lays inward.

STEP 2. Lay the rope from the far side anchor over the left palm. Make one wrap around the palm.

STEP 3. With the wrap thus formed, tie a figure-eight knot around the standing part that leads to the far side anchor.

STEP 4. When dressing the knot down, the tail and the bight must be together.

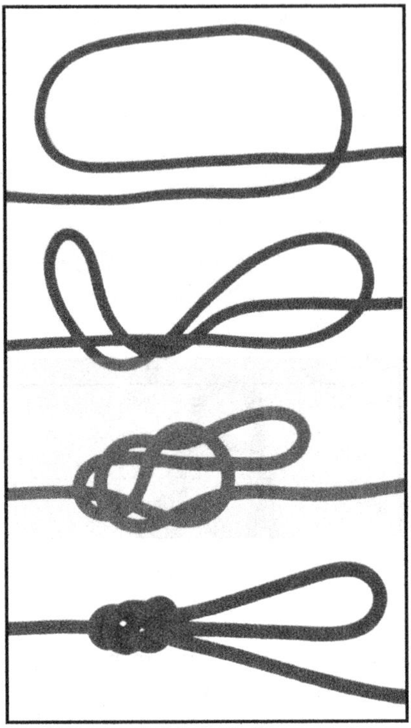

Figure 4-16. Directional figure-eight.

b. **Checkpoints.**

(1) The loop should be large enough to accept a carabiner but no larger than a helmet-size loop.

(2) The tail and bight must be together.

(3) The figure eight is tied tightly.

(4) The bight in the knot faces back toward the near side.

4-18. BOWLINE-ON-A-BIGHT (TWO-LOOP BOWLINE)

The bowline-on-a-bight is used to form two fixed loops in the middle of a rope (Figure 4-17, page 4-20). It is a middle rope knot.

a. **Tying the Knot.**

STEP 1. Form a bight in the rope about twice as long as the finished loops will be.

STEP 2. Tie an overhand knot on a bight.

STEP 3. Hold the overhand knot in the left hand so that the bight is running down and outward.

STEP 4. Grasp the bight with the right hand; fold it back over the overhand knot so that the overhand knot goes through the bight.

STEP 5. From the end (apex) of the bight, follow the bight back to where it forms the cross in the overhand knot. Grasp the two ropes that run down and outward and pull up, forming two loops.

STEP 6. Pull the two ropes out of the overhand knot and dress the knot down.

STEP 7. A final dress is required: grasp the ends of the two fixed loops and pull, spreading them apart to ensure the loops do not slip.

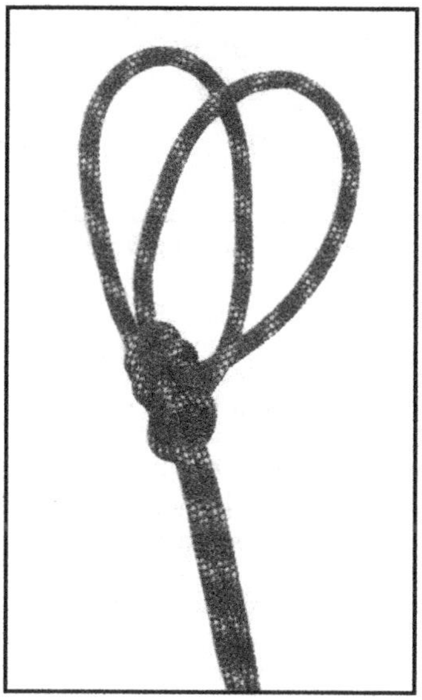

Figure 4-17. Bowline-on-a-bight.

b. **Checkpoints.**

(1) There are two fixed loops that will not slip.

(2) There are no twists in the knot.

(3) A double loop is held in place by a bight.

4-19. TWO-LOOP FIGURE-EIGHT

The two-loop figure-eight is used to form two fixed loops in the middle of a rope (Figure 4-18.) It is a middle rope knot.

a. **Tying the Knot.**

STEP 1. Using a doubled rope, form an 18-inch bight in the left hand with the running end facing to the left.

STEP 2. Grasp the bight with the right hand and make a 360-degree turn around the standing end in a counterclockwise direction.

STEP 3. With the working end, form another bight and place that bight through the loop just formed in the left hand.

STEP 4. Hold the bight with the left hand, and place the original bight (moving toward the left hand) over the knot.

STEP 5. Dress the knot down.

Figure 4-18. Two-loop figure-eight.

b. **Checkpoints.**

(1) There is a double figure-eight knot with two loops that share a common locking bar.

(2) The two loops must be adjustable by means of a common locking bar.

(3) The common locking bar is on the bottom of the double figure-eight knot.

4-20. FIGURE-EIGHT LOOP (FIGURE-EIGHT-ON-A-BIGHT)

The figure-eight loop, also called the figure-eight-on-a-bight, is used to form a fixed loop in a rope (Figure 4-19). It is a middle of the rope knot.

a. **Tying the Knot.**

STEP 1. Form a bight in the rope about as large as the diameter of the desired loop.

STEP 2. With the bight as the working end, form a loop in rope (standing part).

STEP 3. Wrap the working end around the standing part 360 degrees and feed the working end through the loop. Dress the knot tightly.

Figure 4-19. Figure-eight loop.

b. **Checkpoints.**

(1) The loop is the desired size.

(2) The ropes in the loop are parallel and do not cross over each other.

(3) The knot is tightly dressed.

4-21. PRUSIK KNOT

The Prusik knot is used to put a moveable rope on a fixed rope such as a Prusik ascent or a tightening system. This knot can be tied as a middle or end of the rope Prusik. It is a specialty knot.

a. **Tying the Knot.**

(1) ***Middle-of-the-Rope Prusik.*** The middle-of-the-rope Prusik knot can be tied with a short rope to a long rope as follows (Figure 4-20.):

STEP 1. Double the short rope, forming a bight, with the working ends even. Lay it over the long rope so that the closed end of the bight is 12 inches below the long rope and the remaining part of the rope (working ends) is the closest to the climber; spread the working end apart.

STEP 2. Reach down through the 12-inch bight. Pull up both of the working ends and lay them over the long rope. Repeat this process making sure that the working ends pass in the middle of the first two wraps. Now there are four wraps and a locking bar working across them on the long rope.

STEP 3. Dress the wraps and locking bar down to ensure they are tight and not twisted. Tying an overhand knot with both ropes will prevent the knot from slipping during periods of variable tension.

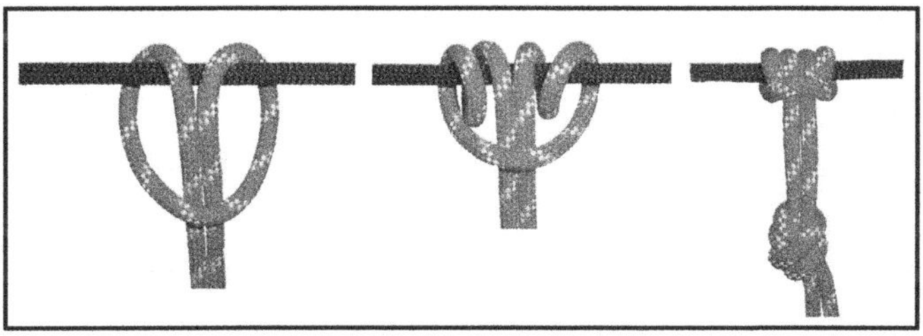

Figure 4-20. Middle-of-the-rope Prusik.

(2) ***End-of-the-Rope Prusik*** (Figure 4-21).

STEP 1. Using an arm's length of rope, and place it over the long rope.

STEP 2. Form a complete round turn in the rope.

STEP 3. Cross over the standing part of the short rope with the working end of the short rope.

STEP 4. Lay the working end under the long rope.

STEP 5. Form a complete round turn in the rope, working back toward the middle of the knot.

STEP 6. There are four wraps and a locking bar running across them on the long rope. Dress the wraps and locking bar down. Ensure they are tight, parallel, and not twisted.

STEP 7. Finish the knot with a bowline to ensure that the Prusik knot will not slip out during periods of varying tension.

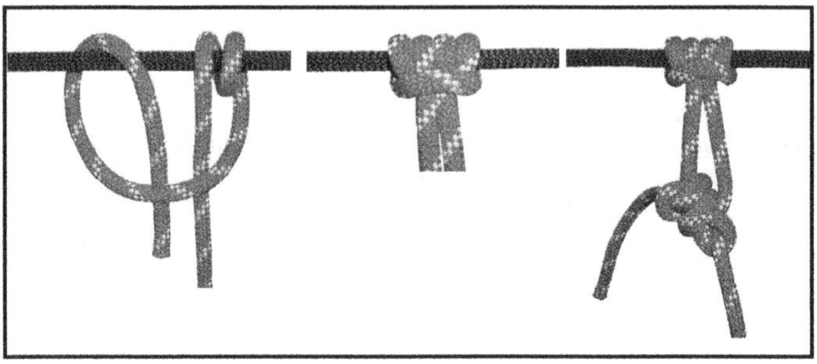

Figure 4-21. End-of-the-rope Prusik knot.

b. **Checkpoints.**

(1) Four wraps with a locking bar.

(2) The locking bar faces the climber.

(3) The knot is tight and dressed down with no ropes twisted or crossed.

(4) Other than a finger Prusik, the knot should contain an overhand or bowline to prevent slipping.

4-22. BACHMAN KNOT

The Bachman knot provides a means of using a makeshift mechanized ascender (Figure 4-22, page 4-24). It is a specialty knot.

a. **Tying the Knot.**

STEP 1. Find the middle of a utility rope and insert it into a carabiner.

STEP 2. Place the carabiner and utility rope next to a long climbing rope.

STEP 3. With the two ropes parallel from the carabiner, make two or more wraps around the climbing rope and through the inside portion of the carabiner.

Note: The rope can be tied into an etrier (stirrup) and used as a Prusik-friction principle ascender.

b. **Checkpoints.**

(1) The bight of the climbing rope is at the top of the carabiner.

(2) The two ropes run parallel without twisting or crossing.

(3) Two or more wraps are made around the long climbing rope and through the inside portion of the carabiner.

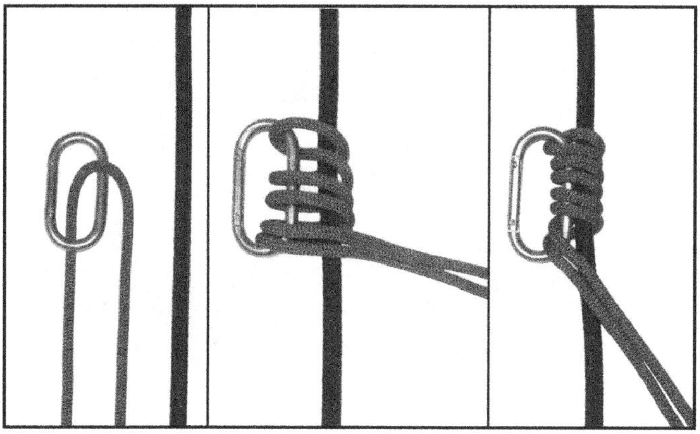

Figure 4-22. Bachman knot.

4-23. BOWLINE-ON-A-COIL

The bowline-on-a-coil is an expedient tie-in used by climbers when a climbing harness is not available (Figure 4-23). It is a specialty knot.

a. **Tying the Knot.**

STEP 1. With the running end, place 3 feet of rope over your right shoulder. The running end is to the back of the body.

STEP 2. Starting at the bottom of your rib cage, wrap the standing part of the rope around your body and down in a clockwise direction four to eight times.

STEP 3. With the standing portion of the rope in your left hand, make a clockwise loop toward the body. The standing portion is on the bottom.

STEP 4. Ensuring the loop does not come uncrossed, bring it up and under the coils between the rope and your body.

STEP 5. Using the standing part, bring a bight up through the loop. Grasp the running end of the rope with the right hand. Pass it through the bight from right to left and back on itself.

STEP 6. Holding the bight loosely, dress the knot down by pulling on the standing end.

STEP 7. Safety the bowline with an overhand around the top, single coil. Then, tie an overhand around all coils, leaving a minimum 4-inch pigtail.

b. **Checkpoints.**

(1) A minimum of four wraps, not crossed, with a bight held in place by a loop.
(2) The loop must be underneath all wraps.
(3) A minimum 4-inch pigtail after the second overhand safety is tied.
(4) Must be centered on the mid-line of the body.

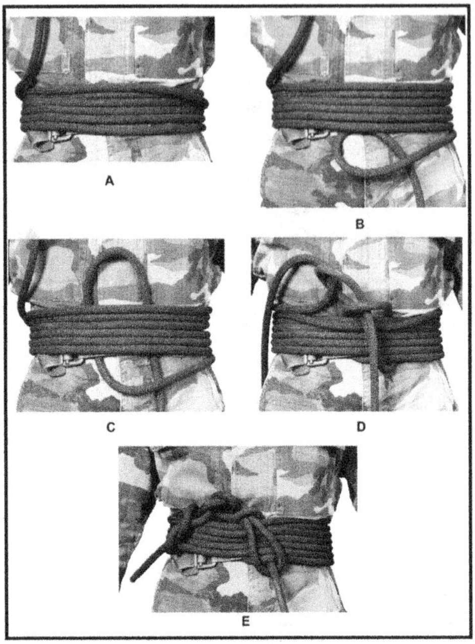

Figure 4-23. Bowline-on-a-coil.

4-24. THREE-LOOP BOWLINE

The three-loop bowline is used to form three fixed loops in the middle of a rope (Figure 4-24, page 4-26). It is used in a self-equalizing anchor system. It is a specialty knot.

a. **Tying the Knot.**

STEP 1. Form an approximate 24-inch bight.

STEP 2. With the right thumb facing toward the body, form a doubled loop in the standing part by turning the wrist clockwise. Lay the loops to the right.

STEP 3. With the right hand, reach down through the loops and pull up a doubled bight from the standing part of the rope.

STEP 4. Place the running end (bight) of the rope (on the left) through the doubled bight from left to right and bring it back on itself. Hold the running end loosely and dress the knot down by pulling on the standing parts.

STEP 5. Safety it off with a doubled overhand knot.

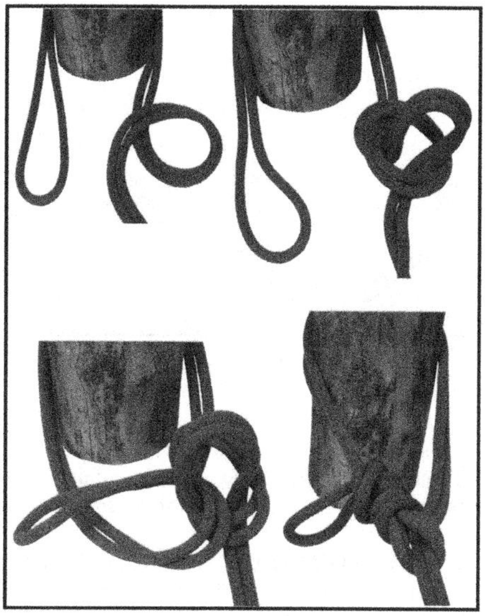

Figure 4-24. Three-loop bowline.

b. **Checkpoints.**

(1) There are two bights held in place by two loops.

(2) The bights form locking bars around the standing parts.

(3) The running end (bight) must be on the inside of the fixed loops.

(4) There is a minimum 4-inch pigtail after the double overhand safety knot is tied.

4-25. FIGURE-EIGHT SLIP KNOT

The figure eight slip knot forms an adjustable bight in a rope (Figure 4-25). It is a specialty knot.

a. **Tying the Knot.**

STEP 1. Form a 12-inch bight in the end of the rope.

STEP 2. Hold the center of the bight in the right hand. Hold the two parallel ropes from the bight in the left hand about 12 inches up the rope.

STEP 3. With the center of the bight in the right hand, twist two complete turns clockwise.

STEP 4. Reach through the bight and grasp the long, standing end of the rope. Pull another bight (from the long standing end) back through the original bight.

STEP 5. Pull down on the short working end of the rope and dress the knot down.

STEP 6. If the knot is to be used in a transport tightening system, take the working end of the rope and form a half hitch around the loop of the figure eight knot.

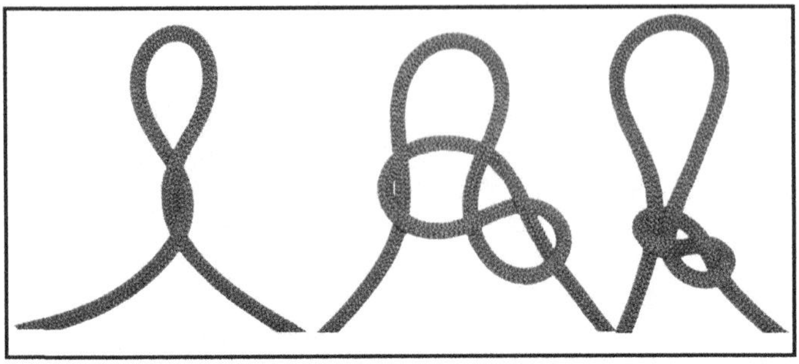

Figure 4-25. Figure-eight slip knot.

b. **Checkpoints.**

(1) The knot is in the shape of a figure eight.

(2) Both ropes of the bight pass through the same loop of the figure eight.

(3) The sliding portion of the rope is the long working end of the rope.

4-26. TRANSPORT KNOT (OVERHAND SLIP KNOT/MULE KNOT)

The transport knot is used to secure the transport tightening system (Figure 4-26, page 4-28). It is simply an overhand slip knot.

a. **Tying the Knot.**

STEP 1. Pass the running end of the rope around the anchor point passing it back under the standing portion (leading to the far side anchor) forming a loop.

STEP 2. Form a bight with the running end of the rope. Pass over the standing portion and down through the loop and dress it down toward the anchor point.

STEP 3. Secure the knot by tying a half hitch around the standing portion with the bight.

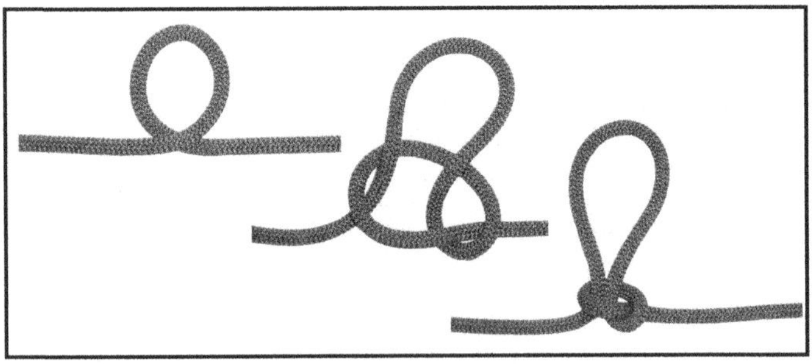

Figure 4-26. Transport knot.

b. **Check Points.**

(1) There is a single overhand slip knot.

(2) The knot is secured using a half hitch on a bight.
(3) The bight is a minimum of 12 inches long.

4-27. KLEIMHIEST KNOT

The Kleimhiest knot provides a moveable, easily adjustable, high-tension knot capable of holding extremely heavy loads while being pulled tight (Figure 4-27). It is a special-purpose knot.

a. **Tying the Knot.**

STEP 1. Using a utility rope or webbing offset the ends by 12 inches. With the ends offset, find the center of the rope and form a bight. Lay the bight over a horizontal rope.

STEP 2. Wrap the tails of the utility rope around the horizontal rope back toward the direction of pull. Wrap at least four complete turns.

STEP 3. With the remaining tails of the utility rope, pass them through the bight (see STEP 1).

STEP 4. Join the two ends of the tail with a joining knot.

STEP 5. Dress the knot down tightly so that all wraps are touching.

Note: Spectra should not be used for the Kleimhiest knot. It has a low melting point and tends to slip .

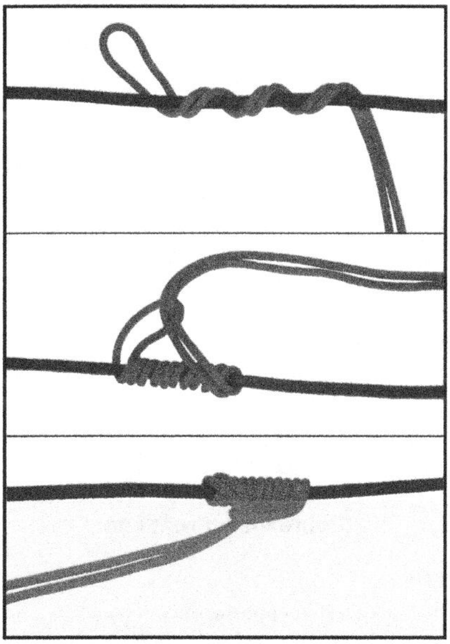

Figure 4-27. Kleimhiest knot.

b. **Checkpoints.**

(1) The bight is opposite the direction of pull.

(2) All wraps are tight and touching.

(3) The ends of the utility rope are properly secured with a joining knot.

4-28. FROST KNOT

The frost knot is used when working with webbing (Figure 4-28, page 4-30). It is used to create the top loop of an etrier. It is a special-purpose knot.

a. **Tying the Knot.**

STEP 1. Lap one end (a bight) of webbing over the other about 10 to 12 inches.

STEP 2. Tie an overhand knot with the newly formed triple-strand webbing; dress tightly.

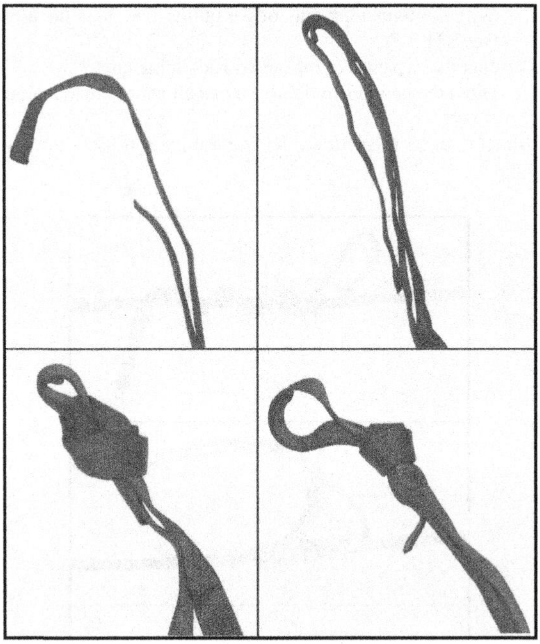

Figure 4-28. Frost knot.

b. **Checkpoints.**

(1) The tails of the webbing run in opposite directions.

(2) Three strands of webbing are formed into a tight overhand knot.

(3) There is a bight and tail exiting the top of the overhand knot.

4-29. GIRTH HITCH

The girth hitch is used to attach a runner to an anchor or piece of equipment (Figure 4-29). It is a special-purpose knot.

a. **Tying the Knot.**

STEP 1: Form a bight.

STEP 2: Bring the runner back through the bight.

STEP 3: Cinch the knot tightly.

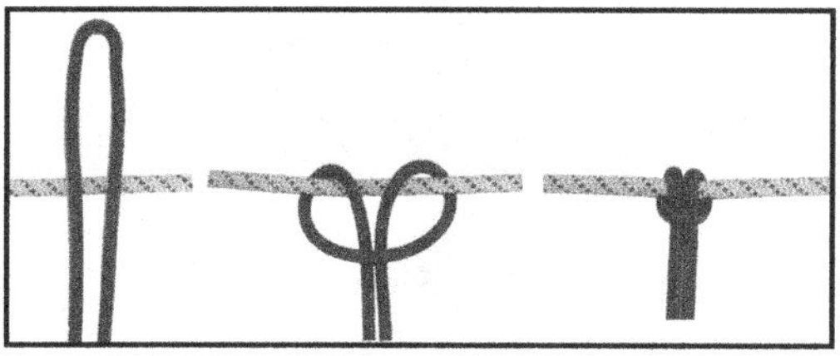

Figure 4-29. Girth hitch.

b. **Checkpoint.**

(1) Two wraps exist with a locking bar running across the wraps.

(2) The knot is dressed tightly.

4-30. MUNTER HITCH

The munter hitch, when used in conjunction with a pear-shaped locking carabiner, is used to form a mechanical belay (Figure 4-30).

a. **Tying the Knot.**

STEP 1. Hold the rope in both hands, palms down about 12 inches apart.

STEP 2. With the right hand, form a loop away from the body toward the left hand. Hold the loop with the left hand.

STEP 3. With the right hand, place the rope that comes from the bottom of the loop over the top of the loop.

STEP 4. Place the bight that has just been formed around the rope into the pear shaped carabiner. Lock the locking mechanism.

b. **Check Points.**

(1) A bight passes through the carabiner, with the closed end around the standing or running part of the rope.

(2) The carabiner is locked.

Figure 4-30. Munter hitch.

4-31. RAPPEL SEAT

The rappel seat is an improvised seat rappel harness made of rope (Figure 4-31, page 4-32). It usually requires a sling rope 14 feet or longer.

a. **Tying the Knot.**

STEP 1. Find the middle of the sling rope and make a bight.

STEP 2. Decide which hand will be used as the brake hand and place the bight on the opposite hip.

STEP 3. Reach around behind and grab a single strand of rope. Bring it around the waist to the front and tie two overhands on the other strand of rope, thus creating a loop around the waist.

STEP 4. Pass the two ends between the legs, ensuring they do not cross.

STEP 5. Pass the two ends up under the loop around the waist, bisecting the pocket flaps on the trousers. Pull up on the ropes, tightening the seat.

STEP 6. From rear to front, pass the two ends through the leg loops creating a half hitch on both hips.

STEP 7. Bring the longer of the two ends across the front to the nonbrake hand hip and secure the two ends with a square knot safetied with overhand knots. Tuck any excess rope in the pocket below the square knot.

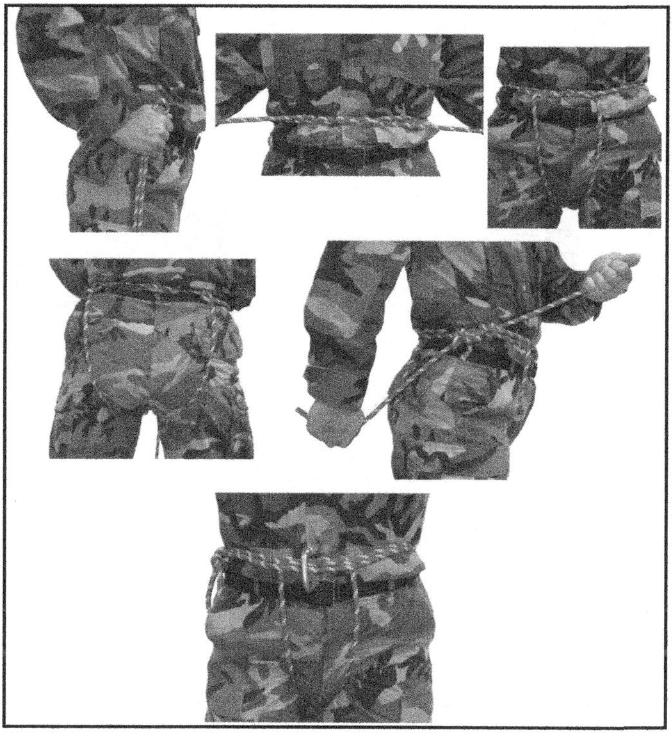

Figure 4-31. Rappel seat.

b. **Check Points.**

(1) There are two overhand knots in the front.

(2) The ropes are not crossed between the legs.

(3) A half hitch is formed on each hip.

(4) Seat is secured with a square knot with overhand safeties on the non-brake hand side.

(5) There is a minimum 4-inch pigtail after the overhand safeties are tied.

4-32. GUARDE KNOT

The guarde knot (ratchet knot, alpine clutch) is a special purpose knot primarily used for hauling systems or rescue (Figure 4-32). The knot works in only one direction and cannot be reversed while under load.

a. **Tying the Knot.**

STEP 1. Place a bight of rope into the two anchored carabiners (works best with two like carabiners, preferably ovals).

STEP 2. Take a loop of rope from the non-load side and place it down into the opposite cararabiner so that the rope comes out between the two carabiners.

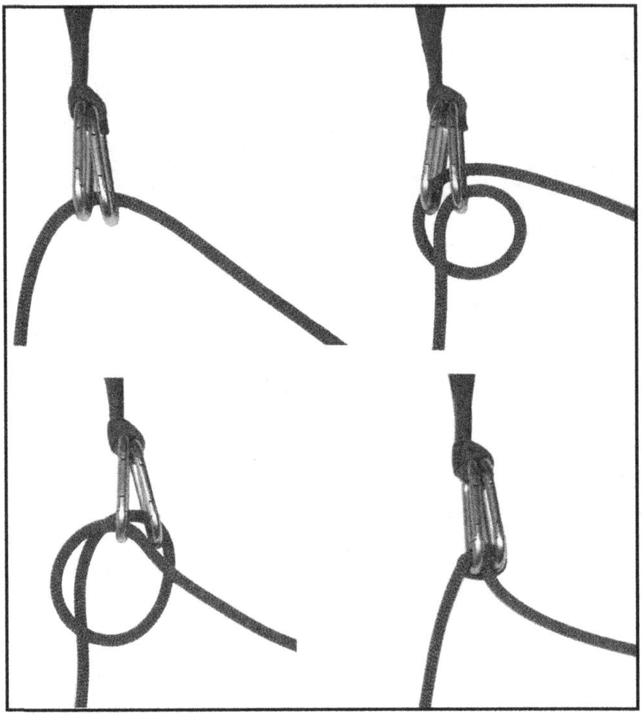

Figure 4-32. Guarde knot.

b. **Check Points.**

(1) When properly dressed, rope can only be pulled in one direction.

(2) The knot will not fail when placed under load.

CHAPTER 5
ANCHORS

This chapter discusses different types of anchors and their application in rope systems and climbing. Proper selection and placement of anchors is a critical skill that requires a great deal of practice. Failure of any system is most likely to occur at the anchor point. If the anchor is not strong enough to support the intended load, it will fail. Failure is usually the result of poor terrain features selected for the anchor point, or the equipment used in rigging the anchor was placed improperly or in insufficient amounts.

When selecting or constructing anchors, always try to make sure the anchor is "bombproof." A bombproof anchor is stronger than any possible load that could be placed on it. An anchor that has more strength than the climbing rope is considered bombproof.

Section I. NATURAL ANCHORS

Natural anchors should be considered for use first. They are usually strong and often simple to construct with minimal use of equipment. Trees, boulders, and other terrain irregularities are already in place and simply require a method of attaching the rope. However, natural anchors should be carefully studied and evaluated for stability and strength before use. Sometimes the climbing rope is tied directly to the anchor, but under most circumstances a sling is attached to the anchor and then the climbing rope is attached to the sling with a carabiner(s). (See paragraph 5-7 for slinging techniques.)

5-1. TREES

Trees are probably the most widely used of all natural anchors depending on the terrain and geographical region (Figure 5-1). However, trees must be carefully checked for suitability.

a. In rocky terrain, trees usually have a shallow root system. This can be checked by pushing or tugging on the tree to see how well it is rooted. Anchoring as low as possible to prevent excess leverage on the tree may be necessary.

b. Use padding on soft, sap producing trees to keep sap off ropes and slings.

Figure 5-1. Trees used as anchors.

5-2. BOULDERS

Boulders and rock nubbins make ideal anchors (Figure 5-2). The rock can be firmly tapped with a piton hammer to ensure it is solid. Sedimentary and other loose rock formations are not stable. Talus and scree fields are an indicator that the rock in the area is not solid. All areas around the rock formation that could cut the rope or sling should be padded.

Figure 5-2. Boulders used as anchors.

5-3. CHOCKSTONES

A chockstone is a rock that is wedged in a crack because the crack narrows downward (Figure 5-3). Chockstones should be checked for strength, security, and crumbling and should always be tested before use. All chockstones must be solid and strong enough to support the load. They must have maximum surface contact and be well tapered with the surrounding rock to remain in position.

a. Chockstones are often directional—they are secure when pulled in one direction but may pop out if pulled in another direction.

b. A creative climber can often make his own chockstone by wedging a rock into position, tying a rope to it, and clipping on a carabiner.

c. Slings should not be wedged between the chockstone and the rock wall since a fall could cut the webbing runner.

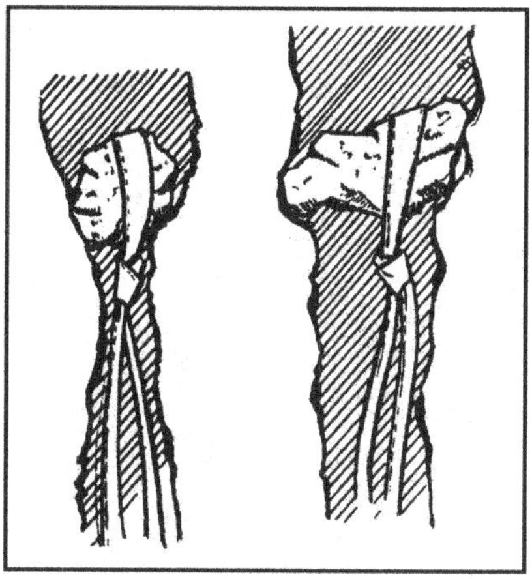

Figure 5-3. Chockstones.

5-4. ROCK PROJECTIONS

Rock projections (sometimes called nubbins) often provide suitable protection (Figure 5-4). These include blocks, flakes, horns, and spikes. If rock projections are used, their firmness is important. They should be checked for cracks or weathering that may impair their firmness. If any of these signs exist, the projection should be avoided.

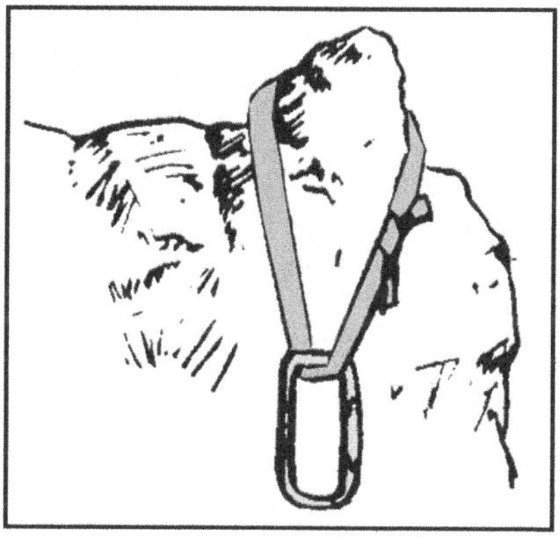

Figure 5-4. Rock projections.

5-5. TUNNELS AND ARCHES

Tunnels and arches are holes formed in solid rock and provide one of the more secure anchor points because they can be pulled in any direction. A sling is threaded through the opening hole and secured with a joining knot or girth hitch. The load-bearing hole must be strong and free of sharp edges (pad if necessary).

5-6. BUSHES AND SHRUBS

If no other suitable anchor is available, the roots of bushes can be used by routing a rope around the bases of several bushes (Figure 5-5). As with trees, the anchoring rope is placed as low as possible to reduce leverage on the anchor. All vegetation should be healthy and well rooted to the ground.

Figure 5-5. Bushes and shrubs.

5-7. SLINGING TECHNIQUES

Three methods are used to attach a sling to a natural anchor—drape, wrap, and girth. Whichever method is used, the knot is set off to the side where it will not interfere with normal carabiner movement. The carabiner gate should face away from the ground and open away from the anchor for easy insertion of the rope. When a locking carabiner cannot be used, two carabiners are used with gates opposed. Correctly opposed gates should open on opposite sides and form an "X" when opened (Figure 5-6).

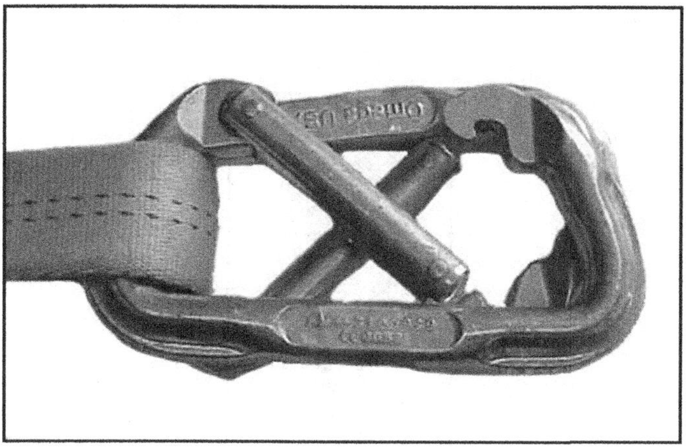

Figure 5-6. Correctly opposed carabiners.

a. **Drape.** Drape the sling over the anchor (Figure 5-7). Untying the sling and routing it around the anchor and then retying is still considered a drape.

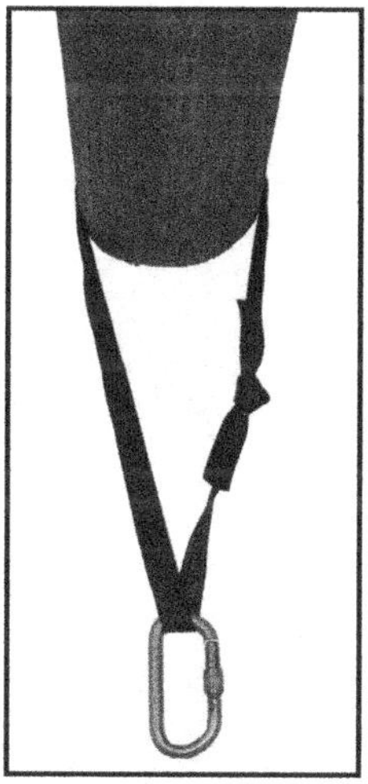

Figure 5-7. Drape.

b. **Wrap.** Wrap the sling around the anchor and connect the two ends together with a carabiner(s) or knot (Figure 5-8).

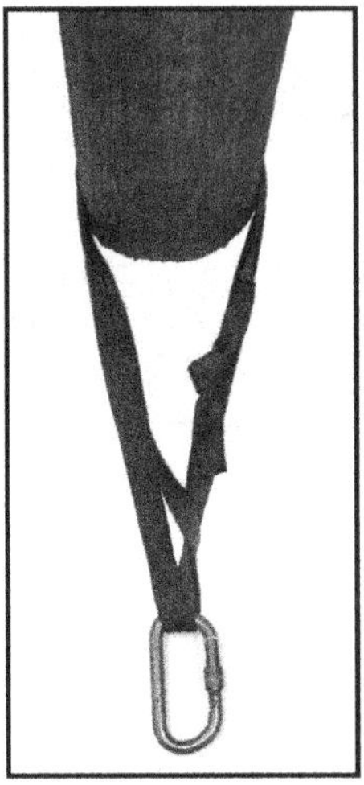

Figure 5-8. Wrap.

c. **Girth.** Tie the sling around the anchor with a girth hitch (Figure 5-9). Although a girth hitch reduces the strength of the sling, it allows the sling to remain in position and not slide on the anchor.

Figure 5-9. Girth.

Section II. ANCHORING WITH THE ROPE

The climbing or installation rope can be tied directly to the anchor using several different techniques. This requires less equipment, but also sacrifices some rope length to tie the anchor. The rope can be tied to the anchor using an appropriate anchor knot such as a bowline or a rerouted figure eight. Round turns can be used to help keep the rope in position on the anchor. A tensionless anchor can be used in high-load installations where tension on the attachment point and knot is undesirable.

5-8. ROPE ANCHOR

When tying the climbing or installation rope around an anchor, the knot should be placed approximately the same distance away from the anchor as the diameter of the anchor (Figure 5-10). The knot shouldn't be placed up against the anchor because this can stress and distort the knot under tension.

Figure 5-10. Rope tied to anchor with anchor knot.

5-9. TENSIONLESS ANCHOR

The tensionless anchor is used to anchor the rope on high-load installations such as bridging and traversing (Figure 5-11, page 5-8). The wraps of the rope around the anchor absorb the tension of the installation and keep the tension off the knot and carabiner. The anchor is usually tied with a minimum of four wraps, more if necessary, to absorb the tension. A smooth anchor may require several wraps, whereas a rough barked tree might only require a few. The rope is wrapped from top to bottom. A fixed loop is placed into the end of the rope and attached loosely back onto the rope with a carabiner.

Figure 5-11. Tensionless anchor.

Section III. ARTIFICIAL ANCHORS

Using artificial anchors becomes necessary when natural anchors are unavailable. The art of choosing and placing good anchors requires a great deal of practice and experience. Artificial anchors are available in many different types such as pitons, chocks, hexcentrics, and SLCDs. Anchor strength varies greatly; the type used depends on the terrain, equipment, and the load to be placed on it.

5-10. DEADMAN

A "deadman" anchor is any solid object buried in the ground and used as an anchor.

a. An object that has a large surface area and some length to it works best. (A hefty timber, such as a railroad tie, would be ideal.) Large boulders can be used, as well as a bundle of smaller tree limbs or poles. As with natural anchors, ensure timbers and tree limbs are not dead or rotting and that boulders are solid. Equipment, such as skis, ice axes, snowshoes, and ruck sacks, can also be used if necessary.

b. In extremely hard, rocky terrain (where digging a trench would be impractical, if not impossible) a variation of the deadman anchor can be constructed by building above the ground. The sling is attached to the anchor, which is set into the ground as deeply as possible. Boulders are then stacked on top of it until the anchor is strong enough for the load. Though normally not as strong as when buried, this method can work well for light-load installations as in anchoring a hand line for a stream crossing.

Note: Artificial anchors, such as pitons and bolts, are not widely accepted for use in all areas because of the scars they leave on the rock and the environment. Often they are left in place and become unnatural, unsightly fixtures in the natural environment. For training planning, local laws and courtesies should be taken into consideration for each area of operation.

5-11. PITONS

Pitons have been in use for over 100 years. Although still available, pitons are not used as often as other types of artificial anchors due primarily to their impact on the environment. Most climbers prefer to use chocks, SLCDs and other artificial anchors rather than pitons because they do not scar the rock and are easier to remove. Eye protection should always be worn when driving a piton into rock.

Note: The proper use and placement of pitons, as with any artificial anchor, should be studied, practiced, and tested while both feet are firmly on the ground and there is no danger of a fall.

a. **Advantages.** Some advantages in using pitons are:

- Depending on type and placement, pitons can support multiple directions of pull.
- Pitons are less complex than other types of artificial anchors.
- Pitons work well in thin cracks where other types of artificial anchors do not.

b. **Disadvantages.** Some disadvantages in using pitons are:

- During military operations, the distinct sound created when hammering pitons is a tactical disadvantage.
- Due to the expansion force of emplacing a piton, the rock could spread apart or break causing an unsafe condition.
- Pitons are more difficult to remove than other types of artificial anchors.
- Pitons leave noticeable scars on the rock.
- Pitons are easily dropped if not tied off when being used.

c. **Piton Placement.** The proper positioning or placement of pitons is critical. (Figure 5-12, page 5-10, shows examples of piton placement.) Usually a properly sized piton for a rock crack will fit one half to two thirds into the crack before being driven with the piton hammer. This helps ensure the depth of the crack is adequate for the size piton selected. As pitons are driven into the rock the pitch or sound that is made will change with each hammer blow, becoming higher pitched as the piton is driven in.

(1) Test the rock for soundness by tapping with the hammer. Driving pitons in soft or rotten rock is not recommended. When this type of rock must be used, clear the loose rock, dirt, and debris from the crack before driving the piton completely in.

(2) While it is being driven, attach the piton to a sling with a carabiner (an old carabiner should be used, if available) so that if the piton is knocked out of the crack, it will not be lost. The greater the resistance overcome while driving the piton, the firmer the anchor will be. The holding power depends on the climber placing the piton in a sound crack, and on the type of rock. The piton should not spread the rock, thereby loosening the emplacement.

Note: Pitons that have rings as attachment points might not display much change in sound as they are driven in as long as the ring moves freely.

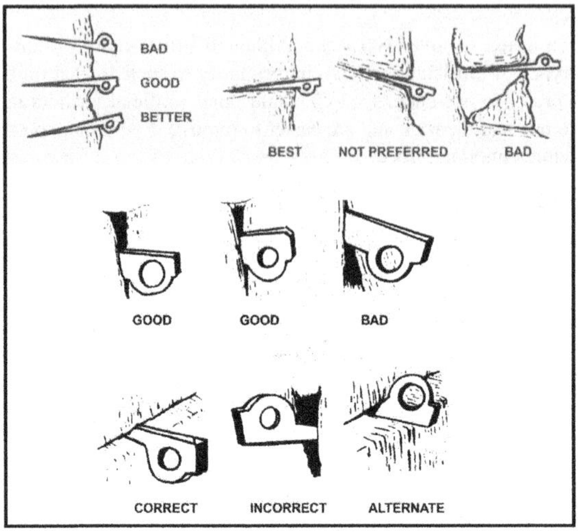

Figure 5-12. Examples of piton placements.

(3) Military mountaineers should practice emplacing pitons using either hand. Sometimes a piton cannot be driven completely into a crack, because the piton is too long. Therefore, it should be tied off using a hero-loop (an endless piece of webbing) (Figure 5-13). Attach this loop to the piton using a girth hitch at the point where the piton enters the rock so that the girth hitch is snug against the rock. Clip a carabiner into the loop.

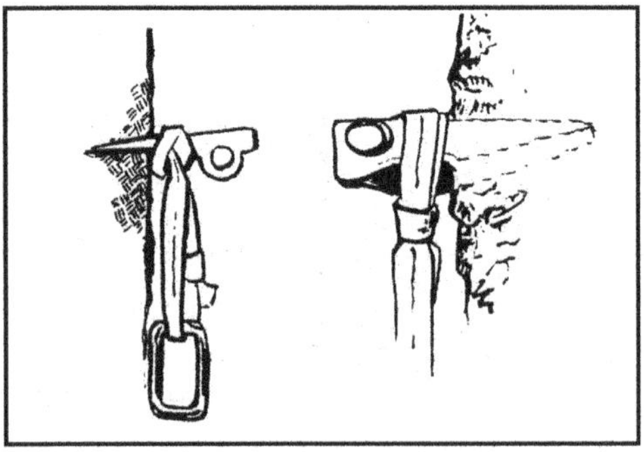

Figure 5-13. Hero-loop.

d. **Testing.** To test pitons pull up about 1 meter of slack in the climbing rope or use a sling. Insert this rope into a carabiner attached to the piton, then grasp the rope at least 1/2 meter from the carabiner. Jerk vigorously upward, downward, to each side, and then outward while observing the piton for movement. Repeat these actions as many times as necessary. Tap the piton to determine if the pitch has changed. If the pitch has changed greatly, drive the piton in as far as possible. If the sound regains its original pitch, the emplacement is probably safe. If the piton shows any sign of moving or if, upon driving it, there is any question of its soundness, drive it into another place. Try to be in a secure position before testing. This procedure is intended for use in testing an omni-directional anchor (one that withstands a pull in any direction). When a directional anchor (pull in one direction) is used, as in most free and direct-aid climbing situations, and when using chocks, concentrate the test in the direction that force will be applied to the anchor.

e. **Removing Pitons.** Attach a carabiner and sling to the piton before removal to eliminate the chance of dropping and losing it. Tap the piton firmly along the axis of the crack in which it is located. Alternate tapping from both sides while applying steady pressure. Pulling out on the attached carabiner eventually removes the piton (Figure 5-14).

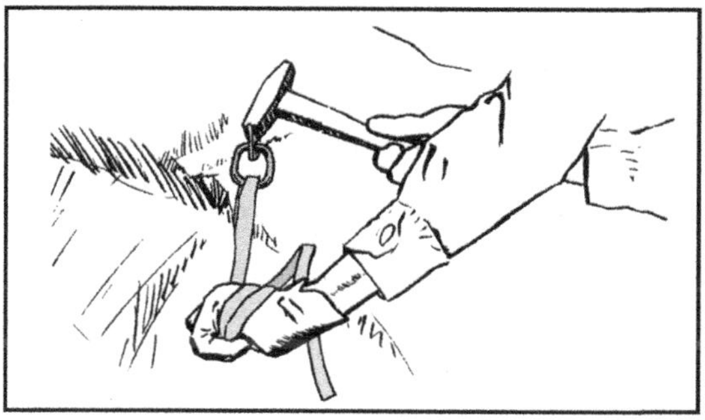

Figure 5-14. Piton removal.

f. **Reusing Pitons.** Soft iron pitons that have been used, removed, and straightened may be reused, but they must be checked for strength. In training areas, pitons already in place should not be trusted since weather loosens them in time. Also, they may have been driven poorly the first time. Before use, test them as described above and drive them again until certain of their soundness.

5-12. CHOCKS

Chock craft has been in use for many decades. A natural chockstone, having fallen and wedged in a crack, provides an excellent anchor point. Sometimes these chockstones are in unstable positions, but can be made into excellent anchors with little adjustment. Chock craft is an art that requires time and technique to master—simple in theory, but complex in practice. Imagination and resourcefulness are key principles to chock craft. The skilled climber must understand the application of mechanical advantage, vectors, and other forces that affect the belay chain in a fall.

a. **Advantages.** The advantages of using chocks are:

- Tactically quiet installation and recovery.
- Usually easy to retrieve and, unless severely damaged, are reusable.
- Light to carry.
- Easy to insert and remove.
- Minimal rock scarring as opposed to pitons.
- Sometimes can be placed where pitons cannot (expanding rock flakes where pitons would further weaken the rock).

b. **Disadvantages.** The disadvantages of using chocks are:

- May not fit in thin cracks, which may accept pitons.
- Often provide only one direction of pull.
- Practice and experience necessary to become proficient in proper placement.

c. **Placement.** The principles of placing chocks are to find a crack with a constriction at some point, place a chock of appropriate size above and behind the constriction, and set the chock by jerking down on the chock loop (Figure 5-15). Maximum surface contact with a tight fit is critical. Chocks are usually good for a single direction of pull.

(1) Avoid cracks that have crumbly (soft) or deteriorating rock, if possible. Some cracks may have loose rock, grass, and dirt, which should be removed before placing the chock. Look for a constriction point in the crack, then select a chock to fit it.

(2) When selecting a chock, choose one that has as much surface area as possible in contact with the rock. A chock resting on one small crystal or point of rock is likely to be unsafe. A chock that sticks partly out of the crack is avoided. Avoid poor protection. Ensure that the chock has a wire or runner long enough; extra ropes, cord, or webbing may be needed to extend the length of the runner.

(3) End weighting of the placement helps to keep the protection in position. A carabiner often provides enough weight

(4) Parallel-sided cracks without constrictions are a problem. Chocks designed to be used in this situation rely on camming principles to remain emplaced. Weighting the emplacement with extra hardware is often necessary to keep the chocks from dropping out.

(a) Emplace the wedge-shaped chock above and behind the constriction; seat it with a sharp downward tug.

(b) Place a camming chock with its narrow side into the crack, then rotate it to the attitude it will assume under load; seat it with a sharp downward tug.

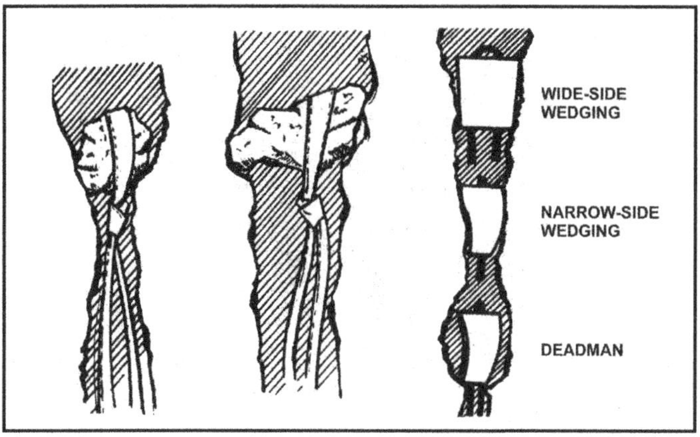

Figure 5-15. Chock placements.

d. **Testing.** After seating a chock, test it to ensure it remains in place. A chock that falls out when the climber moves past it is unsafe and offers no protection. To test it, firmly pull the chock in every anticipated direction of pull. Some chock placements fail in one or more directions; therefore, use pairs of chocks in opposition.

5-13. SPRING-LOADED CAMMING DEVICE

The SLCD offers quick and easy placement of artificial protection. It is well suited in awkward positions and difficult placements, since it can be emplaced with one hand. It can usually be placed quickly and retrieved easily (Figure 5-16, page 5-14).

a. To emplace an SLCD hold the device in either hand like a syringe, pull the retractor bar back, place the device into a crack, and release the retractor bar. The SLCD holds well in parallel-sided hand- and fist-sized cracks. Smaller variations are available for finger-sized cracks.

b. Careful study of the crack should be made before selecting the device for emplacement. It should be placed so that it is aligned in the direction of force applied to it. It should not be placed any deeper than is needed for secure placement, since it may be impossible to reach the extractor bar for removal. An SLCD should be extended with a runner and placed so that the direction of pull is parallel to the shaft; otherwise, it may rotate and pull out. The versions that have a semi-rigid wire cable shaft allow for greater flexibility and usage, without the danger of the shaft snapping off in a fall.

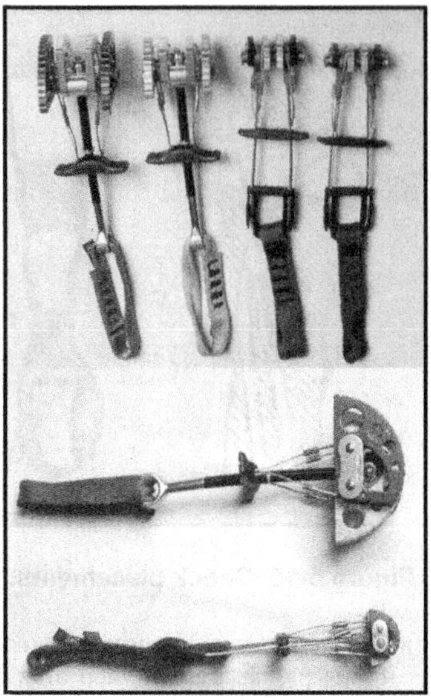

Figure 5-16. SLCD placements.

5-14. BOLTS

Bolts are often used in fixed-rope installations and in aid climbing where cracks are not available.

a. Bolts provide one of the most secure means of establishing protection. The rock should be inspected for evidence of crumbling, flaking, or cracking, and should be tested with a hammer. Emplacing a bolt with a hammer and a hand drill is a time-consuming and difficult process that requires drilling a hole in the rock deeper than the length of the bolt. This normally takes more than 20 minutes for one hole. Electric or even gas-powered drills can be used to greatly shorten drilling time. However, their size and weight can make them difficult to carry on the climbing route.

b. A hanger (carrier) and nut are placed on the bolt, and the bolt is inserted and then driven into the hole. A climber should never hammer on a bolt to test or "improve" it, since this permanently weakens it. Bolts should be used with carriers, carabiners, and runners.

c. When using bolts, the climber uses a piton hammer and hand drill with a masonry bit for drilling holes. Some versions are available in which the sleeve is hammered and turned into the rock (self-drilling), which bores the hole. Split bolts and expanding sleeves are common bolts used to secure hangers and carriers (Figure 5-17). Surgical tubing is useful in blowing dust out of the holes. Nail type bolts are emplaced by driving the nail with a hammer to expand the sleeve against the wall of the drilled hole. Safety glasses should always be worn when emplacing bolts.

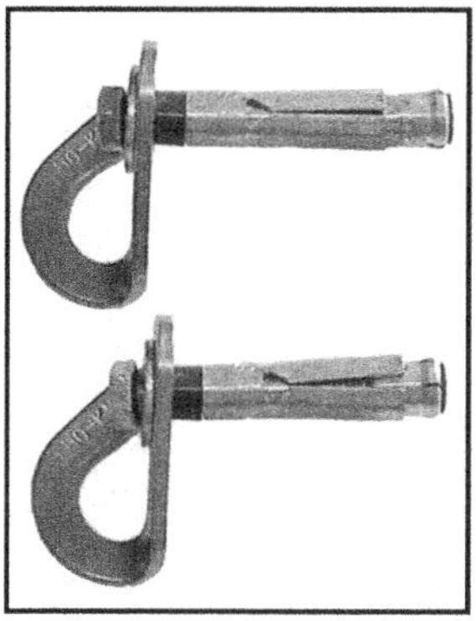

Figure 5-17. Bolt with expanding sleeve.

5-15. EQUALIZING ANCHORS

Equalizing anchors are made up of more than one anchor point joined together so that the intended load is shared equally. This not only provides greater anchor strength, but also adds redundancy or backup because of the multiple points.

a. **Self-equalizing Anchor.** A self-equalizing anchor will maintain an equal load on each individual point as the direction of pull changes (Figure 5-18). This is sometimes used in rappelling when the route must change left or right in the middle of the rappel. A self-equalizing anchor should only be used when necessary because if any one of the individual points fail, the anchor will extend and shock-load the remaining points or even cause complete anchor failure.

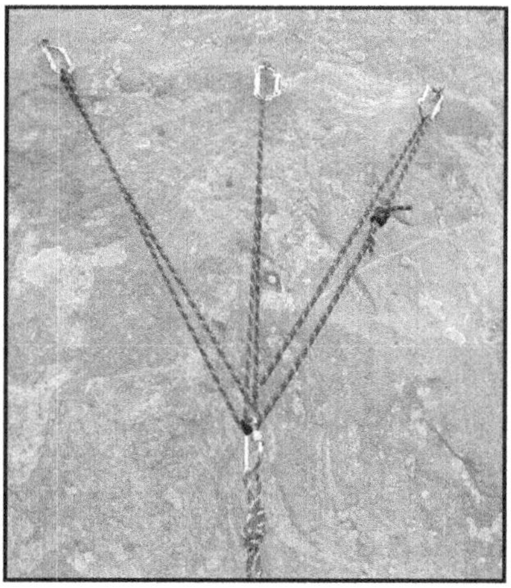

Figure 5-18. Self-equalizing anchors.

b. **Pre-equalized Anchor.** A pre-equalized anchor distributes the load equally to each individual point (Figure 5-19). It is aimed in the direction of the load. A pre-equalized anchor prevents extension and shock-loading of the anchor if an individual point fails. An anchor is pre-equalized by tying an overhand or figure-eight knot in the webbing or sling.

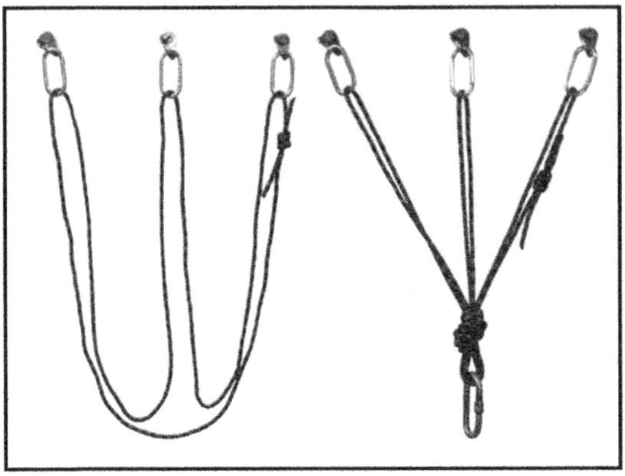

Figure 5-19. Pre-equalized anchor.

Note: When using webbing or slings, the angles of the webbing or slings directly affect the load placed on an anchor. An angle greater than 90 degrees can result in anchor failure (Figure 5-20).

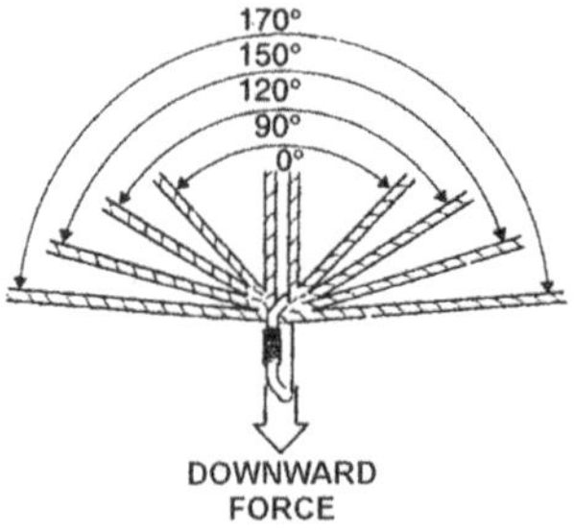

TENSION IN MULTIPLE ANCHOR RIGGING	
ANGLE	RESULTING LEG TENSION*
170°	1150%
150°	200%
120°	100%
90°	70%
0°	50%

**On each leg relative to downward force.*

Figure 5-20. Effects of angles on an anchor.

CHAPTER 6
CLIMBING

A steep rock face is a terrain feature that can be avoided most of the time through prior planning and good route selection. Rock climbing can be time consuming, especially for a larger unit with a heavy combat load. It can leave the climbing party totally exposed to weather, terrain hazards, and the enemy for the length of the climb.

Sometimes steep rock cannot be avoided. Climbing relatively short sections of steep rock (one or two pitches) may prove quicker and safer than using alternate routes. A steep rock route would normally be considered an unlikely avenue of approach and, therefore, might be weakly defended or not defended at all.

All personnel in a unit preparing for deployment to mountainous terrain should be trained in the basics of climbing. Forward observers, reconnaissance personnel, and security teams are a few examples of small units who may require rock climbing skills to gain their vantage points in mountainous terrain. Select personnel demonstrating the highest degree of skill and experience should be trained in roped climbing techniques. These personnel will have the job of picking and "fixing" the route for the rest of the unit.

Rock climbing has evolved into a specialized "sport" with a wide range of varying techniques and styles. This chapter focuses on the basics most applicable to military operations.

Section I. CLIMBING FUNDAMENTALS

A variety of refined techniques are used to climb different types of rock formations. The foundation for all of these styles is the art of climbing. Climbing technique stresses climbing with the weight centered over the feet, using the hands primarily for balance. It can be thought of as a combination of the balanced movement required to walk a tightrope and the technique used to ascend a ladder. No mountaineering equipment is required; however, the climbing technique is also used in roped climbing.

6-1. ROUTE SELECTION

The experienced climber has learned to climb with the "eyes." Even before getting on the rock, the climber studies all possible routes, or "lines," to the top looking for cracks, ledges, nubbins, and other irregularities in the rock that will be used for footholds and handholds, taking note of any larger ledges or benches for resting places. When picking the line, he mentally climbs the route, rehearsing the step-by-step sequence of movements that will be required to do the climb, ensuring himself that the route has an adequate number of holds and the difficulty of the climb will be well within the limit of his ability.

6-2. TERRAIN SELECTION FOR TRAINING

Route selection for military climbing involves picking the easiest and quickest possible line for all personnel to follow. However, climbing skill and experience can only be developed by increasing the length and difficulty of routes as training progresses. In the training environment, beginning lessons in climbing should be performed CLOSE to the ground on lower-angled rock with plenty of holds for the hands and feet. Personnel not climbing can act as "spotters" for those climbing. In later lessons, a "top-rope" belay can be used for safety, allowing the individual to increase the length and difficulty of the climb under the protection of the climbing rope.

6-3. PREPARATION

In preparation for climbing, the boot soles should be dry and clean. A small stick can be used to clean out dirt and small rocks that might be caught between the lugs of the boot sole. If the soles are wet or damp, dry them off by stomping and rubbing the soles on clean, dry rock. All jewelry should be removed from the fingers. Watches and bracelets can interfere with hand placements and may become damaged if worn while climbing. Helmets should be worn to protect the head from injury if an object, such as a rock or climbing gear, falls from climbers above. Most climbing helmets are not designed to provide protection from impact to the head if the wearer falls, but will provide a minimal amount of protection if a climber comes in contact with the rock during climbing.

CAUTION

Rings can become caught on rock facial features and or lodged into cracks, which could cause injuries during a slip or fall.

6-4. SPOTTING

Spotting is a technique used to add a level of safety to climbing without a rope. A second man stands below and just outside of the climbers fall path and helps (spots) the climber to land safely if he should fall. Spotting is only applicable if the climber is not going above the spotters head on the rock. Beyond that height a roped climbing should be conducted. If an individual climbs beyond the effective range of the spotter(s), he has climbed TOO HIGH for his own safety. The duties of the spotter are to help prevent the falling climber from impacting the head and or spine, help the climber land feet first, and reduce the impact of a fall.

CAUTION

The spotter should not catch the climber against the rock because additional injuries could result. If the spotter pushes the falling climber into the rock, deep abrasions of the skin or knee may occur. Ankle joints could be twisted by the fall if the climber's foot remained high on the rock. The spotter might be required to fully support the weight of the climber causing injury to the spotter.

6-5. CLIMBING TECHNIQUE

Climbing involves linking together a series of movements based on foot and hand placement, weight shift, and movement. When this series of movements is combined correctly, a smooth climbing technique results. This technique reduces excess force on the limbs, helping to minimize fatigue. The basic principle is based on the five body parts described here.

a. **Five Body Parts.** The five body parts used for climbing are the right hand, left hand, right foot, left foot, and body (trunk). The basic principle to achieve smooth climbing is to move only one of the five body parts at a time. The trunk is not moved in conjunction with a foot or in conjunction with a hand, a hand is not moved in conjunction with a foot, and so on. Following this simple technique forces both legs to do all the lifting simultaneously.

b. **Stance or Body Position.** Body position is probably the single most important element to good technique. A relaxed, comfortable stance is essential. (Figure 6-1 shows a correct climbing stance, and Figure 6-2, page 6-4, shows an incorrect stance.) The body should be in a near vertical or erect stance with the weight centered over the feet. Leaning in towards the rock will cause the feet to push outward, away from the rock, resulting in a loss of friction between the boot sole and rock surface. The legs are straight and the heels are kept low to reduce fatigue. Bent legs and tense muscles tire quickly. If strained for too long, tense muscles may vibrate uncontrollably. This vibration, known as "Elvis-ing" or "sewing-machine leg" can be cured by straightening the leg, lowering the heel, or moving on to a more restful position. The hands are used to maintain balance. Keeping the hands between waist and shoulder level will reduce arm fatigue.

Figure 6-1. Correct climbing stance—balanced over both feet.

Figure 6-2. Incorrect stance—stretched out.

(1) Whenever possible, three points of contact are maintained with the rock. Proper positioning of the hips and shoulders is critical. When using two footholds and one handhold, the hips and shoulders should be centered over both feet. In most cases, as the climbing progresses, the body is resting on one foot with two handholds for balance. The hips and shoulders must be centered over the support foot to maintain balance, allowing the "free" foot to maneuver.

(2) The angle or steepness of the rock also determines how far away from the rock the hips and shoulders should be. On low-angle slopes, the hips are moved out away from the rock to keep the body in balance with the weight over the feet. The shoulders can be moved closer to the rock to reach handholds. On steep rock, the hips are pushed closer to the rock. The shoulders are moved away from the rock by arching the back. The body is still in balance over the feet and the eyes can see where the hands need to go. Sometimes, when footholds are small, the hips are moved back to increase friction between the foot and the rock. This is normally done on quick, intermediate holds. It should be avoided in the rest position as it places more weight on the arms and hands. When weight must be placed on handholds, the arms should be kept straight to reduce fatigue. Again, flexed muscles tire quickly.

c. **Climbing Sequence.** The steps defined below provide a complete sequence of events to move the entire body on the rock. These are the basic steps to follow for a smooth climbing technique. Performing these steps in this exact order will not always be necessary because the nature of the route will dictate the availability of hand and foot placements. The basic steps are weight, shift, and movement (movement being either the foot, hand, or body). (A typical climbing sequence is shown in Figure 6-3, pages 6-6 through 6-8.)

STEP ONE: Shift the weight from both feet to one foot. This will allow lifting of one foot with no effect on the stance.

STEP TWO:	Lift the unweighted foot and place it in a new location, within one to two feet of the starting position, with no effect on body position or balance (higher placement will result in a potentially higher lift for the legs to make, creating more stress, and is called a high step) The trunk does not move during foot movement.
STEP THREE:	Shift the weight onto both feet. (Repeat steps 1 through 3 for remaining foot.)
STEP FOUR:	Lift the body into a new stance with both legs.
STEP FIVE:	Move one hand to a new position between waist and head height. During this movement, the trunk should be completely balanced in position and the removed hand should have no effect on stability.
STEP SIX:	Move the remaining hand as in Step 5.

Now the entire body is in a new position and ready to start the process again. Following these steps will prevent lifting with the hands and arms, which are used to maintain stance and balance. If both legs are bent, leg extension can be performed as soon as one foot has been moved. Hand movements can be delayed until numerous foot movements have been made, which not only creates shorter lifts with the legs, but may allow a better choice for the next hand movements because the reach will have increased.

Figure 6-3. Typical climbing sequence.

Figure 6-3. Typical climbing sequence (continued).

RIGHT HAND MOVED

Figure 6-3. Typical climbing sequence (continued).

(1) Many climbers will move more than one body part at a time, usually resulting in lifting the body with one leg or one leg and both arms. This type of lifting is inefficient, requiring one leg to perform the work of two or using the arms to lift the body. Proper climbing technique is lifting the body with the legs, not the arms, because the legs are much stronger.

(2) When the angle of the rock increases, these movements become more critical. Holding or pulling the body into the rock with the arms and hands may be necessary as the angle increases (this is still not lifting with the arms). Many climbing routes have angles greater than ninety degrees (overhanging) and the arms are used to support partial body weight. The same technique applies even at those angles.

(3) The climber should avoid moving on the knees and elbows. Other than being uncomfortable, even painful, to rest on, these bony portions of the limbs offer little friction and "feel" on the rock.

6-6. SAFETY PRECAUTIONS

The following safety precautions should be observed when rock climbing.

a. While ascending a seldom or never traveled route, you may encounter precariously perched rocks. If the rock will endanger your second, it may be possible to remove it from the route and trundle it, tossing it down. This is extremely dangerous to climbers below and should not be attempted unless you are absolutely sure no men are below. If you are not sure that the flight path is clear, do not do it. Never dislodge loose rocks carelessly. Should a rock become loose accidentally, immediately shout the warning "ROCK" to alert climbers below. Upon hearing the warning, personnel should seek immediate cover behind any rock bulges or overhangs available, or flatten themselves against the rock to minimize exposure.

b. Should a climber fall, he should do his utmost to maintain control and not panic. If on a low-angle climb, he may be able to arrest his own fall by staying in contact with the rock, grasping for any possible hold available. He should shout the warning "FALLING" to alert personnel below.

CAUTION

Grasping at the rock in a fall can result in serious injuries to the upper body. If conducting a roped climb, let the rope provide protection.

c. When climbing close to the ground and without a rope, a spotter can be used for safety. The duties of the spotter are to ensure the falling climber does not impact the head or spine, and to reduce the impact of a fall.

d. Avoid climbing directly above or below other climbers (with the exception of spotters). When personnel must climb at the same time, following the same line, a fixed rope should be installed.

e. Avoid climbing with gloves on because of the decreased "feel" for the rock. The use of gloves in the training environment is especially discouraged, while their use in the mountains is often mandatory when it is cold. A thin polypropylene or wool glove is best for rock climbing, although heavier cotton or leather work gloves are often used for belaying.

f. Be extremely careful when climbing on wet or moss-covered rock; friction on holds is greatly reduced.

g. Avoid grasping small vegetation for handholds; the root systems can be shallow and will usually not support much weight.

6-7. MARGIN OF SAFETY

Besides observing the standard safety precautions, the climber can avoid catastrophe by climbing with a wide margin of safety. The margin of safety is a protective buffer the climber places between himself and potential climbing hazards. Both subjective (personnel-related) and objective (environmental) hazards must be considered when applying the margin of safety. The leader must apply the margin of safety taking into account the strengths and weaknesses of the entire team or unit.

a. When climbing, the climber increases his margin of safety by selecting routes that are well within the limit of his ability. When leading a group of climbers, he selects a route well within the ability of the weakest member.

b. When the rock is wet, or when climbing in other adverse weather conditions, the climber's ability is reduced and routes are selected accordingly. When the climbing becomes difficult or exposed, the climber knows to use the protection of the climbing rope and belays. A lead climber increases his margin of safety by placing protection along the route to limit the length of a potential fall.

Section II. USE OF HOLDS

The climber should check each hold before use. This may simply be a quick, visual inspection if he knows the rock to be solid. When in doubt, he should grab and tug on the hold to test it for soundness BEFORE depending on it. Sometimes, a hold that appears weak can actually be solid as long as minimal force is applied to it, or the force is applied in a direction that strengthens it. A loose nubbin might not be strong enough to support the climber's weight, but it may serve as an adequate handhold. Be especially careful when climbing on weathered, sedimentary-type rock.

6-8. CLIMBING WITH THE FEET

"Climb with the feet and use the hands for balance" is extremely important to remember. In the early learning stages of climbing, most individuals will rely heavily on the arms, forgetting to use the feet properly. It is true that solid handholds and a firm grip are needed in some combination techniques; however, even the most strenuous techniques require good footwork and a quick return to a balanced position over one or both feet. Failure to climb any route, easy or difficult, is usually the result of poor footwork.

a. The beginning climber will have a natural tendency to look up for handholds. Try to keep the hands low and train your eyes to look down for footholds. Even the smallest irregularity in the rock can support the climber once the foot is positioned properly and weight is committed to it.

b. The foot remains on the rock as a result of friction. Maximum friction is obtained from a correct stance over a properly positioned foot. The following describes a few ways the foot can be positioned on the rock to maximize friction.

(1) ***Maximum Sole Contact.*** The principle of using full sole contact, as in mountain walking, also applies in climbing. Maximum friction is obtained by placing as much of the boot sole on the rock as possible. Also, the leg muscles can relax the most when the entire foot is placed on the rock. (Figure 6-4 shows examples of maximum and minimum sole contact.)

(a) Smooth, low-angled rock (slab) and rock containing large "bucket" holds and ledges are typical formations where the entire boot sole should be used.

(b) On some large holds, like bucket holds that extend deep into the rock, the entire foot cannot be used. The climber may not be able to achieve a balanced position if the foot is stuck too far underneath a bulge in the rock. In this case, placing only part of the foot on the hold may allow the climber to achieve a balanced stance. The key is to use as much of the boot sole as possible. Remember to keep the heels low to reduce strain on the lower leg muscles.

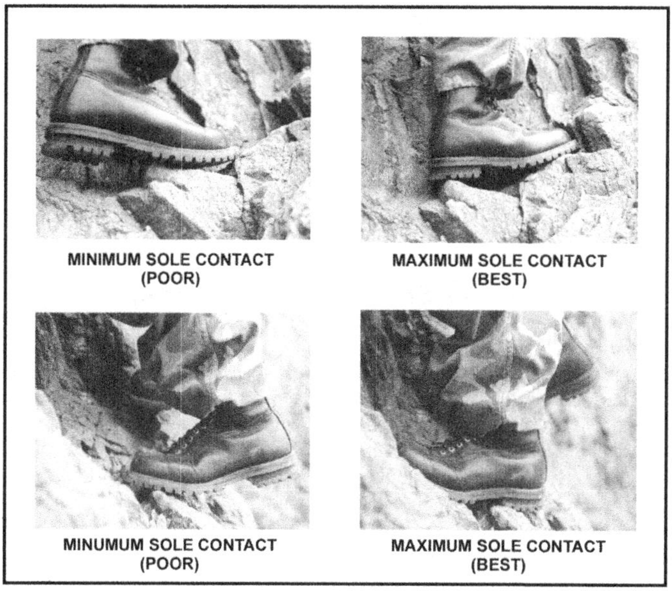

Figure 6-4. Examples of maximum and minimum sole contact.

(2) ***Edging.*** The edging technique is used where horizontal crack systems and other irregularities in the rock form small, well-defined ledges. The edge of the boot sole is placed on the ledge for the foothold. Usually, the inside edge of the boot or the edge area around the toes is used. Whenever possible, turn the foot sideways and use the entire inside edge of the boot. Again, more sole contact equals more friction and the legs can rest more when the heel is on the rock. (Figure 6-5, page 6-12, shows examples of the edging technique.)

(a) On smaller holds, edging with the front of the boot, or toe, may be used. Use of the toe is most tiring because the heel is off the rock and the toes support the climber's weight. Remember to keep the heel low to reduce fatigue. Curling and stiffening the toes in the boot increases support on the hold. A stronger position is usually obtained on small ledges by turning the foot at about a 45-degree angle, using the strength of the big toe and the ball of the foot.

(b) Effective edging on small ledges requires stiff-soled footwear. The stiffer the sole, the better the edging capability. Typical mountain boots worn by the US military have a relatively flexible lugged sole and, therefore, edging ability on smaller holds will be somewhat limited.

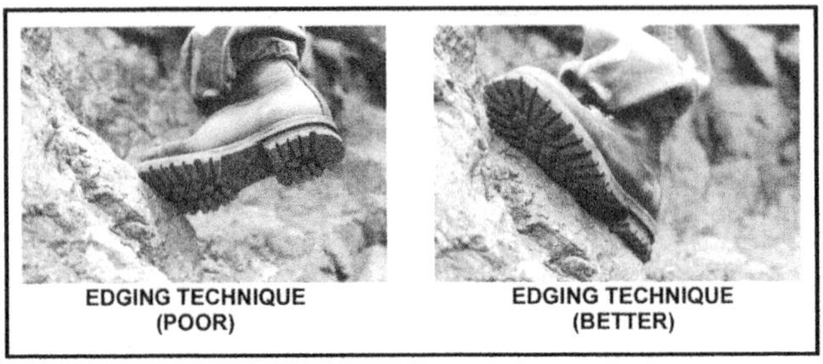

Figure 6-5. Examples of edging technique.

(3) ***Smearing.*** When footholds are too small to use a good edging technique, the ball of the foot can be "smeared" over the hold. The smearing technique requires the boot to adhere to the rock by deformation of the sole and by friction. Rock climbing shoes are specifically designed to maximize friction for smearing; some athletic shoes also work well. The Army mountain boot, with its softer sole, usually works better for smearing than for edging. Rounded, down-sloping ledges and low-angled slab rock often require good smearing technique. (Figure 6-6 shows examples of the smearing technique.)

(a) Effective smearing requires maximum friction between the foot and the rock. Cover as much of the hold as possible with the ball of the foot. Keeping the heel low will not only reduce muscle strain, but will increase the amount of surface contact between the foot and the rock.

(b) Sometimes flexing the ankles and knees slightly will place the climber's weight more directly over the ball of the foot and increase friction; however, this is more tiring and should only be used for quick, intermediate holds. The leg should be kept straight whenever possible.

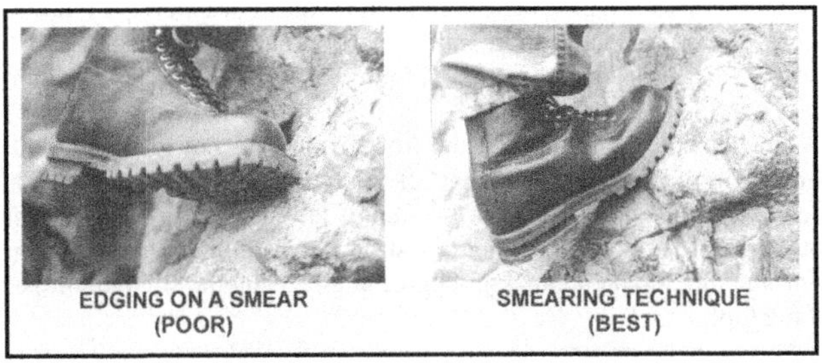

Figure 6-6. Examples of the smearing technique.

(4) ***Jamming.*** The jamming technique works on the same principal as chock placement. The foot is set into a crack in such a way that it "jams" into place, resisting a downward pull. The jamming technique is a specialized skill used to climb vertical or

near vertical cracks when no other holds are available on the rock face. The technique is not limited to just wedging the feet; fingers, hands, arms, even the entire leg or body are all used in the jamming technique, depending on the size of the crack. Jam holds are described in this text to broaden the range of climbing skills. Jamming holds can be used in a crack while other hand/foot holds are used on the face of the rock. Many cracks will have facial features, such as edges, pockets, and so on, inside and within reach. Always look or feel for easier to use features. (Figure 6-7 shows examples of jamming.)

(a) The foot can be jammed in a crack in different ways. It can be inserted above a constriction and set into the narrow portion, or it can be placed in the crack and turned, like a camming device, until it locks in place tight enough to support the climber's weight. Aside from these two basic ideas, the possibilities are endless. The toes, ball of the foot, or the entire foot can be used. Try to use as much of the foot as possible for maximum surface contact. Some positions are more tiring, and even more painful on the foot, than others. Practice jamming the foot in various ways to see what offers the most secure, restful position.

(b) Some foot jams may be difficult to remove once weight has been committed to them, especially if a stiffer sole boot is used. The foot is less likely to get stuck when it is twisted or "cammed" into position. When removing the boot from a crack, reverse the way it was placed to prevent further constriction.

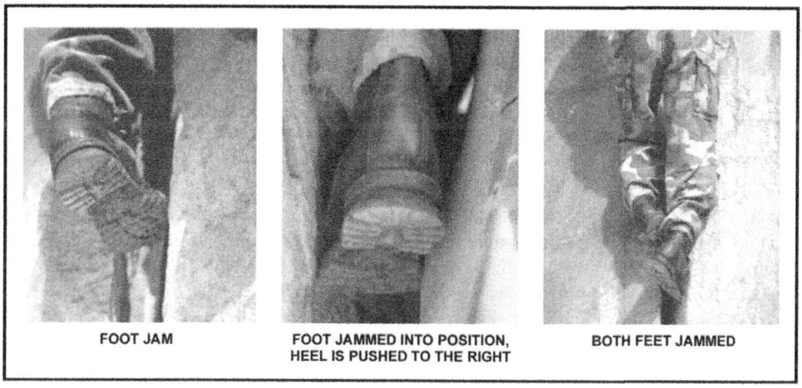

Figure 6-7. Examples of jamming.

6-9. USING THE HANDS

The hands can be placed on the rock in many ways. Exactly how and where to position the hands and arms depends on what holds are available, and what configuration will best support the current stance as well as the movement to the next stance. Selecting handholds between waist and shoulder level helps in different ways. Circulation in the arms and hands is best when the arms are kept low. Secondly, the climber has less tendency to "hang" on his arms when the handholds are at shoulder level and below. Both of these contribute to a relaxed stance and reduce fatigue in the hands and arms.

a. As the individual climbs, he continually repositions his hands and arms to keep the body in balance, with the weight centered over the feet. On lower-angled rock, he may simply need to place the hands up against the rock and extend the arm to maintain balance; just like using an ice ax as a third point of contact in mountain walking. Sometimes, he will be able to push directly down on a large hold with the palm of the hand. More often though, he will need to "grip" the rock in some fashion and then push or pull against the hold to maintain balance.

b. As stated earlier, the beginner will undoubtedly place too much weight on the hands and arms. If we think of ourselves climbing a ladder, our body weight is on our legs. Our hands grip, and our arms pull on each rung only enough to maintain our balance and footing on the ladder. Ideally, this is the amount of grip and pull that should be used in climbing. Of course, as the size and availability of holds decreases, and the steepness of the rock approaches the vertical, the grip must be stronger and more weight might be placed on the arms and handholds for brief moments. The key is to move quickly from the smaller, intermediate holds to the larger holds where the weight can be placed back on the feet allowing the hands and arms to relax. The following describes some of the basic handholds and how the hand can be positioned to maximize grip on smaller holds.

(1) ***Push Holds.*** Push holds rely on the friction created when the hand is pushed against the rock. Most often a climber will use a push hold by applying "downward pressure" on a ledge or nubbin. This is fine, and works well; however, the climber should not limit his use of push holds to the application of down pressure. Pushing sideways, and on occasion, even upward on less obvious holds can prove quite secure. Push holds often work best when used in combination with other holds. Pushing in opposite directions and "push-pull" combinations are excellent techniques. (Figure 6-8 shows examples of push holds.)

(a) An effective push hold does not necessarily require the use of the entire hand. On smaller holds, the side of the palm, the fingers, or the thumb may be all that is needed to support the stance. Some holds may not feel secure when the hand is initially placed on them. The hold may improve or weaken during the movement. The key is to try and select a hold that will improve as the climber moves past it.

(b) Most push holds do not require much grip; however, friction might be increased by taking advantage of any rough surfaces or irregularities in the rock. Sometimes the strength of the hold can be increased by squeezing, or "pinching," the rock between the thumb and fingers (see paragraph on pinch holds).

Figure 6-8. Examples of push holds.

(2) ***Pull Holds.*** Pull holds, also called "cling holds," which are grasped and pulled upon, are probably the most widely used holds in climbing. Grip plays more of a role in a pull hold, and, therefore, it normally feels more secure to the climber than a push hold. Because of this increased feeling of security, pull holds are often overworked. These are the holds the climber has a tendency to hang from. Most pull holds do not require great strength, just good technique. Avoid the "death grip" syndrome by climbing with the feet. (Figure 6-9, page 6-16, shows examples of pull holds.)

(a) Like push holds, pressure on a pull hold can be applied straight down, sideways, or upward. Again, these are the holds the climber tends to stretch and reach for, creating an unbalanced stance. Remember to try and keep the hands between waist and shoulder level, making use of intermediate holds instead of reaching for those above the head.

(b) Pulling sideways on vertical cracks can be very secure. There is less tendency to hang from "side-clings" and the hands naturally remain lower. The thumb can often push against one side of the crack, in opposition to the pull by the fingers, creating a stronger hold. Both hands can also be placed in the same crack, with the hands pulling in opposite directions. The number of possible combinations is limited only by the imagination and experience of the climber.

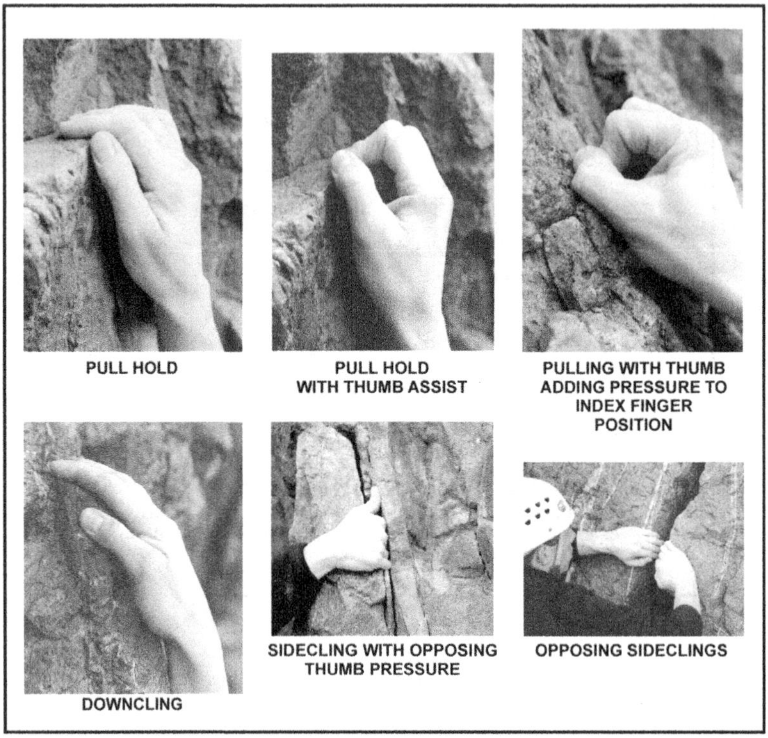

Figure 6-9. Examples of pull holds.

(c) Friction and strength of a pull hold can be increased by the way the hand grips the rock. Normally, the grip is stronger when the fingers are closed together; however, sometimes more friction is obtained by spreading the fingers apart and placing them between irregularities on the rock surface. On small holds, grip can often be improved by bending the fingers upward, forcing the palm of the hand to push against the rock. This helps to hold the finger tips in place and reduces muscle strain in the hand. Keeping the forearm up against the rock also allows the arm and hand muscles to relax more.

(d) Another technique that helps to strengthen a cling hold for a downward pull is to press the thumb against the side of the index finger, or place it on top of the index finger and press down. This hand configuration, known as a "ring grip," works well on smaller holds.

(3) ***Pinch Holds.*** Sometimes a small nubbin or protrusion in the rock can be "squeezed" between the thumb and fingers. This technique is called a pinch hold. Friction is applied by increasing the grip on the rock. Pinch holds are often overlooked by the

novice climber because they feel insecure at first and cannot be relied upon to support much body weight. If the climber has his weight over his feet properly, the pinch hold will work well in providing balance. The pinch hold can also be used as a gripping technique for push holds and pull holds. (Figure 6-10 shows examples of pinch holds.)

PINCH HOLD | PINCH HOLD | LARGE PINCH HOLD

Figure 6-10. Examples of pinch holds.

(4) ***Jam Holds.*** Like foot jams, the fingers and hands can be wedged or cammed into a crack so they resist a downward or outward pull. Jamming with the fingers and hands can be painful and may cause minor cuts and abrasions to tender skin. Cotton tape can be used to protect the fingertips, knuckles, and the back of the hand; however, prolonged jamming technique requiring hand taping should be avoided. Tape also adds friction to the hand in jammed position. (Figure 6-11, page 6-18, shows examples of jam holds.)

(a) The hand can be placed in a crack a number of ways. Sometimes an open hand can be inserted and wedged into a narrower portion of the crack. Other times a clenched fist will provide the necessary grip. Friction can be created by applying cross pressure between the fingers and the back of the hand. Another technique for vertical cracks is to place the hand in the crack with the thumb pointed either up or down. The hand is then clenched as much as possible. When the arm is straightened, it will twist the hand and tend to cam it into place. This combination of clenching and camming usually produces the most friction, and the most secure hand jam in vertical cracks.

(b) In smaller cracks, only the fingers will fit. Use as many fingers as the crack will allow. The fingers can sometimes be stacked in some configuration to increase friction. The thumb is usually kept outside the crack in finger jams and pressed against the rock to increase friction or create cross pressure. In vertical cracks it is best to insert the fingers with the thumb pointing down to make use of the natural camming action of the fingers that occurs when the arm is twisted towards a normal position.

(c) Jamming technique for large cracks, or "off widths," requiring the use of arm, leg, and body jams, is another technique. To jam or cam an arm, leg, or body into an off width, the principle is the same as for fingers, hands, or feet—you are making the jammed appendage "fatter" by folding or twisting it inside the crack. For off widths, you may place your entire arm inside the crack with the arm folded and the palm pointing outward. The leg can be used, from the calf to the thigh, and flexed to fit the crack. Routes requiring this type of climbing should be avoided as the equipment normally used for protection might not be large enough to protect larger cracks and openings. However, sometimes a narrower section may be deeper in the crack allowing the use of "normal" size protection.

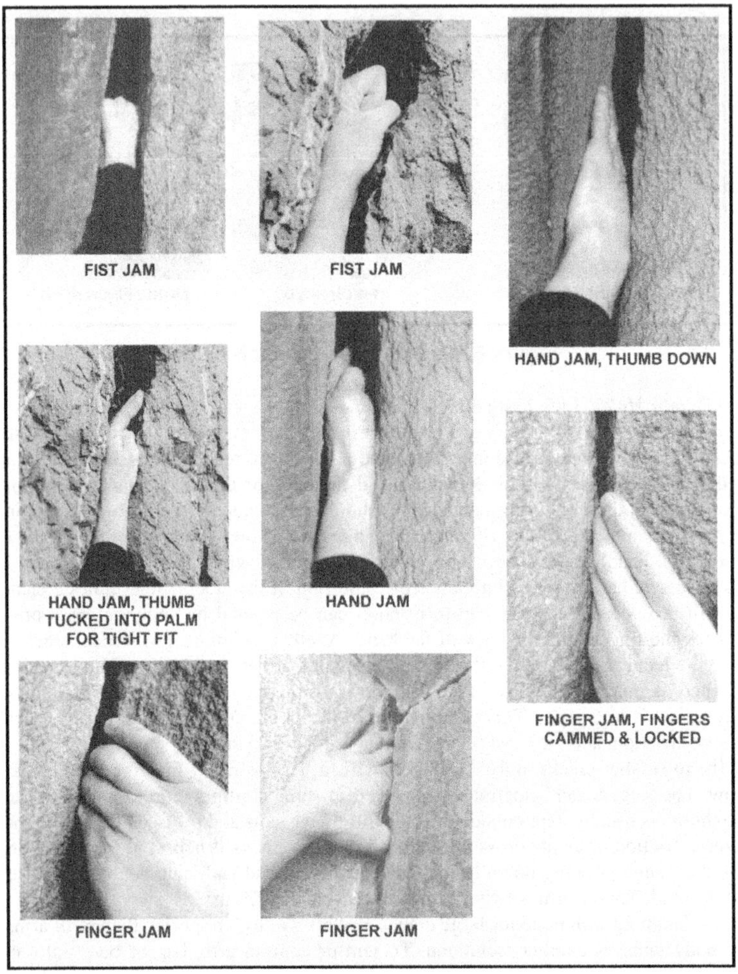

Figure 6-11. Examples of jam holds.

6-10. COMBINATION TECHNIQUES

The positions and holds previously discussed are the basics and the ones most common to climbing. From these fundamentals, numerous combination techniques are possible. As the climber gains experience, he will learn more ways to position the hands, feet, and body in relation to the holds available; however, he should always strive to climb with his weight on his feet from a balanced stance.

a. Sometimes, even on an easy route, the climber may come upon a section of the rock that defies the basic principles of climbing. Short of turning back, the only alternative is to figure out some combination technique that will work. Many of these type problems require the hands and feet to work in opposition to one another. Most will place more weight on the hands and arms than is desirable, and some will put the climber in an "out of balance" position. To make the move, the climber may have to "break the rules" momentarily. This is not a problem and is done quite frequently by experienced climbers. The key to using these type of combination techniques is to plan and execute them deliberately, without lunging or groping for holds, yet quickly, before the hands, arms, or other body parts tire. Still, most of these maneuvers require good technique more than great strength, though a certain degree of hand and arm strength certainly helps.

b. Combination possibilities are endless. The following is a brief description of some of the more common techniques.

(1) ***Change Step.*** The change step, or hop step, can be used when the climber needs to change position of the feet. It is commonly used when traversing to avoid crossing the feet, which might put the climber in an awkward position. To prevent an off balance situation, two solid handholds should be used. The climber simply places his weight on his handholds while he repositions the feet. He often does this with a quick "hop," replacing the lead foot with the trail foot on the same hold. Keeping the forearms against the rock during the maneuver takes some of the strain off the hands, while at the same time strengthening the grip on the holds.

(2) ***Mantling.*** Mantling is a technique that can be used when the distance between the holds increases and there are no immediate places to move the hands or feet. It does require a ledge (mantle) or projection in the rock that the climber can press straight down upon. (Figure 6-12, page 6-20, shows the mantling sequence.)

(a) When the ledge is above head height, mantling begins with pull holds, usually "hooking" both hands over the ledge. The climber pulls himself up until his head is above the hands, where the pull holds become push holds. He elevates himself until the arms are straight and he can lock the elbows to relax the muscles. Rotating the hands inward during the transition to push holds helps to place the palms more securely on the ledge. Once the arms are locked, a foot can be raised and placed on the ledge. The climber may have to remove one hand to make room for the foot. Mantling can be fairly strenuous; however, most individuals should be able to support their weight, momentarily, on one arm if they keep it straight and locked. With the foot on the ledge, weight can be taken off the arms and the climber can grasp the holds that were previously out of reach. Once balanced over the foot, he can stand up on the ledge and plan his next move.

(b) Pure mantling uses arm strength to raise the body; however, the climber can often smear the balls of the feet against the rock and "walk" the feet up during the maneuver to take some of the weight off the arms. Sometimes edges will be available for short steps in the process.

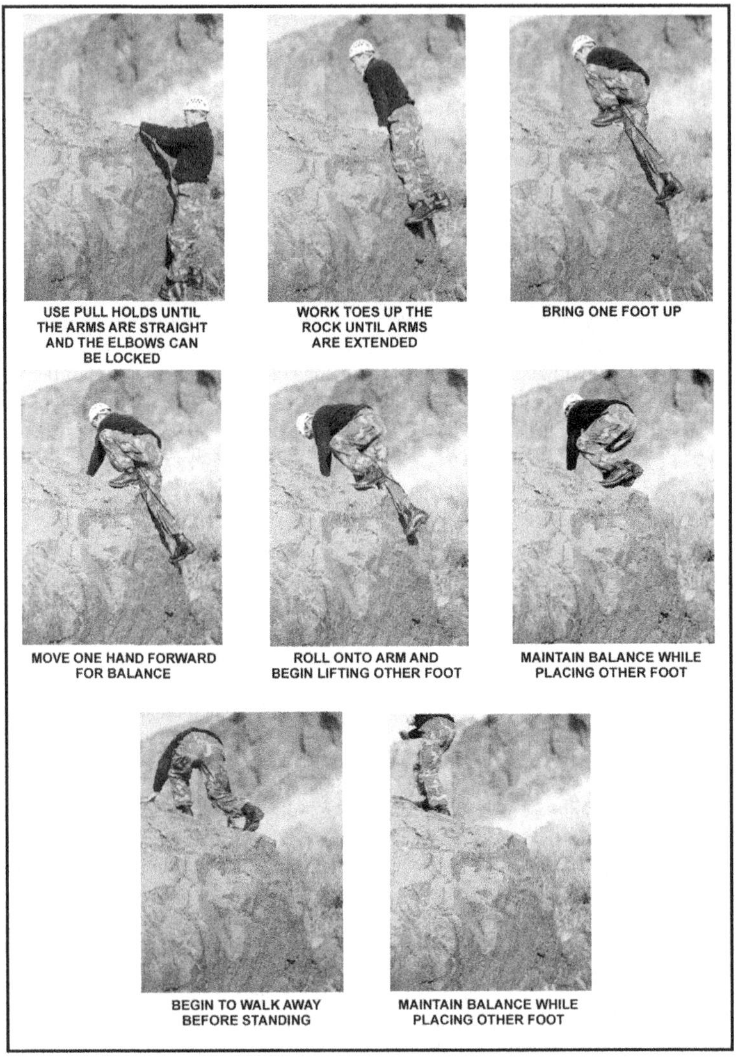

Figure 6-12. Mantling sequence.

(3) ***Undercling.*** An "undercling" is a classic example of handholds and footholds working in opposition (Figure 6-13). It is commonly used in places where the rock

projects outward, forming a bulge or small overhang. Underclings can be used in the tops of buckets, also. The hands are placed "palms-up" underneath the bulge, applying an upward pull. Increasing this upward pull creates a counterforce, or body tension, which applies more weight and friction to the footholds. The arms and legs should be kept as straight as possible to reduce fatigue. The climber can often lean back slightly in the undercling position, enabling him to see above the overhang better and search for the next hold.

Figure 6-13. Undercling.

(4) ***Lieback.*** The "lieback" is another good example of the hands working in opposition to the feet. The technique is often used in a vertical or diagonal crack separating two rock faces that come together at, more or less, a right angle (commonly referred to as a dihedral). The crack edge closest to the body is used for handholds while the feet are pressed against the other edge. The climber bends at the waist, putting the body into an L-shaped position. Leaning away from the crack on two pull holds, body tension creates friction between the feet and the hands. The feet must be kept relatively high to maintain weight, creating maximum friction between the sole and the rock surface. Either full sole contact or the smearing technique can be used, whichever seems to produce the most friction.

(a) The climber ascends a dihedral by alternately shuffling the hands and feet upward. The lieback technique can be extremely tiring, especially when the dihedral is near vertical. If the hands and arms tire out before completing the sequence, the climber will likely fall. The arms should be kept straight throughout the entire maneuver so the climber's weight is pulling against bones and ligaments, rather than muscle. The legs should be straightened whenever possible.

(b) Placing protection in a lieback is especially tiring. Look for edges or pockets for the feet in the crack or on the face for a better position to place protection from, or for a rest position. Often, a lieback can be avoided with closer examination of the available face features. The lieback can be used alternately with the jamming technique, or vice versa, for variation or to get past a section of a crack with difficult or nonexistent jam possibilities. The lieback can sometimes be used as a face maneuver (Figure 6-14).

Figure 6-14. Lieback on a face.

(5) ***Stemming.*** When the feet work in opposition from a relatively wide stance, the maneuver is known as stemming. The stemming technique can sometimes be used on faces, as well as in a dihedral in the absence of solid handholds for the lieback (Figure 6-15).

Figure 6-15. Stemming on a face.

(a) The classic example of stemming is when used in combination with two opposing push holds in wide, parallel cracks, known as chimneys. Chimneys are cracks in which the walls are at least 1 foot apart and just big enough to squeeze the body into. Friction is created by pushing outward with the hands and feet on each side of the crack. The climber ascends the chimney by alternately moving the hands and feet up the crack (Figure 6-16). Applying pressure with the back and bottom is usually necessary in wider chimneys. Usually, full sole contact of the shoes will provide the most friction, although smearing may work best in some instances. Chimneys that do not allow a full stemming position can be negotiated using the arms, legs, or body as an integral contact point. This technique will often feel more secure since there is more body to rock contact.

Figure 6-16. Chimney sequence.

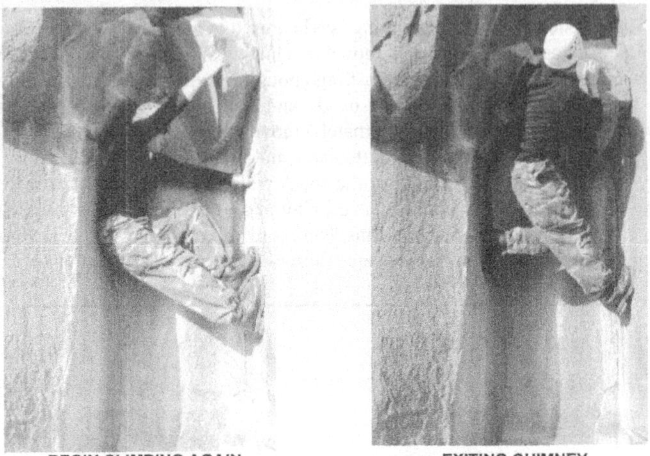

Figure 6-16. Chimney sequence (continued).

(b) The climber can sometimes rest by placing both feet on the same side of the crack, forcing the body against the opposing wall. The feet must be kept relatively high up under the body so the force is directed sideways against the walls of the crack. The arms should be straightened with the elbows locked whenever possible to reduce muscle strain. The climber must ensure that the crack does not widen beyond the climbable width before committing to the maneuver. Remember to look for face features inside chimneys for more security in the climb.

(c) Routes requiring this type of climbing should be avoided as the equipment normally used for protection might not be large enough to protect chimneys. However, face features, or a much narrower crack in one or both corners, may sometimes be found deeper in the chimney allowing the use of normal size protection.

(6) ***Slab Technique.*** A slab is a relatively smooth, low-angled rock formation that requires a slightly modified climbing technique (Figure 6-17). Since slab rock normally contains few, if any holds, the technique requires maximum friction and perfect balance over the feet.

(a) On lower-angled slab, the climber can often stand erect and climb using full sole contact and other mountain walking techniques. On steeper slab, the climber will need to apply good smearing technique. Often, maximum friction cannot be attained on steeper slab from an erect stance. The climber will have to flex the ankles and knees so his weight is placed more directly over the balls of the feet. He may then have to bend at the waist to place the hands on the rock, while keeping the hips over his feet.

(b) The climber must pay attention to any changes in slope angle and adjust his body accordingly. Even the slightest change in the position of the hips over the feet can mean the difference between a good grip or a quick slip. The climber should also take advantage of any rough surfaces, or other irregularities in the rock he can place his hands or feet on, to increase friction.

Figure 6-17. Slab technique.

(7) ***Down Climbing.*** Descending steep rock is normally performed using a roped method; however, the climber may at some point be required to down climb a route. Even if climbing ropes and related equipment are on hand, down climbing easier terrain is often quicker than taking the time to rig a rappel point. Also, a climber might find himself confronted with difficulties part way up a route that exceed his climbing ability, or the abilities of others to follow. Whatever the case may be, down climbing is a skill well worth practicing.

CAUTIONS

1. Down climbing can inadvertently lead into an unforeseen dangerous position on a descent. When in doubt, use a roped descent.
2. Down climbing is accomplished at a difficulty level well below the ability of the climber. When in doubt, use a roped descent.

(a) On easier terrain, the climber can face outward, away from the rock, enabling him to see the route better and descend quickly. As the steepness and difficulty increase, he can often turn sideways, still having a good view of the descent route, but being better able to use the hands and feet on the holds available. On the steepest terrain, the climber will have to face the rock and down climb using good climbing techniques.

(b) Down climbing is usually more difficult than ascending a given route. Some holds will be less visible when down climbing, and slips are more likely to occur. The climber must often lean well away from the rock to look for holds and plan his movements. More weight is placed on the arms and handholds at times to accomplish this, as well as to help lower the climber to the next foothold. Hands should be moved to holds as low as waist level to give the climber more range of movement with each step. If the handholds are too high, he may have trouble reaching the next foothold. The climber must be careful not to overextend himself, forcing a release of his handholds before reaching the next foothold.

CAUTION

Do not drop from good handholds to a standing position. A bad landing could lead to injured ankles or a fall beyond the planned landing area.

(c) Descending slab formations can be especially tricky. The generally lower angle of slab rock may give the climber a false sense of security, and a tendency to move too quickly. Down climbing must be slow and deliberate, as in ascending, to maintain perfect balance and weight distribution over the feet. On lower-angle slab the climber may be able to stand more or less erect, facing outward or sideways, and descend using good flat foot technique. The climber should avoid the tendency to move faster, which can lead to uncontrollable speed.

(d) On steeper slab, the climber will normally face the rock and down climb, using the same smearing technique as for ascending. An alternate method for descending slab is to face away from the rock in a "crab" position (Figure 6-18). Weight is still concentrated over the feet, but may be shifted partly onto the hands to increase overall friction. The climber is able to maintain full sole contact with the rock and see the entire descent route. Allowing the buttocks to "drag behind" on the rock will decrease the actual weight on the footholds, reducing friction, and leading to the likelihood of a slip. Facing the rock, and down-climbing with good smearing technique, is usually best on steeper slab.

Figure 6-18. Descending slab in the crab position.

Section III. ROPED CLIMBING

When the angle, length, and difficulty of the proposed climbing route surpasses the ability of the climbers' safety margin (possibly on class 4 and usually on class 5 terrain), ropes must be used to proceed. Roped climbing is only safe if accomplished correctly. Reading this manual does not constitute skill with ropes—much training and practice is necessary. Many aspects of roped climbing take time to understand and learn. Ropes are normally not used in training until the basic principles of climbing are covered.

Note: A rope is completely useless for climbing unless the climber knows how to use it safely.

6-11. TYING-IN TO THE CLIMBING ROPE

Over the years, climbers have developed many different knots and procedures for tying-in to the climbing rope. Some of the older methods of tying directly into the rope require minimal equipment and are relatively easy to inspect; however, they offer little support to the climber, may induce further injuries, and may even lead to strangulation in a severe fall. A severe fall, where the climber might fall 20 feet or more and be left dangling on the end of the rope, is highly unlikely in most instances, especially for most personnel involved in military climbing. Tying directly into the rope is perfectly safe for many roped party climbs used in training on lower-angled rock. All climbers should know how to properly tie the rope around the waist in case a climbing harness is unavailable.

6-12. PRESEWN HARNESSES

Although improvised harnesses are made from readily available materials and take little space in the pack or pocket, presewn harnesses provide other aspects that should be considered. No assembly is required, which reduces preparation time for roped movement. All presewn harnesses provide a range of adjustability. These harnesses have a fixed buckle that, when used correctly, will not fail before the nylon materials connected to it. However, specialized equipment, such as a presewn harness, reduce the flexibility of gear. Presewn harness are bulky, also.

a. **Seat Harness.** Many presewn seat harnesses are available with many different qualities separating them, including cost.

(1) The most notable difference will be the amount and placement of padding. The more padding the higher the price and the more comfort. Gear loops sewn into the waist belt on the sides and in the back are a common feature and are usually strong enough to hold quite a few carabiners and or protection. The gear loops will vary in number from one model/manufacturer to another.

(2) Although most presewn seat harnesses have a permanently attached belay loop connecting the waist belt and the leg loops, the climbing rope should be run around the waist belt and leg loop connector. The presewn belay loop adds another link to the chain of possible failure points and only gives one point of security whereas running the rope through the waist belt and leg loop connector provides two points of contact.

(3) If more than two men will be on the rope, connect the middle position(s) to the rope with a carabiner routed the same as stated in the previous paragraph.

(4) Many manufactured seat harnesses will have a presewn loop of webbing on the rear. Although this loop is much stronger than the gear loops, it is not for a belay anchor. It is a quick attachment point to haul an additional rope.

b. **Chest Harness.** The chest harness will provide an additional connecting point for the rope, usually in the form of a carabiner loop to attach a carabiner and rope to. This type of additional connection will provide a comfortable hanging position on the rope, but otherwise provides no additional protection from injury during a fall (if the seat harness is fitted correctly).

(1) A chest harness will help the climber remain upright on the rope during rappelling or ascending a fixed rope, especially while wearing a heavy pack. (If rappelling or ascending long or multiple pitches, let the pack hang on a drop cord below the feet and attached to the harness tie-in point.)

(2) The presewn chest harnesses available commercially will invariably offer more comfort or performance features, such as padding, gear loops, or ease of adjustment, than an improvised chest harness.

c. **Full-Body Harness.** Full-body harnesses incorporate a chest and seat harness into one assembly. This is the safest harness to use as it relocates the tie-in point higher, at the chest, reducing the chance of an inverted position when hanging on the rope. This is especially helpful when moving on ropes with heavy packs. A full-body harness only affects the body position when hanging on the rope and will not prevent head injury in a fall.

CAUTION

This type of harness does not prevent the climber from falling head first. Body position during a fall is affected only by the forces that generated the fall, and this type of harness promotes an upright position only when hanging on the rope from the attachment point.

6-13. IMPROVISED HARNESSES

Without the use of a manufactured harness, many methods are still available for attaching oneself to a rope. Harnesses can be improvised using rope or webbing and knots.

a. **Swami Belt.** The swami belt is a simple, belt-only harness created by wrapping rope or webbing around the waistline and securing the ends. One-inch webbing will provide more comfort. Although an effective swami belt can be assembled with a minimum of one wrap, at least two wraps are recommended for comfort, usually with approximately ten feet of material. The ends are secured with an appropriate knot.

b. **Bowline-on-a-Coil.** Traditionally, the standard method for attaching oneself to the climbing rope was with a bowline-on-a-coil around the waist. The extra wraps distribute the force of a fall over a larger area of the torso than a single bowline would, and help prevent the rope from riding up over the rib cage and under the armpits. The knot must be tied snugly around the narrow part of the waist, just above the bony portions of the hips (pelvis). Avoid crossing the wraps by keeping them spread over the waist area. "Sucking in the gut" a bit when making the wraps will ensure a snug fit.

(1) The bowline-on-a-coil can be used to tie-in to the end of the rope (Figure 6-19). The end man should have a minimum of four wraps around the waist before completing the knot.

Figure 6-19. Tying-in with a bowline-on-a-coil.

(2) The bowline-on-a-coil is a safe and effective method for attaching to the rope when the terrain is low-angled, WITHOUT THE POSSIBILITY OF A SEVERE FALL. When the terrain becomes steeper, a fall will generate more force on the climber and this will be felt through the coils of this type of attachment. A hard fall will cause the coils to ride up against the ribs. In a severe fall, any tie-in around the waist only could place a "shock load" on the climber's lower back. Even in a relatively short fall, if the climber ends up suspended in mid-air and unable to regain footing on the rock, the rope around the waist can easily cut off circulation and breathing in a relatively short time.

(3) The climbing harness distributes the force of a fall over the entire pelvic region, like a parachute harness. Every climber should know how to tie some sort of improvised climbing harness from sling material. A safe, and comfortable, seat/chest combination harness can be tied from one-inch tubular nylon.

c. **Improvised Seat Harness.** A seat harness can be tied from a length of webbing approximately 25 feet long (Figure 6-20, page 6-30).

(1) Locate the center of the rope. Off to one side, tie two fixed loops approximately 6 inches apart (overhand loops). Adjust the size of the loops so they fit snugly around the thigh. The loops are tied into the sling "off center" so the remaining ends are different lengths. The short end should be approximately 4 feet long (4 to 5 feet for larger individuals).

(2) Slip the leg loops over the feet and up to the crotch, with the knots to the front. Make one complete wrap around the waist with the short end, wrapping to the outside, and hold it in place on the hip. Keep the webbing flat and free of twists when wrapping.

(3) Make two to three wraps around the waist with the long end in the opposite direction (wrapping to the outside), binding down on the short end to hold it in place. Grasping both ends, adjust the waist wraps to a snug fit. Connect the ends with the

appropriate knot between the front and one side so you will be able to see what you are doing.

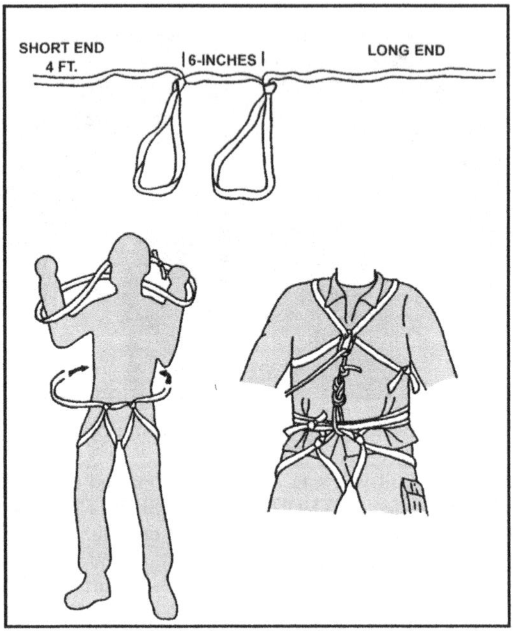

Figure 6-20. Improvised seat and chest harness.

d. **Improvised Chest Harness.** The chest harness can be tied from rope or webbing, but remember that with webbing, wider is better and will be more comfortable when you load this harness. Remember as you tie this harness that the remaining ends will need to be secured so choose the best length. Approximately 6 to 10 feet usually works.

(1) Tie the ends of the webbing together with the appropriate knot, making a sling 3 to 4 feet long.

(2) Put a single twist into the sling, forming two loops.

(3) Place an arm through each loop formed by the twist, just as you would put on a jacket, and drape the sling over the shoulders. The twist, or cross, in the sling should be in the middle of the back.

(4) Join the two loops at the chest with a carabiner. The water knot should be set off to either side for easy inspection (if a pack is to be worn, the knot will be uncomfortable if it gets between the body and the pack). The chest harness should fit just loose enough to allow necessary clothing and not to restrict breathing or circulation. Adjust the size of the sling if necessary.

e. **Improvised Full-Body Harness.** Full-body harnesses incorporate a chest and seat harness into one assembly.

(1) The full-body harness is the safest harness because it relocates the tie-in point higher, at the chest, reducing the chance of an inverted hanging position on the rope. This is especially helpful when moving on ropes with heavy packs. A full-body harness affects the body position only when hanging on the rope.

CAUTION

A full-body harness does not prevent falling head first; body position in a fall is caused by the forces that caused the fall.

(2) Although running the rope through the carabiner of the chest harness does, in effect, create a type of full-body harness, it is not a true full-body harness until the chest harness and the seat harness are connected as one piece. A true full-body harness can be improvised by connecting the chest harness to the seat harness, but not by just tying the rope into both—the two harnesses must be "fixed" as one harness. Fix them together with a short loop of webbing or rope so that the climbing rope can be connected directly to the chest harness and your weight is supported by the seat harness through the connecting material.

f. **Attaching the Rope to the Improvised Harness.** The attachment of the climbing rope to the harness is a CRITICAL LINK. The strength of the rope means nothing if it is attached poorly, or incorrectly, and comes off the harness in a fall. The climber ties the end of the climbing rope to the seat harness with an appropriate knot. If using a chest harness, the standing part of the rope is then clipped into the chest harness carabiner. The seat harness absorbs the main force of the fall, and the chest harness helps keep the body upright.

CAUTION

The knot must be tied around all the waist wraps and the 6-inch length of webbing between the leg loops.

(1) A middleman must create a fixed loop to tie in to. A rethreaded figure-eight loop tied on a doubled rope or the three loop bowline can be used. If using the three loop bowline, ensure the end, or third loop formed in the knot, is secured around the tie-in loops with an overhand knot. The standing part of the rope going to the lead climber is clipped into the chest harness carabiner.

Note: The climbing rope is not clipped into the chest harness when belaying.

(2) The choice of whether to tie-in with a bowline-on-a-coil or into a climbing harness depends entirely on the climber's judgment, and possibly the equipment

available. A good rule of thumb is: "Wear a climbing harness when the potential for severe falls exists and for all travel over snow-covered glaciers because of the crevasse fall hazard."

(3) Under certain conditions many climbers prefer to attach the rope to the seat harness with a locking carabiner, rather than tying the rope to it. This is a common practice for moderate snow and ice climbing, and especially for glacier travel where wet and frozen knots become difficult to untie.

> **CAUTION**
> Because the carabiner gate may be broken or opened by protruding rocks during a fall, tie the rope directly to the harness for maximum safety.

Section IV. BELAY TECHNIQUES

Tying-in to the climbing rope and moving as a member of a rope team increases the climber's margin of safety on difficult, exposed terrain. In some instances, such as when traveling over snow-covered glaciers, rope team members can often move at the same time, relying on the security of a tight rope and "team arrest" techniques to halt a fall by any one member. On steep terrain, however, simultaneous movement only helps to ensure that if one climber falls, he will jerk the other rope team members off the slope. For the climbing rope to be of any value on steep rock climbs, the rope team must incorporate "belays" into the movement.

Belaying is a method of managing the rope in such a way that, if one person falls, the fall can be halted or "arrested" by another rope team member (belayer). One person climbs at a time, while being belayed from above or below by another. The belayer manipulates the rope so that friction, or a "brake," can be applied to halt a fall. Belay techniques are also used to control the descent of personnel and equipment on fixed rope installations, and for additional safety on rappels and stream crossings.

Belaying is a skill that requires practice to develop proficiency. Setting up a belay may at first appear confusing to the beginner, but with practice, the procedure should become "second nature." If confronted with a peculiar problem during the setup of a belay, try to use common sense and apply the basic principles stressed throughout this text. Remember the following key points:

- Select the best possible terrain features for the position and use terrain to your advantage.
- Use a well braced, sitting position whenever possible.
- Aim and anchor the belay for all possible load directions.
- Follow the "minimum" rule for belay anchors—2 for a downward pull, 1 for an upward pull.
- Ensure anchor attachments are aligned, independent, and snug.
- Stack the rope properly.
- Choose a belay technique appropriate for the climbing.

- Use a guide carabiner for rope control in all body belays.
- Ensure anchor attachments, guide carabiner (if applicable), and rope running to the climber are all on the guidehand side.
- The brake hand remains on the rope when belaying.

CAUTION
Never remove the brake hand from the rope while belaying. If the brake hand is removed, there is no belay.

- Ensure you are satisfied with your position before giving the command "BELAY ON."
- The belay remains in place until the climber gives the command "OFF BELAY."

CAUTION
The belay remains in place the from the time the belayer commands *"BELAY ON"* until the climber commands *"OFF BELAY."*

6-14. PROCEDURE FOR MANAGING THE ROPE

A number of different belay techniques are used in modern climbing, ranging from the basic "body belays" to the various "mechanical belays," which incorporate some type of friction device.

a. Whether the rope is wrapped around the body, or run through a friction device, the rope management procedure is basically the same. The belayer must be able to perform three basic functions: manipulate the rope to give the climber slack during movement, take up rope to remove excess slack, and apply the brake to halt a fall.

b. The belayer must be able to perform all three functions while maintaining "total control" of the rope at all times. Total control means the brake hand is NEVER removed from the rope. When giving slack, the rope simply slides through the grasp of the brake hand, at times being fed to the climber with the other "feeling" or guide hand. Taking up rope, however, requires a certain technique to ensure the brake hand remains on the rope at all times. The following procedure describes how to take up excess rope and apply the brake in a basic body belay.

(1) Grasping the rope with both hands, place it behind the back and around the hips. The hand on the section of rope between the belayer and the climber would be the guide hand. The other hand is the brake hand.

(2) Take in rope with the brake hand until the arm is fully extended. The guide hand can also help to pull in the rope (Figure 6-21, step 1, page 6-34).

(3) Holding the rope in the brake hand, slide the guide hand out, extending the arm so the guide hand is father away from the body than the brake hand (Figure 6-21, step 2).

(4) Grasp both parts of the rope, to the front of the brake hand, with the guide hand (Figure 6-21, step 3).

(5) Slide the brake hand back towards the body (Figure 6-21, step 4).

(6) Repeat step 5 of Figure 6-21. The brake can be applied at any moment during the procedure. It is applied by wrapping the rope around the front of the hips while increasing grip with the brake hand (Figure 6-21, step 6).

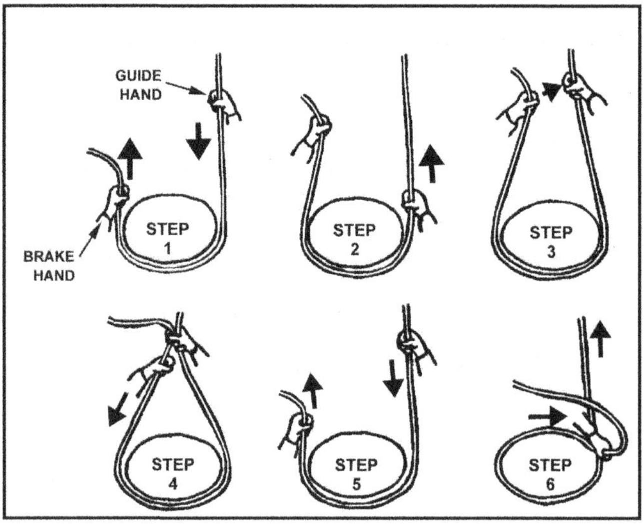

Figure 6-21. Managing the rope.

6-15. CHOOSING A BELAY TECHNIQUE

The climber may choose from a variety of belay techniques. A method that works well in one situation may not be the best choice in another. The choice between body belays and mechanical belays depends largely on equipment available, what the climber feels most comfortable with, and the amount of load, or fall force, the belay may have to absorb. The following describes a few of the more widely used techniques, and the ones most applicable to military mountaineering.

a. **Body Belay.** The basic body belay is the most widely used technique on moderate terrain. It uses friction between the rope and the clothed body as the rope is pressured across the clothing. It is the simplest belay, requiring no special equipment, and should be the first technique learned by all climbers. A body belay gives the belayer the greatest "feel" for the climber, letting him know when to give slack or take up rope. Rope management in a body belay is quick and easy, especially for beginners, and is effective in snow and ice climbing when ropes often become wet, stiff, and frozen. The body belay, in its various forms, will hold low to moderate impact falls well. It has been known to arrest some severe falls, although probably not without inflicting great pain on the belayer.

CAUTION
The belayer must ensure he is wearing adequate clothing to protect his body from rope burns when using a body belay. Heavy duty cotton or leather work gloves can also be worn to protect the hands.

(1) ***Sitting Body Belay.*** The sitting body belay is the preferred position and is usually the most secure (Figure 6-22). The belayer sits facing the direction where the force of a fall will likely come from, using terrain to his advantage, and attempts to brace both feet against the rock to support his position. It is best to sit in a slight depression, placing the buttocks lower than the feet, and straightening the legs for maximum support. When perfectly aligned, the rope running to the climber will pass between the belayer's feet, and both legs will equally absorb the force of a fall. Sometimes, the belayer may not be able to sit facing the direction he would like, or both feet cannot be braced well. The leg on the "guide hand" side should then point towards the load, bracing the foot on the rock when possible. The belayer can also "straddle" a large tree or rock nubbin for support, as long as the object is solid enough to sustain the possible load.

Figure 6-22. Sitting body belay.

(2) ***Standing Body Belay.*** The standing body belay is used on smaller ledges where there is no room for the belayer to sit (Figure 6-23). What appears at first to be a fairly unstable position can actually be quite secure when belay anchors are placed at or above shoulder height to support the stance when the force will be downward.

Figure 6-23. Standing body belay.

(a) For a body belay to work effectively, the belayer must ensure that the rope runs around the hips properly, and remains there under load when applying the brake. The rope should run around the narrow portion of the pelvic girdle, just below the bony high points of the hips. If the rope runs too high, the force of a fall could injure the belayer's midsection and lower rib cage. If the rope runs too low, the load may pull the rope below the buttocks, dumping the belayer out of position. It is also possible for a strong upward or downward pull to strip the rope away from the belayer, rendering the belay useless.

(b) To prevent any of these possibilities from happening, the belay rope is clipped into a carabiner attached to the guide hand side of the seat harness (or bowline-on-a-coil). This "guide carabiner" helps keep the rope in place around the hips and prevents loss of control in upward or downward loads (Figure 6-24).

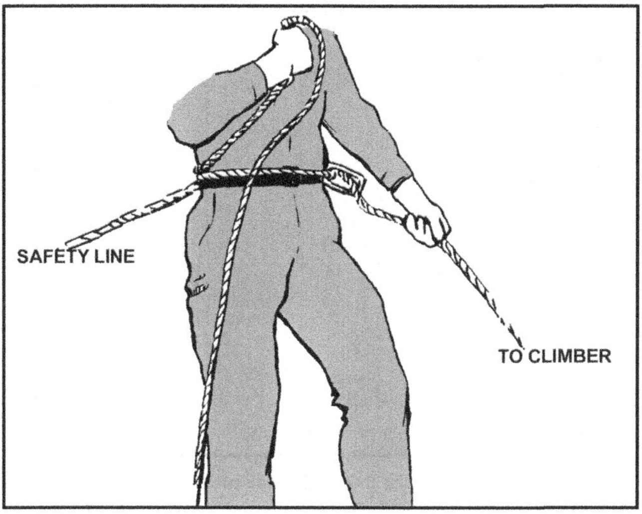

Figure 6-24. Guide carabiner for rope control in a body belay.

b. **Mechanical Belay.** A mechanical belay **must** be used whenever there is potential for the lead climber to take a severe fall. The holding power of a belay device is vastly superior to any body belay under high loads. However, rope management in a mechanical belay is more difficult to master and requires more practice. For the most part, the basic body belay should be totally adequate on a typical military route, as routes used during military operations should be the easiest to negotiate.

(1) ***Munter Hitch.*** The Munter hitch is an excellent mechanical belay technique and requires only a rope and a carabiner (Figure 6-25, page 6-38). The Munter is actually a two-way friction hitch. The Munter hitch will flip back and forth through the carabiner as the belayer switches from giving slack to taking up rope. The carabiner must be large enough, and of the proper design, to allow this function. The locking pear-shaped carabiner, or pearabiner, is designed for the Munter hitch.

(a) The Munter hitch works exceptionally well as a lowering belay off the anchor. As a climbing belay, the carabiner should be attached to the front of the belayer's seat harness. The hitch is tied by forming a loop and a bight in the rope, attaching both to the carabiner. It's fairly easy to place the bight on the carabiner backwards, which forms an obvious, useless hitch. Put some tension on the Munter to ensure it is formed correctly, as depicted in the following illustrations.

(b) The Munter hitch will automatically "lock-up" under load as the brake hand grips the rope. The brake is increased by pulling the slack rope away from the body, towards the load. The belayer must be aware that flipping the hitch DOES NOT change the function of the hands. The hand on the rope running to the climber, or load, is always the guide hand.

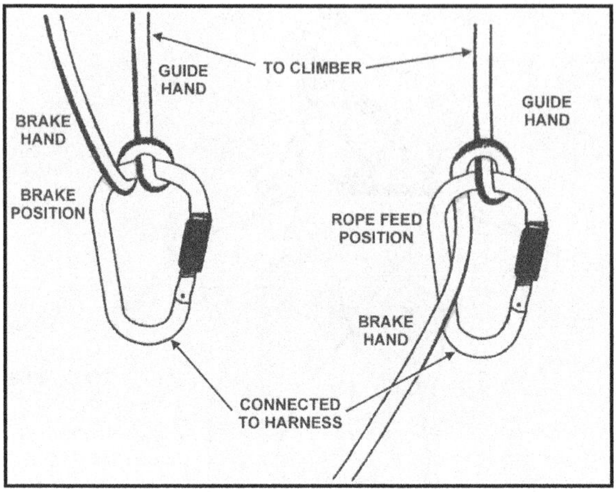

Figure 6-25. Munter hitch.

(2) ***Figure-Eight Device.*** The figure-eight device is a versatile piece of equipment and, though developed as a rappel device, has become widely accepted as an effective mechanical belay device (Figure 6-26). The advantage of any mechanical belay is friction required to halt a fall is applied on the rope through the device, rather than around the belayer's body. The device itself provides rope control for upward and downward pulls and excellent friction for halting severe falls. The main principle behind the figure-eight device in belay mode is the friction developing on the rope as it reaches and exceeds the 90-degree angle between the rope entering the device and leaving the device. As a belay device, the figure-eight works well for both belayed climbing and for lowering personnel and equipment on fixed-rope installations.

(a) As a climbing belay, a bight placed into the climbing rope is run through the "small eye" of the device and attached to a locking carabiner at the front of the belayer's seat harness. A short, small diameter safety rope is used to connect the "large eye" of the figure eight to the locking carabiner for control of the device. The guide hand is placed on the rope running to the climber. Rope management is performed as in a body belay. The brake is applied by pulling the slack rope in the brake hand towards the body, locking the rope between the device and the carabiner.

(b) As a lowering belay, the device is normally attached directly to the anchor with the rope routed as in rappelling.

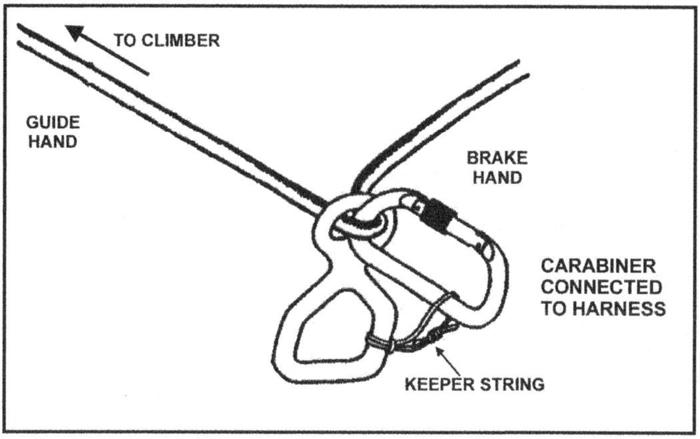

Figure 6-26. Figure-eight device.

Note: Some figure-eight descenders should not be used as belay devices due to their construction and design. Always refer to manufacturer's specifications and directions before use.

(3) ***Mechanical Camming Device.*** The mechanical camming device has an internal camming action that begins locking the rope in place as friction is increased. Unlike the other devices, the mechanical camming device can stop a falling climber without any input from the belayer. A few other devices perform similarly to this, but have no moving parts. Some limitations to these type devices are minimum and maximum rope diameters.

(4) ***Other Mechanical Belay Devices.*** There are many other commercially available mechanical belay devices. Most of these work with the same rope movement direction and the same braking principle. The air traffic controller (ATC), slotted plate, and other tube devices are made in many different shapes. These all work on the same principle as the figure-eight device—friction increases on the rope as it reaches and exceeds the 90-degree angle between the rope entering the device and leaving the device.

6-16. ESTABLISHING A BELAY

A belay can be established using either a direct or indirect connection. Each type has advantages and disadvantages. The choice will depend on the intended use of the belay.

a. **Direct Belay.** The direct belay removes any possible forces from the belayer and places this force completely on the anchor. Used often for rescue installations or to bring a second climber up to a new belay position in conjunction with the Munter hitch, the belay can be placed above the belayer's stance, creating a comfortable position and ease of applying the brake. Also, if the second falls or weights the rope, the belayer is not locked into a position. Direct belays provide no shock-absorbing properties from the belayer's attachment to the system as does the indirect belay; therefore, the belayer is apt to pay closer attention to the belaying process.

b. **Indirect Belay.** An indirect belay, the most commonly used, uses a belay device attached to the belayer's harness. This type of belay provides dynamic shock or weight absorption by the belayer if the climber falls or weights the rope, which reduces the direct force on the anchor and prevents a severe shock load to the anchor.

6-17. SETTING UP A BELAY

In rock climbing, climbers must sometimes make do with marginal protection placements along a route, but belay positions must be made as "bombproof" as possible. Additionally, the belayer must set up the belay in relation to where the fall force will come from and pay strict attention to proper rope management for the belay to be effective. All belay positions are established with the anchor connection to the front of the harness. If the belay is correctly established, the belayer will feel little or no force if the climber falls or has to rest on the rope. Regardless of the actual belay technique used, five basic steps are required to set up a sound belay.

a. **Select Position and Stance.** Once the climbing line is picked, the belayer selects his position. It's best if the position is off to the side of the actual line, putting the belayer out of the direct path of a potential fall or any rocks kicked loose by the climber. The position should allow the belayer to maintain a comfortable, relaxed stance, as he could be in the position for a fairly long time. Large ledges that allow a well braced, sitting stance are preferred. Look for belay positions close to bombproof natural anchors. The position must at least allow for solid artificial placements.

b. **Aim the Belay.** With the belay position selected, the belay must now be "aimed." The belayer determines where the rope leading to the climber will run and the direction the force of a fall will likely come from. When a lead climber begins placing protection, the fall force on the belayer will be in some upward direction, and in line with the first protection placement. If this placement fails under load, the force on the belay could be straight down again. The belayer must aim his belay for all possible load directions, adjusting his position or stance when necessary. The belay can be aimed through an anchor placement to immediately establish an upward pull; however, the belayer must always be prepared for the more severe downward fall force in the event intermediate protection placements fail.

c. **Anchor the Belay.** For a climbing belay to be considered bombproof, the belayer must be attached to a solid anchor capable of withstanding the highest possible fall force. A solid natural anchor would be ideal, but more often the belayer will have to place pitons or chocks. A single artificial placement should never be considered adequate for anchoring a belay (except at ground level). Multiple anchor points capable of supporting both upward and downward pulls should be placed. The rule of thumb is to place two anchors for a downward pull and one anchor for an upward pull as a MINIMUM. The following key points also apply to anchoring belays.

(1) Each anchor must be placed in line with the direction of pull it is intended to support.

(2) Each anchor attachment must be rigged "independently" so a failure of one will not shock load remaining placements or cause the belayer to be pulled out of position.

(3) The attachment between the anchor and the belayer must be snug to support the stance. Both belayer's stance and belay anchors should absorb the force of a fall.

(4) It is best for the anchors to be placed relatively close to the belayer with short attachments. If the climber has to be tied-off in an emergency, say after a severe fall, the belayer can attach a Prusik sling to the climbing rope, reach back, and connect the sling to one of the anchors. The load can be placed on the Prusik and the belayer can come out of the system to render help.

(5) The belayer can use either a portion of the climbing rope or slings of the appropriate length to connect himself to the anchors. It's best to use the climbing rope whenever possible, saving the slings for the climb. The rope is attached using either figure eight loops or clove hitches. Clove hitches have the advantage of being easily adjusted. If the belayer has to change his stance at some point, he can reach back with the guide hand and adjust the length of the attachment through the clove hitch as needed.

(6) The anchor attachments should also help prevent the force of a fall from "rotating" the belayer out of position. To accomplish this, the climbing rope must pass around the "guide-hand side" of the body to the anchors. Sling attachments are connected to the belayer's seat harness (or bowline-on-a-coil) on the guide-hand side.

(7) Arrangement of rope and sling attachments may vary according to the number and location of placements. Follow the guidelines set forth and remember the key points for belay anchors; "in line", "independent", and "snug". Figure 6-27 shows an example of a common arrangement, attaching the rope to the two "downward" anchors and a sling to the "upward" anchor. Note how the rope is connected from one of the anchors back to the belayer. This is not mandatory, but often helps "line-up" the second attachment.

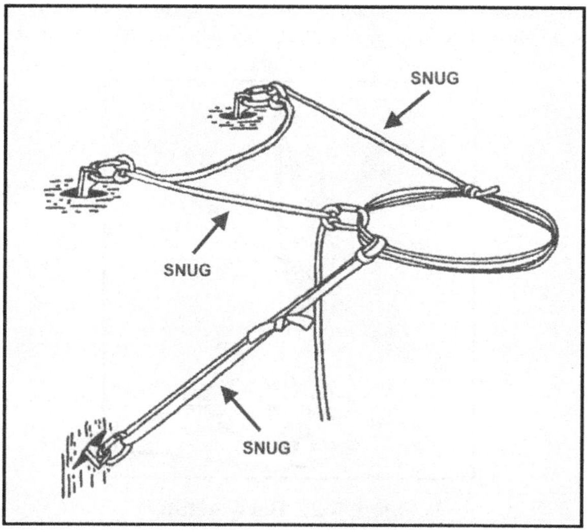

Figure 6-27. Anchoring a belay.

d. **Stack the Rope.** Once the belayer is anchored into position, he must stack the rope to ensure it is free of twists and tangles that might hinder rope management in the belay. The rope should be stacked on the ground, or on the ledge, where it will not get caught in cracks or nubbins as it is fed out to the climber.

(1) On small ledges, the rope can be stacked on top of the anchor attachments if there is no other place to lay it, but make sure to stack it carefully so it won't tangle with the anchored portion of the rope or other slings. The belayer must also ensure that the rope will not get tangled around his legs or other body parts as it "feeds" out.

(2) The rope should never be allowed to hang down over the ledge. If it gets caught in the rock below the position, the belayer may have to tie-off the climber and come out of the belay to free the rope; a time-consuming and unnecessary task. The final point to remember is the rope must be stacked "from the belayer's end" so the rope running to the climber comes off the "top" of the stacked pile.

e. **Attach the Belay.** The final step of the procedure is to attach the belay. With the rope properly stacked, the belayer takes the rope coming off the top of the pile, removes any slack between himself and the climber, and applies the actual belay technique. If using a body belay, ensure the rope is clipped into the guide carabiner.

(1) The belayer should make one quick, final inspection of his belay. If the belay is set up correctly, the anchor attachments, guide carabiner if applicable, and the rope running to the climber will all be on the "guide hand" side, which is normally closest to the rock (Figure 6-28). If the climber takes a fall, the force, if any, should not have any negative effect on the belayer's involvement in the system. The brake hand is out away from the slope where it won't be jammed between the body and the rock. The guide hand can be placed on the rock to help support the stance when applying the brake.

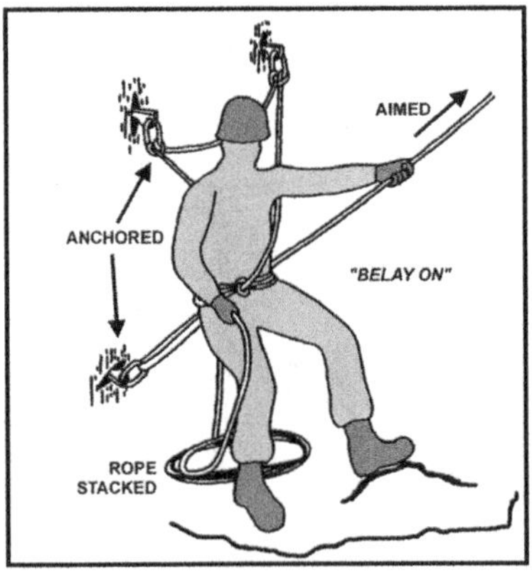

Figure 6-28. Belay setup.

(2) When the belayer is satisfied with his position, he gives the signal, "BELAY ON!". When belaying the "second", the same procedure is used to set up the belay. Unless the belay is aimed for an upward pull, the fall force is of course downward and the

belayer is usually facing away from the rock, the exception being a hanging belay on a vertical face. If the rope runs straight down to the climber and the anchors are directly behind the position, the belayer may choose to brake with the hand he feels most comfortable with. Anchor attachments, guide carabiner, and rope running to the climber through the guide hand must still be aligned on the same side to prevent the belayer from being rotated out of position, unless the belayer is using an improvised harness and the anchor attachment is at the rear.

6-18. TOP-ROPE BELAY

A "top-rope" is a belay setup used in training to protect a climber while climbing on longer, exposed routes. A solid, bombproof anchor is required at the top of the pitch. The belayer is positioned either on the ground with the rope running through the top anchor and back to the climber, or at the top at the anchor. The belayer takes in rope as the climber proceeds up the rock. If this is accomplished with the belayer at the bottom, the instructor can watch the belayer while he coaches the climber through the movements.

CAUTION

Do not use a body belay for top-rope climbing. The rope will burn the belayer if the climber has to be lowered.

Section V. CLIMBING COMMANDS

Communication is often difficult during a climb. As the distance between climber and belayer increases, it becomes harder to distinguish one word from another and the shortest sentence may be heard as nothing more than jumbled syllables. A series of standard voice commands were developed over the years to signal the essential rope management functions in a belayed climb. Each command is concise and sounds a bit different from another to reduce the risk of a misunderstanding between climber and belayer. They must be pronounced clearly and loudly so they can be heard and understood in the worst conditions.

6-19. VERBAL COMMANDS

Table 6-1, page 6-44, lists standard rope commands and their meanings in sequence as they would normally be used on a typical climb. (Note how the critical "BELAY" commands are reversed so they sound different and will not be confused.)

BELAYER	CLIMBER	MEANING/ACTION TAKEN
"BELAY ON"		The belay is on; you may climb when ready; the rope will be managed as needed.
	"CLIMBING" (as a courtesy)	I am ready to climb.
"CLIMB" (as a courtesy)		Proceed, and again, the rope will be managed as necessary.
"ROCK"	"ROCK"	**PROTECT YOURSELF FROM FALLING OBJECTS.** Signal will be echoed by all climbers in the area. If multipitch climbing, ensure climbers below hear.
	"TAKE ROPE"	Take in excess rope between us without pulling me off the route. Belayer takes in rope.
	"SLACK"	Release all braking/tension on the rope so I can have slack without pulling the rope. Belayer removes brake/tension.
	"TENSION"	Take all the slack, apply brake, and hold me. My weight will be on the rope. Belayer removes slack and applies brake.
	"FALLING"	I am falling. Belayer applies brake to arrest the fall.
"TWEN-TY-FIVE"		You have approximately 25 feet of rope left. Start looking for the next belay position. Climber selects a belay position.
"FIF-TEEN"		You have approximately 15 feet of rope left. Start looking for the next belay position. Climber selects a belay position within the next few feet.
"FIVE"	Set up the belay.	You have 5 feet of rope left. Set up the belay position. You have no more rope. Climber sets up the belay.
Removes the belay, remains anchored. Prepares to climb.	"OFF BELAY"	I have finished climbing and I am anchored. You may remove the belay. Belayer removes the belay and, remaining anchored, prepares to climb.

Table 6-1. Rope commands.

6-20. ROPE TUG COMMANDS

Sometimes the loudest scream cannot be heard when the climber and belayer are far apart. This is especially true in windy conditions, or when the climber is around a corner, above an overhang, or at the back of a ledge. It may be necessary to use a series of "tugs" on the rope in place of the standard voice commands. To avoid any possible confusion with interpretation of multiple rope tug commands, use only one.

a. While a lead climb is in progress, the most important command is "BELAY ON." This command is given only by the climber when the climber is anchored and is prepared for the second to begin climbing. With the issue of this command, the second knows the climber is anchored and the second prepares to climb.

b. For a rope tug command, the leader issues three distinct tugs on the rope AFTER anchoring and putting the second on belay. This is the signal for "BELAY ON" and signals the second to climb when ready. The new belayer keeps slack out of the rope.

Section VI. ROPED CLIMBING METHODS

In military mountaineering, the primary mission of a roped climbing team is to "fix" a route with some type of rope installation to assist movement of less trained personnel in the unit. This duty falls upon the most experienced climbers in the unit, usually working in two- or three-man groups or teams called assault climbing teams. Even if the climbing is for another purpose, roped climbing should be performed whenever the terrain becomes difficult and exposed.

6-21. TOP-ROPED CLIMBING

Top-roped climbing is used for training purposes only. This method of climbing is not used for movement due to the necessity of pre-placing anchors at the top of a climb. If you can easily access the top of a climb, you can easily avoid the climb itself.

a. For training, top-roped climbing is valuable because it allows climbers to attempt climbs above their skill level and or to hone present skills without the risk of a fall. Top-roped climbing may be used to increase the stamina of a climber training to climb longer routes as well as for a climber practicing protection placements.

b. The belayer is positioned either at the base of a climb with the rope running through the top anchor and back to the climber or at the top at the anchor. The belayer takes in rope as the climber moves up the rock, giving the climber the same protection as a belay from above. If this is accomplished with the belayer at the bottom, the instructor is able to keep an eye on the belayer while he coaches the climber through the movements.

6-22. LEAD CLIMBING

A lead climb consists of a belayer, a leader or climber, rope(s), and webbing or hardware used to establish anchors or protect the climb. As he climbs the route, the leader emplaces "intermediate" anchors, and the climbing rope is connected to these anchors with a carabiner. These "intermediate" anchors protect the climber against a fall-thus the term "protecting the climb."

Note: Intermediate anchors are commonly referred to as "protection," "pro," "pieces," "pieces of pro," "pro placements," and so on. For standardization within this publication, these specific anchors will be referred to as "protection;" anchors established for other purposes, such as rappel points, belays, or other rope installations, will be referred to as "anchors."

CAUTION

During all lead climbing, each climber in the team is either anchored or being belayed.

a. Lead climbing with two climbers is the preferred combination for movement on technically difficult terrain. Two climbers are at least twice as fast as three climbers, and are efficient for installing a "fixed rope," probably the most widely used rope installation in the mountains. A group of three climbers are typically used on moderate snow, ice, and snow-covered glaciers where the rope team can often move at the same time, stopping occasionally to set up belays on particularly difficult sections. A group or team of three climbers is sometimes used in rock climbing because of an odd number of personnel, a shortage of ropes (such as six climbers and only two ropes), or to protect and assist an individual who has little or no experience in climbing and belaying. Whichever technique is chosen, a standard roped climbing procedure is used for maximum speed and safety.

b. When the difficulty of the climbing is within the "leading ability" of both climbers, valuable time can be saved by "swinging leads." This is normally the most efficient method for climbing multipitch routes. The second finishes cleaning the first pitch and continues climbing, taking on the role of lead climber. Unless he requires equipment from the other rack or desires a break, he can climb past the belay and immediately begin leading. The belayer simply adjusts his position, re-aiming the belay once the new leader begins placing protection. Swinging leads, or "leap frogging," should be planned before starting the climb so the leader knows to anchor the upper belay for both upward and downward pulls during the setup.

c. The procedures for conducting a lead climb with a group of two are relatively simple. The most experienced individual is the "lead" climber or leader, and is responsible for selecting the route. The leader must ensure the route is well within his ability and the ability of the second. The lead climber carries most of the climbing equipment in order to place protection along the route and set up the next belay. The leader must also ensure that the second has the necessary equipment, such as a piton hammer, nut tool, etc., to remove any protection that the leader may place.

(1) The leader is responsible for emplacing protection frequently enough and in such a manner that, in the event that either the leader or the second should fall, the fall will be neither long enough nor hard enough to result in injury. The leader must also ensure that the rope is routed in a way that will allow it to run freely through the protection placements, thus minimizing friction, or "rope drag".

(2) The other member of the climbing team, the belayer (sometimes referred to as the "second"), is responsible for belaying the leader, removing the belay anchor, and retrieving the protection placed by the leader between belay positions (also called "cleaning the pitch").

(3) Before the climb starts, the second will normally set up the first belay while the leader is arranging his rack. When the belay is ready, the belayer signals, "BELAY ON", affirming that the belay is "on" and the rope will be managed as necessary. When the leader is ready, he double checks the belay. The leader can then signal, "CLIMBING", only as a courtesy, to let the belayer know he is ready to move. The belayer can reply with "CLIMB", again, only as a courtesy, reaffirming that the belay is "on" and the rope will be managed as necessary. The leader then begins climbing.

(4) While belaying, the second must pay close attention to the climber's every move, ensuring that the rope runs free and does not inhibit the climber's movements. If he cannot see the climber, he must "feel" the climber through the rope. Unless told otherwise by the climber, the belayer can slowly give slack on the rope as the climber proceeds on the route. The belayer should keep just enough slack in the rope so the climber does not have to pull it through the belay. If the climber wants a tighter rope, it can be called for. If the belayer notices too much slack developing in the rope, the excess

rope should be taken in quickly. It is the belayer's responsibility to manage the rope, whether by sight or feel, until the climber tells him otherwise.

(5) As the leader protects the climb, slack will sometimes be needed to place the rope through the carabiner (clipping), in a piece of protection above the tie-in point on the leaders harness. In this situation, the leader gives the command "SLACK" and the belayer gives slack, (if more slack is needed the command will be repeated). The leader is able to pull a bight of rope above the tie-in point and clip it into the carabiner in the protection above. When the leader has completed the connection, or the clip, the command "TAKE ROPE" is given by the leader and the belayer takes in the remaining slack.

(6) The leader continues on the route until either a designated belay location is reached or he is at the end of or near the end of the rope. At this position, the leader sets an anchor, connects to the anchor and signals "OFF BELAY". The belayer prepares to climb by removing all but at least one of his anchors and secures the remaining equipment. The belayer remains attached to at least one anchor until the command "BELAY ON" is given.

d. When the leader selects a particular route, he must also determine how much, and what types, of equipment might be required to safely negotiate the route. The selected equipment must be carried by the leader. The leader must carry enough equipment to safely protect the route, additional anchors for the next belay, and any other items to be carried individually such as rucksacks or individual weapons.

(1) The leader will assemble, or "rack," the necessary equipment onto his harness or onto slings around the head and shoulder. A typical leader "rack" consists of:

- Six to eight small wired stoppers on a carabiner.
- Four to six medium to large wired stoppers on a carabiner.
- Assorted hexentrics, each on a separate carabiner.
- SLCDs of required size, each on a separate carabiner.
- Five to ten standard length runners, with two carabiners on each.
- Two to three double length runners, with two carabiners on each.
- Extra carabiners.
- Nut tool.

Note: The route chosen will dictate, to some degree, the necessary equipment. Members of a climbing team may need to consolidate gear to climb a particular route.

(2) The belayer and the leader both should carry many duplicate items while climbing.

- Short Prusik sling.
- Long Prusik sling.
- Cordellette.
- 10 feet of 1-inch webbing.
- 20 feet of 1-inch webbing.
- Belay device (a combination belay/rappel device is multifunctional).
- Rappel device (a combination belay/rappel device is multifunctional).
- Large locking carabiner (pear shape carabiners are multifunctional).

- Extra carabiners.
- Nut tool (if stoppers are carried).

Note: If using an over the shoulder gear sling, place the items in order from smallest to the front and largest to the rear.

e. Leading a difficult pitch is the most hazardous task in roped climbing. The lead climber may be exposed to potentially long, hard falls and must exercise keen judgment in route selection, placement of protection, and routing of the climbing rope through the protection. The leader should try to keep the climbing line as direct as possible to the next belay to allow the rope to run smoothly through the protection with minimal friction. Protection should be placed whenever the leader feels he needs it, and BEFORE moving past a difficult section.

CAUTION

The climber must remember he will fall twice the distance from his last piece of protection before the rope can even begin to stop him.

(1) ***Placing Protection.*** Generally, protection is placed from one stable position to the next. The anchor should be placed as high as possible to reduce the potential fall distance between placements. If the climbing is difficult, protection should be placed more frequently. If the climbing becomes easier, protection can be placed farther apart, saving hardware for difficult sections. On some routes an extended diagonal or horizontal movement, known as a traverse, is required. As the leader begins this type of move, he must consider the second's safety as well as his own. The potential fall of the second will result in a pendulum swing if protection is not adequate to prevent this. The danger comes from any objects in the swinging path of the second.

CAUTION

Leader should place protection prior to, during, and upon completion of any traverse. This will minimize the potential swing, or pendulum, for both the leader and second if either should fall.

(2) ***Correct Clipping Technique.*** Once an anchor is placed, the climber "clips" the rope into the carabiner (Figure 6-29, page 6-50). As a carabiner hangs from the protection, the rope can be routed through the carabiner in two possible ways. One way will allow the rope to run smoothly as the climber moves past the placement; the other way will often create a dangerous situation in which the rope could become "unclipped" from the carabiner if the leader were to fall on this piece of protection. In addition, a

series of incorrectly clipped carabiners may contribute to rope drag. When placing protection, the leader must ensure the carabiner on the protection does not hang with the carabiner gate facing the rock; when placing protection in a crack ensure the carabiner gate is not facing into the crack.

- Grasp the rope with either hand with the thumb pointing down the rope towards the belayer
- Pull enough rope to reach the carabiner with a bight
- Note the direction the carabiner is hanging from the protection
- Place the bight into the carabiner so that, when released, the rope does not cause the carabiner to twist.

(a) If the route changes direction, clipping the carabiner will require a little more thought. Once leaving that piece of protection, the rope may force the carabiner to twist if not correctly clipped. If the clip is made correctly, a rotation of the clipped carabiner to ensure that the gate is not resting against the rock may be all that is necessary.

CAUTION

Ensure the carabiner gate is not resting against a protrusion or crack edge in the rock surface; the rock may cause the gate to open.

(b) Once the rope is clipped into the carabiner, the climber should check to see that it is routed correctly by pulling on the rope in the direction it will travel when the climber moves past that position.

(c) Another potential hazard peculiar to leading should be eliminated before the climber continues. The carabiner is attached to the anchor or runner with the gate facing away from the rock and opening down for easy insertion of the rope. However, in a leader fall, it is possible for the rope to run back over the carabiner as the climber falls below the placement. If the carabiner is left with the gate facing the direction of the route there is a chance that the rope will open the gate and unclip itself entirely from the placement. To prevent this possibility, the climber should ensure that after the clip has been made, the gate is facing away from the direction of the route. There are two ways to accomplish this: determine which direction the gate will face before the protection or runner is placed or once clipped, rotate the carabiner upwards 180 degrees. This problem is more apt to occur if bent gate carabiners are used. Straight gate ovals or "Ds" are less likely to have this problem and are stronger and are highly recommended. Bent gate carabiners are easier to clip the rope into and are used mostly on routes with bolts preplaced for protection. Bent gate carabiners are not recommended for many climbing situations.

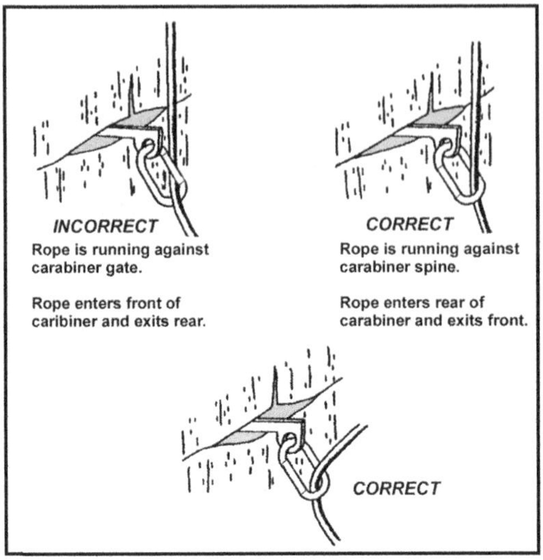

Figure 6-29. Clipping on to protection.

(3) ***Reducing Rope Drag; Using Runners***. No matter how direct the route, the climber will often encounter problems with "rope drag" through the protection positions. The friction created by rope drag will increase to some degree every time the rope passes through a carabiner, or anchor. It will increase dramatically if the rope begins to "zigzag" as it travels through the carabiners. To prevent this, the placements should be positioned so the rope creates a smooth, almost straight line as it passes through the carabiners (Figure 6-30). Minimal rope drag is an inconvenience; severe rope drag may actually pull the climber off balance, inducing a fall.

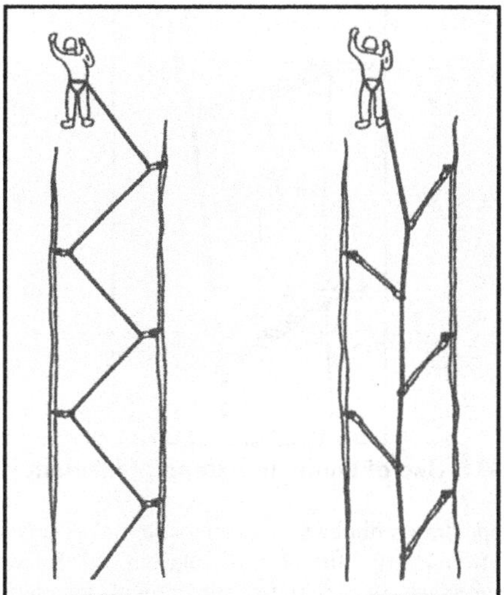

Figure 6-30. Use of slings on protection.

> **CAUTION**
> Rope drag can cause confusion when belaying the second or follower up to a new belay position. Rope drag can be mistaken for the climber, causing the belayer to not take in the necessary slack in the rope and possibly resulting in a serious fall.

(a) If it is not possible to place all the protection so the carabiners form a straight line as the rope moves through, you should "extend" the protection (Figure 6-31, page 6-52). Do this by attaching an appropriate length sling, or runner, to the protection to extend the rope connection in the necessary direction. The runner is attached to the protection's carabiner while the rope is clipped into a carabiner at the other end of the runner. Extending placements with runners will allow the climber to vary the route slightly while the rope continues to run in a relatively straight line.

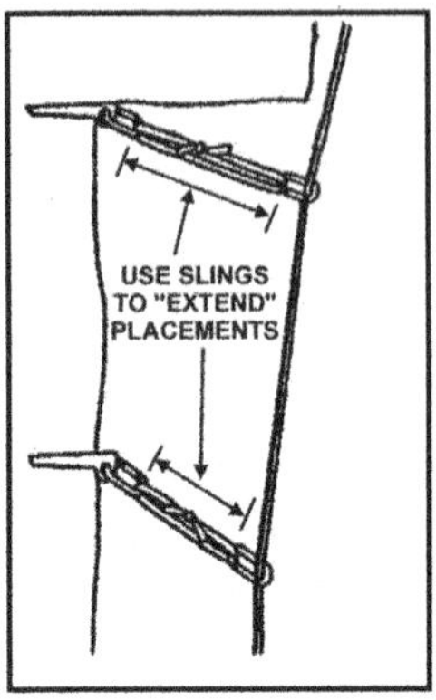

Figure 6-31. Use of slings to extend placement positions.

(b) Not only is rope drag a hindrance, it can cause undue movement of protection as the rope tightens between any "out of line" placements. Rope drag through chock placements can be dangerous. As the climber moves above the placements, an outward or upward pull from rope drag may cause correctly set chocks to pop out, even when used "actively". Most all chocks placed for leader protection should be extended with a runner, even if the line is direct to eliminate the possibility of movement.

(c) Wired chocks are especially prone to wiggling loose as the rope pulls on the stiff cable attachment. All wired chocks used for leader protection should be extended to reduce the chance of the rope pulling them out (Figure 6-32). Some of the larger chocks, such as roped Hexentrics and Tri-Cams, have longer slings pre-attached that will normally serve as an adequate runner for the placement. Chocks with smaller sling attachments must often be extended with a runner. Many of today's chocks are manufactured with pre-sewn webbing installed instead of cable.

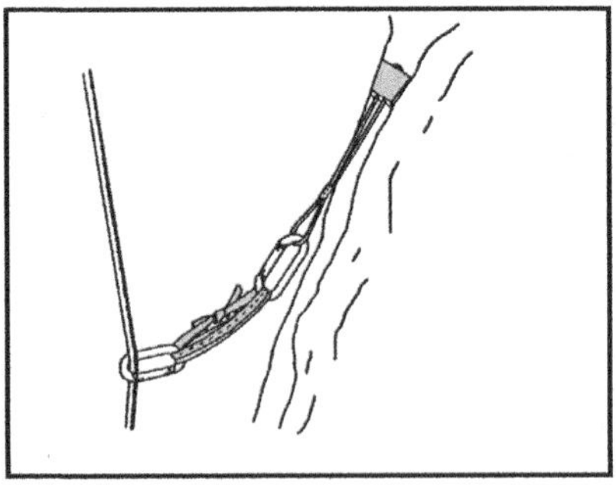

Figure 6-32. Use of sling on a wired stopper.

(d) When a correctly placed piton is used for protection, it will normally not be affected by rope drag. A correctly placed piton is generally a multi-directional anchor, therefore, rope drag through pitons will usually only affect the leader's movements but will continue to protect as expected.

(e) Rope drag will quite often move SLCDs out of position, or "walk" them deeper into the crack than initially placed, resulting in difficult removal or inability to remove them at all. Furthermore, most cases of SLCD movement result in the SLCD moving to a position that does not provide protection in the correct direction or no protection at all due to the lobes being at different angles from those at the original position.

Note: Any placement extended with a runner will increase the distance of a potential fall by the actual length of the sling. Try to use the shortest runners possible, ensuring they are long enough to function properly.

f. Belaying the follower is similar to belaying a top-roped climb in that the follower is not able to fall any farther than rope stretch will allow. This does not imply there is no danger in following. Sharp rocks, rock fall, and inadequately protected traverses can result in damage to equipment or injury to the second.

g. Following, or seconding, a leader has a variety of responsibilities. The second has to issue commands to the leader, as well as follow the leader's commands. Once the lead climber reaches a good belay position, he immediately establishes an anchor and connects to it. When this is completed he can signal "OFF BELAY" to the belayer. The second can now remove the leader's belay and prepare to climb. The second must remain attached to at least one of the original anchors while the leader is preparing the next belay position. The removed materials and hardware can be organized and secured on the second's rack in preparation to climb.

(1) When the leader has established the new belay position and is ready to belay the follower, the "new" belayer signals "BELAY ON." The second, now the climber, removes any remaining anchor hardware/materials and completes any final preparations. The belayer maintains tension on the rope, unless otherwise directed, while the final preparations are taking place, since removal of these remaining anchors can introduce slack into the rope. When the second is ready, he can, as a courtesy, signal "CLIMBING," and the leader can, again as a courtesy, reply with "CLIMB."

(2) Upon signaling "BELAY ON," the belayer must remove and keep all slack from the rope. (This is especially important as in many situations the belayer cannot see the follower. A long pitch induces weight and sometimes "drag" on the rope and the belayer above will have difficulty distinguishing these from a rope with no slack.)

h. When removing protection, the man cleaning the pitch should rack it properly to facilitate the exchange and or arrangement of equipment at the end of the pitch. When removing the protection, or "cleaning the pitch", SLCDs or chocks may be left attached to the rope to prevent loss if they are accidentally dropped during removal. If necessary, the hardware can remain on the rope until the second reaches a more secure stance. If removing a piton, the rope should be unclipped from the piton to avoid the possibility of damaging the rope with a hammer strike.

(1) The second may need to place full body weight on the rope to facilitate use of both hands for protection removal by giving the command "TENSION." The second

must also ensure that he does not climb faster than the rope is being taken in by the belayer. If too much slack develops, he should signal "TAKE ROPE" and wait until the excess is removed before continuing the climb. Once the second completes the pitch, he should immediately connect to the anchor. Once secured, he can signal "OFF BELAY." The leader removes the belay, while remaining attached to an anchor. The equipment is exchanged or organized in preparation for the next pitch or climb.

(2) When the difficulty of the climbing is within the "leading ability" of both climbers, valuable time can be saved by "swinging leads." This is normally the most efficient method for climbing multi-pitch routes. The second finishes cleaning the first pitch and continues climbing, taking on the role of lead climber. Unless he requires equipment from the belayer or desires a break, he can climb past the belay and immediately begin leading. The belayer simply adjusts his position, re-aiming the belay once the new leader begins placing protection. Swinging leads, or "leap frogging," should be planned before starting the climb so the leader knows to anchor the upper belay for both upward and downward pulls during the setup.

6-23. AID CLIMBING

When a route is too difficult to free climb and is unavoidable, if the correct equipment is available you might aid climb the route. Aid climbing consists of placing protection and putting full body weight on the piece. This allows you to hang solely on the protection you place, giving you the ability to ascend more difficult routes than you can free climb. Clean aid consists of using SLCDs and chocks, and is the simplest form of aid climbing.

a. **Equipment**. Aid climbing can be accomplished with various types of protection. Regardless of the type of protection used, the method of aid climbing is the same. In addition to the equipment for free climbing, other specialized equipment will be needed.

(1) ***Pitons***. Pitons are used the same as for free climbing. Most piton placements will require the use of both hands. Piton usage will usually leave a scar in the rock just by virtue of the hardness of the piton and the force required to set it with a hammer. Swinging a hammer to place pitons will lead to climber fatigue sooner than clean aid. Since pitons are multidirectional, the strength of a well-placed piton is more secure than most clean aid protection. Consider other forms of protection when noise could be hazardous to tactics.

(2) ***Bolts***. Bolts are used when no other protection will work. They are a more permanent form of protection and more time is needed to place them. Placing bolts creates more noise whether drilled by hand or by motorized drill. Bolts used in climbing are a multi-part expanding system pounded into predrilled holes and then tightened to the desired torque with a wrench or other tool. Bolts are used in many ways in climbing today. The most common use is with a hanger attached and placed for anchors in face climbing. However, bolts can be used for aid climbing, with or without the hanger.

(a) Placing bolts for aid climbing takes much more time than using pitons or clean aid. Bolting for aid climbing consists of consecutive bolts about 2 feet apart. Drilling a deep enough hole takes approximately thirty minutes with a hand drill and up to two minutes with a powered hammer drill. A lot of time and work is expended in a short distance no matter how the hole is drilled. (The weight of a powered hammer drill becomes an issue in itself.) Noise will also be a factor in both applications. A constant pounding with a hammer on the hand drill or the motorized pounding of the powered drill may alert the enemy to the position. The typical climbing bolt/hanger combination normally is left in the hole where it was placed.

(b) Other items that can be used instead of the bolt/hanger combination are the removable and reusable "spring-loaded removable bolts" such as rivets (hex head threaded bolts sized to fit tightly into the hole and pounded in with a hammer), split-shaft rivets, and some piton sizes that can be pounded into the holes. When using rivets or bolts without a hanger, place a loop of cable over the head and onto the shaft of the rivet or bolt and attach a carabiner to the other end of the loop (a stopper with the chock slid back will suffice). Rivet hangers are available that slide onto the rivet or bolt after it is placed and are easily removed for reuse. Easy removal means a slight loss of security while in use.

(3) ***SLCDs.*** SLCDs are used the same as for free climbing, although in aid climbing, full body weight is applied to the SLCD as soon as it is placed.

(4) ***Chocks.*** Chocks are used the same as for free climbing, although in aid climbing, weight is applied to the chock as soon as it is placed.

(5) ***Daisy Chains.*** Daisy chains are tied or presewn loops of webbing with small tied or presewn loops approximately every two inches. The small loops are just large enough for two or three carabiners. Two daisy chains should be girth-hitched to the tie-in point in the harness.

(6) ***Etriers (or Aiders).*** Etriers (aiders) are tied or presewn webbing loops with four to six tied or presewn internal loops, or steps, approximately every 12 inches. The internal loops are large enough to easily place one booted foot into. At least two etriers (aiders) should be connected by carabiner to the free ends of the daisy chains.

(7) ***Fifi Hook.*** A fifi hook is a small, smooth-surfaced hook strong enough for body weight. The fifi hook should be girth-hitched to the tie-in point in the harness and is used in the small loops of the daisy chain. A carabiner can be used in place of the fifi hook, although the fifi hook is simpler and adequate.

(8) ***Ascenders.*** Ascenders are mechanical devices that will move easily in one direction on the rope, but will lock in place if pushed or pulled the other direction. (Prusiks can be used but are more difficult than ascenders.)

b. **Technique.** The belay will be the same as in normal lead climbing and the rope will be routed through the protection the same way also. The big difference is the movement up the rock. With the daisy chains, aiders, and fifi hook attached to the rope tie-in point of the harness as stated above, and secured temporarily to a gear loop or gear sling, the climb continues as follows:

(1) The leader places the first piece of protection as high as can safely be reached and attaches the appropriate sling/carabiner

(2) Attach one daisy chain/aider group to the newly placed protection

(3) Clip the rope into the protection, (the same as for normal lead climbing)

(4) Insure the protection is sound by weighting it gradually; place both feet, one at a time, into the steps in the aider, secure your balance by grasping the top of the aider with your hands.

(5) When both feet are in the aider, move up the steps until your waist is no higher than the top of the aider.

(6) Place the fifi hook (or substituted carabiner) into the loop of the daisy chain closest to the daisy chain/aider carabiner, this effectively shortens the daisy chain;

maintain tension on the daisy chain as the hook can fall out of the daisy chain loop if it is unweighted.

Note: Moving the waist higher than the top of the aider is possible, but this creates a potential for a fall to occur even though you are on the aider and "hooked" close to the protection with the daisy chain. As the daisy chain tie-in point on the harness moves above the top of the aider, you are no longer supported from above by the daisy chain, you are now standing above your support. From this height, the fifi hook can easily fall out of the daisy chain loop if it is unweighted. If this happens, you could fall the full length of the daisy chain resulting in a static fall on the last piece of protection placed.

(7) Release one hand from the aider and place the next piece of protection, again, as high as you can comfortably reach; if using pitons or bolts you may need both hands free- "lean" backwards slowly, and rest your upper body on the daisy chain that you have "shortened" with the fifi hook

(8) Clip the rope into the protection

(9) Attach the other daisy chain/aider group to the next piece of protection

(10) Repeat entire process until climb is finished

c. **Seconding.** When the pitch is completed, the belayer will need to ascend the route. To ascend the route, use ascenders instead of Prusiks, ascenders are much faster and safer than Prusiks. Attach each ascender to a daisy chain/aider group with carabiners. To adjust the maximum reach/height of the ascenders on the rope, adjust the effective length of the daisy chains with a carabiner the same as with the fifi hook; the typical height will be enough to hold the attached ascender in the hand at nose level. When adjusted to the correct height, the arms need not support much body weight. If the ascender is too high, you will have difficulty reaching and maintaining a grip on the handle.

(1) Unlike lead climbing, there will be a continuous load on the rope during the cleaning of the route, this would normally increase the difficulty of removing protection. To make this easier, as you approach the protection on the ascenders, move the ascenders, one at a time, above the piece. When your weight is on the rope above the piece, you can easily unclip and remove the protection.

CAUTION

If both ascenders should fail while ascending the pitch, a serious fall could result. To prevent this possibility, *tie-in short* on the rope every 10-20 feet by tying a figure eight loop and clipping it into the harness with a separate locking carabiner as soon as the ascent is started. After ascending another 20 feet, repeat this procedure. Do not unclip the previous figure eight until the new knot is attached to another locking carabiner. Clear each knot as you unclip it.

Notes: 1. Ensure the loops formed by the short tie-ins do not catch on anything below as you ascend.

2. If the nature of the rock will cause the "hanging loop" of rope, formed by tying in at the end of the rope, to get caught as you move upward, do not tie into the end of the rope.

(2) Seconding an aid pitch can be done in a similar fashion as seconding free-climbed pitches. The second can be belayed from above as the second "climbs" the protection. However, the rope is unclipped from the protection before the aider/daisy chain is attached.

d. **Seconding Through a Traverse.** While leading an aid traverse, the climber is hanging on the protection placed in front of the current position. If the second were to clean the section by hanging on the rope while cleaning, the protection will be pulled in more than one direction, possibly resulting in the protection failing. To make this safer and easier, the second should hang on the protection just as the leader did. As the second moves to the beginning of the traverse, one ascender/daisy chain/aider group is removed from the rope and clipped to the protection with a carabiner, (keep the ascenders attached to the daisy chain/aider group for convenience when the traverse ends). The second will negotiate the traverse by leapfrogging the daisy chain/aider groups on the next protection just as the leader did. Cleaning is accomplished by removing the protection as it is passed when all weight is removed from it. This is in effect a self-belay. The second maintains a shorter safety tie-in on the rope than for vertical movement to reduce the possibility of a lengthy pendulum if the protection should pull before intended.

e. **Clean Aid Climbing**. Clean aid climbing consists of using protection placed without a hammer or drill involvement: chocks, SLCDs, hooks, and other protection placed easily by hand. This type of aid climbing will normally leave no trace of the climb when completed. When climbing the aiders on clean aid protection, ensure the protection does not "move" from it's original position.

(1) Hooks are any device that rests on the rock surface without a camming or gripping action. Hooks are just what the name implies, a curved piece of hard steel with a hole in one end for webbing attachment. The hook blade shape will vary from one model to another, some have curved or notched "blades" to better fit a certain crystal shape on a face placement. These types of devices due to their passive application, are only secure while weighted by the climber.

(2) Some featureless sections of rock can be negotiated with hook use, although bolts can be used. Hook usage is faster and quieter but the margin of safety is not there unless hooks are alternated with more active forms of protection. If the last twenty foot section of a route is negotiated with hooks, a forty foot fall could result.

6-24. THREE-MAN CLIMBING TEAM

Often times a movement on steep terrain will require a team of more than two climbers, which involves more difficulties. A four-man team (or more) more than doubles the difficulty found in three men climbing together. A four-man team should be broken down into two groups of two unless prevented by a severe lack of gear.

a. Given one rope, a three-man team is at a disadvantage on a steep, belayed climb. It takes at least twice as long to climb an average length pitch because of the third climber and the extra belaying required. The distance between belay positions will be halved if only one rope is used because one climber must tie in at the middle of the rope. Two ropes are recommended for a team of three climbers.

Note: Time and complications will increase when a three-man team uses only one rope. For example: a 100-foot climb with a 150-foot rope would normally require two belays for two climbers; a 100-foot climb with a 150-foot rope would require six belays for three climbers.

b. At times a three-man climb may be unavoidable and personnel should be familiar with the procedure. Although a team of three may choose from many different methods, only two are described below. If the climb is only one pitch, the methods will vary.

CAUTION

When climbing with a team of three, protected traverses will require additional time. The equipment used to protect the traverse must be left in place to protect both the second and third climbers.

(1) The first method can be used when the belay positions are not large enough for three men. If using one rope, two climbers tie in at each end and the other at the mid point. When using two ropes, the second will tie in at one end of both ropes, and the other two climbers will each tie in to the other ends. The most experienced individual is the leader, or number 1 climber. The second, or number 2 climber, is the stronger of the remaining two and will be the belayer for both number 1 and number 3. Number 3 will be the last to climb. Although the number 3 climber does no belaying in this method, each climber should be skilled in the belay techniques required. The sequence for this method (in one pitch increments) is as follows (repeated until the climb is complete):

(a) Number 1 ascends belayed by number 2. Number 2 belays the leader up the first pitch while number 3 is simply anchored to the rock for security (unless starting off at ground level) and manages the rope between himself and number 2. When the leader completes the pitch, he sets up the next belay and belays number 2 up.

(b) Number 2 ascends belayed by number 1, and cleans the route (except for traverses). Number 2 returns the hardware to the leader and belays him up the next pitch. When the leader completes this pitch, he again sets up a new belay. When number 2 receives "OFF BELAY" from the leader, he changes ropes and puts number 3 on belay. He should not have to change anchor attachments because the position was already aimed for a downward as well as an upward pull when he belayed the leader.

(c) Number 3 ascends belayed by number 2. When number 3 receives "BELAY ON," he removes his anchor and climbs to number 2's position. When the pitch is completed he secures himself to one of number 2's belay anchors. When number 1's belay is ready, he brings up number 2 while number 3 remains anchored for security. Number 2 again cleans the pitch and the procedure is continued until the climb is completed.

(d) In this method, number 3 performs no belay function. He climbs when told to do so by number 2. When number 3 is not climbing, he remains anchored to the rock for security. The standard rope commands are used; however, the number 2 climber may include the trailing climber's name or number in the commands to avoid confusion as to who should be climbing.

(d) Normally, only one climber would be climbing at a time; however, the number 3 climber could ascend a fixed rope to number 2's belay position using proper ascending technique, with no effect on the other two members of the team. This would save time for a team of three, since number 2 would not have to belay number 3 and could be either belaying number 1 to the next belay or climbing to number 1. If number 3 is to ascend a fixed rope to the next belay position, the rope will be loaded with number 3's weight, and positioned directly off the anchors established for the belay. The rope should be located so it does not contact any sharp edges. The rope to the ascending number 3 could be secured to a separate anchor, but this would require additional time and gear.

(2) The second method uses either two ropes or a doubled rope, and number 2 and number 3 climb simultaneously. This requires either a special belay device that accepts two ropes, such as the tuber type, or with two Munter hitches. The ropes must travel through the belay device(s) without affecting each other.

(a) As the leader climbs the pitch, he will trail a second rope or will be tied in with a figure eight in the middle of a doubled rope. The leader reaches the next belay position and establishes the anchor and then places both remaining climbers on belay. One remaining climber will start the ascent toward the leader and the other will start when a gap of at least 10 feet is created between the two climbers. The belayer will have to remain alert for differences in rope movement and the climbers will have to climb at the same speed. One of the "second" climbers also cleans the pitch.

(b) Having at least two experienced climbers in this team will also save time. The belayer will have additional requirements to meet as opposed to having just one second. The possible force on the anchor will be twice that of one second. The second that is not cleaning the pitch can climb off route, but staying on route will usually prevent a possible swing if stance is not maintained.

CHAPTER 7

ROPE INSTALLATIONS

Obstacles on the battlefield today are inevitable. They can limit the battlefield and, even worse, prevent a unit from accomplishing its mission. However, with highly skilled personnel trained on rope installations, leaders can be assured that even a unit with limited mountain skills and experience will be able to successfully move and operate in terrain that would otherwise have been impassable.

Section I. FIXED ROPE

A fixed rope is a rope anchored in place to assist soldiers in movement over difficult terrain. Its simplest form is a rope tied off on the top of steep terrain. As terrain becomes steeper or more difficult, fixed rope systems may require intermediate anchors along the route. Moving on a fixed rope requires minimal equipment. The use of harnesses, ascenders, and other technical gear makes fixed rope movement easier, faster, and safer, but adds to total mission weight.

7-1. INSTALLATION

To install a fixed rope, two experienced climbers rope up for a roped climb. The leader must have the necessary equipment to rig the anchor at the top of the pitch. Although leader protection is usually not needed on a typical slope, additional hardware can be brought along and placed at the leader's discretion. The second will establish a belay if protection is being placed. Otherwise, he will stack and manage the rope. He ensures the rope runs smoothly up the slope and does not get tangled as the climber ascends. Upon reaching the end of the pitch, the leader will establish the top anchor. Once the anchor is rigged, the leader will take up any remaining slack between himself and the second. He will anchor the installation rope and remain tied into the rope. The second unties from his end of the rope and begins to climb. If the leader placed protection, the second will clean the pitch on his way up.

7-2. UTILIZATION

All personnel using the fixed rope grasp the rope with the palm downward and use it for assistance as they ascend the slope (Figure 7-1, page 7-2). An individual can easily prevent a long fall by attaching himself to the rope with a sling using a friction knot (for example, Prusik, autoblock). The knot is slid along the rope as the individual ascends. If the climber slips and loses control of the rope, the friction knot will grab the rope and arrest the fall. The friction knot used in this manner is referred to as a self-belay (Figure 7-2, page 7-2).

Figure 7-1. Using a fixed rope.

Figure 7-2. Using a self-belay.

7-3. RETRIEVAL

If the fixed rope is to be used on the descent, it can be left in place and recovered after the last rappel. If not, the last climber will tie into the rope and be belayed from above. The climber now can easily free the rope if it gets caught on anything as it is taken up from the belayer.

7-4. FIXED ROPE WITH INTERMEDIATE ANCHORS

Whenever the route varies from the fall line of the slope, the fixed rope must be anchored at intermediate anchor points (Figure 7-3). Intermediate anchor points should also be used on any long routes that exceed the length of a single rope. The use of intermediate anchor points creates independent sections and allows for changes in direction from one section to the next. The independent sections allow for more personnel to move on the fixed rope. This type of fixed rope is commonly used along exposed ridges and narrow mountain passes.

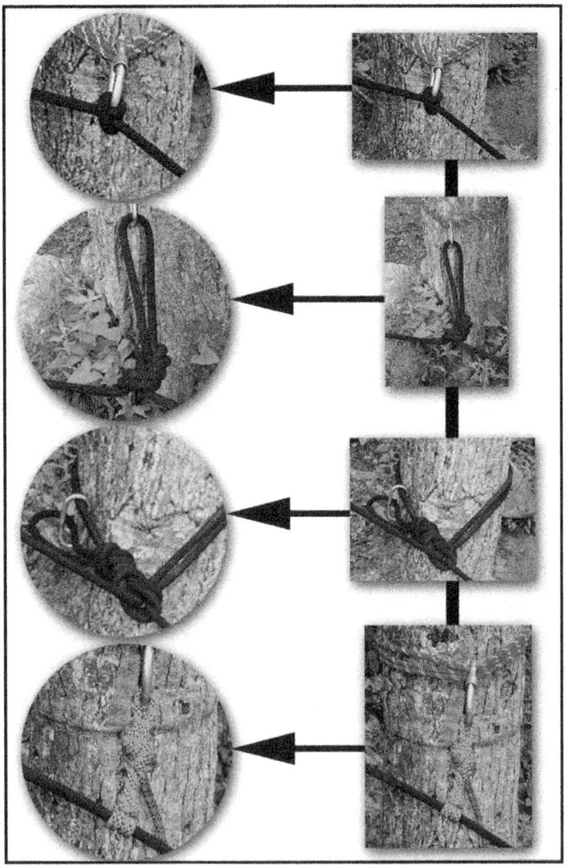

Figure 7-3. Fixed rope with intermediate anchors.

a. **Installation.** Two experienced climbers prepare for a roped climb. The leader will carry a typical rack with enough hardware to place an adequate number of intermediate anchor points. The second sets up a standard belay. The route they select must have the following characteristics:

- Most suitable location, ease of negotiation, avoids obstacles.
- Availability of anchors (natural and artificial).
- Area is safe from falling rock or ice.
- Tactical considerations are met.
- A rope routed between knee and chest height (waist high preferred).
- Rope crossovers should be avoided.

As the leader climbs the route, he will place the anchors and route the climbing rope as in a typical roped climb. The leader makes use of any available natural anchors.

Note: Sling attachments should be kept as short and snug as possible to ensure that a load on the fixed rope below the anchor is placed only on that anchor. This will prevent one section from affecting another section.

(1) The leader places an anchor at all points where a change of direction occurs. He also makes every attempt to route the rope so personnel will not have to cross back and forth over the rope between sections.

(2) When the leader reaches the end of the pitch, he temporarily anchors the rope. He should use a sling to anchor himself if there is any chance of slipping and falling. He then takes up any excess slack, and attaches the rope to the anchor.

Note: Enough slack must be left in the rope so the second can tie the knots necessary to fix the rope.

(3) The second unties from the rope and anchors it at the bottom. He attaches himself to the rope with a sling using a friction knot to create a self-belay. The self-belay will protect the second as he climbs and fixes the rope to the intermediate anchor points. When he reaches an anchor point, he unclips the climbing rope so he can advance the self-belay beyond the anchor point. He then takes the slack out of the section below the anchor point. He ensures that the fixed rope will be approximately knee to chest level as climbers negotiate the installation. He then attaches the rope using an anchor knot (for example, middle-of-the-rope clove hitch, double figure-eight). The second then moves to next anchor point and repeats the process.

(4) If a long runner is to be used at any anchor point, the second should adjust the section below it so the runner is oriented in the direction where the load or pull on the anchor will come from. This will help isolate the section.

(5) The sections are normally adjusted fairly snug between anchor points. A slack section may be necessary to move around obstacles in the route or large bulges in the terrain. If clove hitches are used, adjusting the clove hitches at each end of the section can leave any amount of slack.

(6) A middle-of-the-rope Prusik safetied with a figure eight may be used when utility ropes are available. These are used to adjust the rope height (either higher or lower).

(7) In addition to the fixed rope, the second could anchor etriers to be used as footholds.

(8) When the second reaches the end of the pitch, the rope is removed from the top anchor and the remaining slack is removed from the last section. The rope is reattached to the anchor. If additional fixed rope is required the procedure is repeated using another rope. The second will tie the ropes together before anchoring the next section, creating one continuous fixed rope.

b. **Utilization.** Personnel should be attached to the fixed rope during movement for safety reasons.

(1) If a self-belay is desired, a harness should be worn. A friction knot will be tied to the installation rope using a short sling. The sling will then be attached to the harness. Another short sling will be used as a safety line. One end of the sling will be attached to the harness and the other will have a carabiner inserted. This safety line is also attached to the fixed rope during movement. Once the climber reaches an anchor point, he removes his safety line and attaches it to the anchor or attaches it to the next section of rope. He will then untie the friction knot and tie another friction knot beyond the anchor point. The use of a mechanical ascender in the place of the friction knot could greatly speed up movement.

(2) There will be many situations where a self-belay may not be required. In these situations an individual may attach himself to the fixed rope using only a safety line. The individual will tie into the middle of a sling rope approximately 12 feet long. Fixed loops are tied into the running ends and a carabiner is attached into each of the fixed loops. The individual now has two points of attachments to the fixed rope. Upon reaching an anchor point, one safety line is removed and advanced beyond the anchor point onto the next section. Then the next safety is removed and placed on the next section. This way the individual is always secured to the fixed rope at all times.

(3) Personnel will move one at a time per section during the entire movement. Once an individual changes over to the next section he signals the next man to climb. When descending on the fixed rope, personnel can down climb using the installation for assistance. Another option would be to descend using a hasty rappel.

c. **Retrieval.** When the installation is retrieved, the next to last man on the system will untie the knots at the intermediate anchor points and reclips the rope as he ascends. He will be attached to rope using a self-belay. Once he reaches the top of the pitch, the rope should be running the same as when the leader initially placed it. The last man will untie the rope from the bottom anchor and tie into the rope. He will the clean the pitch as he climbs while being belayed from above.

Section II. RAPPELLING

When an individual or group must descend a vertical surface quickly, a rappel may be performed. Rappelling is a quick method of descent but it is extremely dangerous. These dangers include anchor failure, equipment failure, and individual error. Anchors in a mountainous environment should be selected carefully. Great care must be taken to load the anchor slowly and to ensure that no excessive stress is placed on the anchor. To ensure this, bounding rappels should be prohibited, and only walk down rappels used. Constant vigilance to every detail will guarantee a safe descent every time.

7-5. SELECTION OF A RAPPEL POINT

The selection of the rappel point depends on factors such as mission, cover, route, anchor points, and edge composition (loose or jagged rocks). There must be good anchors (primary and secondary). The anchor point should be above the rappeller's departure point. Suitable loading and off-loading platforms should be available.

7-6. INSTALLATION OF THE RAPPEL POINT

A rappel lane should have equal tension between all anchor points by establishing primary and secondary anchor points. The rappel rope should not extend if one anchor point fails. The following methods of establishing an anchor can be performed with a single or double rope. A double rope application should be used when possible for safety purposes.

a. If a rappel lane is less than half the rope length, the climber may apply one of the following techniques:

(1) Double the rope and tie a three-loop bowline around the primary anchor to include the primary anchor inside two loops and enough rope in the third loop to run to the secondary anchor (another three-loop bowline secured with an overhand knot).

(2) Bowline secured with an overhand knot (or any appropriate anchor knot).

(3) Double the rope and establish a self-equalizing anchor system with a three-loop bowline or any other appropriate anchor knot

b. If a rappel lane is greater than half the rope length, the climber may apply one of the following techniques:

(1) Use two ropes. With both ropes, tie a round turn anchor bowline around a primary anchor point. Take the remaining rope (the tail from the primary anchor bowline) and tie another round turn anchor bowline to a secondary anchor point. The secondary anchor point should be in a direct line behind the primary anchor point. The anchor can be either natural or artificial.

(2) Use two ropes. Establish a multi-point anchor system using a bowline on a bight or any other appropriate anchor knot.

c. Situations may arise where, due to the length of the rappel, the rappel rope cannot be tied to the anchor (if the rope is used to tie the knots, it will be too short to accomplish the rappel). The following techniques can be used:

(1) When using a natural anchor, tie a sling rope, piece of webbing, or another rope around the anchor using proper techniques for slinging natural anchors. The rappel rope will have a fixed loop tied in one end, which is attached to the anchor created.

(2) When using an artificial anchor, tie off a sling rope, piece of webbing, runner, or another rope to form a loop. Use this loop to create an equalizing or pre-equalized anchor, to which the rappel rope will be attached.

7-7. OPERATION OF THE RAPPEL POINT

Due to the inherent dangers of rappelling, special care must be taken to ensure a safe and successful descent.

a. **Communication.** Climbers at the top of a rappel point must be able to communicate with those at the bottom. During a tactical rappel, radios, hand signals, and rope signals are considered. For training situations use the commands shown in Table 7-1.

COMMAND	GIVEN BY	MEANING
LANE NUMBER ___, ON RAPPEL	Rappeller	I am ready to begin rappelling.
LANE NUMBER ___, ON BELAY	Belayer	I am on belay and you may begin your rappel.
LANE NUMBER ___, OFF RAPPEL	Rappeller	I have completed the rappel, cleared the rappel lane, and am off the rope.
LANE NUMBER ___, OFF BELAY	Belayer	I am off belay.

Table 7-1. Rappel commands.

Notes: 1. In a training environment, the lane number must be understood.

2. In a tactical situation, a series of tugs on the rope may be substituted for the oral commands to maintain noise discipline. The number of tugs used to indicate each command is IAW the unit SOP.

b. **Duties and Responsibilities.**

(1) Duties of the rappel point commander are as follows:

- Ensures that the anchors are sound and the knots are properly tied.
- Ensures that loose rock and debris are cleared from the loading platform.
- Allows only one man on the loading platform at a time and ensures that the rappel point is run orderly.
- Ensures that each man is properly prepared for the particular rappel: gloves on, sleeves down, helmet with chin strap fastened, gear prepared properly, and rappel seat and knots correct (if required). He also ensures that the rappeller is hooked up to the rope correctly and is aware of the proper braking position.
- Ensures that the proper signals or commands are used.
- Dispatches each man down the rope.
- Is the last man down the rope.

(2) Duties of the first rappeller down are as follows:

- Selects a smooth route, for the rope, that is clear of sharp rocks.
- Conducts a self-belay.
- Clears the route, placing loose rocks far enough back on ledges to be out of the way, which the rope may dislodge.
- Ensures the rope reaches the bottom or is at a place from which additional rappels can be made.
- Ensures that the rope will run freely around the rappel point when pulled from below.
- Clears the rappel lane by straightening all twists and tangles from the ropes.
- Belays subsequent rappellers down the rope or monitors subsequent belayers
- Takes charge of personnel as they arrive at the bottom (off-loading platform).

Note: A rappeller is always belayed from the bottom, except for the first man down. The first man belays himself down the rope using a self-belay attached to his rappel seat, which is hooked to the rappel rope with a friction knot. As the first man rappels down the rope, he "walks" the friction knot down with him.

(3) Each rappeller down clears the ropes, and shouts, "Off rappel," (if the tactical situation permits). After the rope is cleared and the rappeller is off rappel, he acts as the belayer for next rappeller.

(4) Soldiers wear gloves for all types of rappels to protect their hands from rope burns.

(5) Rappellers descend in a smooth, controlled manner.

(6) The body forms an L-shape with the feet shoulder-width apart, legs straight, and buttocks parallel to the ground. When carrying equipment or additional weight, a modified L-shape is used with the legs slightly lower than the buttocks to compensate for the additional weight. The rappeller's back is straight. He looks over the brake shoulder. The guide hand is extended on the rope with the elbow extended and locked. The rope slides freely through the guide hand. The guide hand is used to adjust equipment and assist balance during descent. The rappeller grasps the rope firmly with the brake hand and places it in the brake position. Releasing tension on the rope and moving the brake hand regulates the rate of descent. The rappeller never lets go of the ropes with his brake hand until the rappel is complete.

c. **Tying Off During the Rappel.** It may be necessary to stop during descent. This can be accomplished by passing the rope around the body and placing three or more wraps around the guide-hand-side leg, or by tying off using the appropriate knot for the rappel device.

7-8. RECOVERY OF THE RAPPEL POINT

After almost all personnel have descended, only two personnel will remain at the top of the rappel point. They will be responsible for establishing a retrievable rappel.

a. **Establishing the Retrievable Rappel.** To set up a retrievable rappel point, a climber must apply one of the following methods:

(1) Double the rope when the rappel is less than half the total length of the rope. Place the rope, with the bight formed by the midpoint, around the primary anchor. Join the tails of the rappel rope and throw the rope over the cliff. Tie a clove hitch around a carabiner, just below the anchor point, with the locking bar outside the carabiner away from the gate opening end and facing uphill. Snap the opposite standing portion into the carabiner. When the rappeller reaches the bottom, he pulls on that portion of the rope to which the carabiner is secured to allow the rope to slide around the anchor point.

(2) When the length of the rappel is greater than half the length of the rope used, join two ropes around the anchor point with an appropriate joining knot (except the square knot). Adjust the joining knot so that it is away from the anchor. Tie a clove hitch around a carabiner just below the anchor point with the locking bar outside the carabiner away from the gate opening end and facing uphill. Snap the opposite standing portion into the carabiner. Upon completion of the rappel, pull the rope to which the carabiner is secured to allow the rope to slide around the anchor point.

Notes: 1. When setting up a retrievable rappel, use only a primary point; care is taken in selecting the point.
2. Ensure the soldiers have a safety line when approaching the rappel point, with only the rappeller going near the edge.

b. **Retrieving the Rappel Rope.** The next to last rappeller will descend the lane, removing any twists, and routes the rope for easiest retrieval. Once he reaches the end of the rappel, he tests the rope for retrieval. If the rappel is retrievable, the last man will rappel down. Once he is off rappel, he pulls the lane down.

7-9. TYPES OF RAPPELS

During military mountaineering operations, many types of rappels may be used. The following paragraphs describe some these rappels.

a. **Hasty Rappel** (Figure 7-4). The hasty rappel is used only on moderate pitches. Its main advantage is that it is easier and faster than other methods. Gloves are worn to prevent rope burns.

(1) Facing slightly sideways to the anchor, the rappeller places the ropes horizontally across his back. The hand nearest to the anchor is his guide hand, and the other is the brake hand.

(2) To stop, the rappeller brings his brake hand across in front of his body locking the rope. At the same time, he turns to face up toward the anchor point.

Figure 7-4. Hasty rappel.

b. **Body Rappel** (Figure 7-5). The rappeller faces the anchor point and straddles the rope. He then pulls the rope from behind, and runs it around either hip, diagonally across the chest, and back over the opposite shoulder. From there, the rope runs to the brake hand, which is on the same side of the hip that the rope crosses (for example, the right hip to the left shoulder to the right hand). The rappeller leads with the brake hand down and faces slightly sideways. The foot corresponding to the brake hand precedes the guide hand at all times. The rappeller keeps the guide hand on the rope above him to guide himself--not to brake himself. He must lean out at a sharp angle to the rock. He keeps his legs spread well apart and relatively straight for lateral stability, and his back straight to reduce friction. The BDU collar is turned up to prevent rope burns on the neck. Gloves are worn, and other clothing may be used to pad the shoulders and buttocks. To brake, the rappeller leans back and faces directly toward the rock area so his feet are horizontal to the ground.

Figure 7-5. Body rappel.

Notes: 1. Hasty rappels and body rappels are not used on pitches that have overhangs; feet must maintain surface contact.

2. Hasty rappels and body rappels are not belayed from below.

c. **Seat-Hip Rappel** (Figure 7-6). The seat rappel differs from the body rappel in that the friction is absorbed by a carabiner that is inserted in a sling rope seat and fastened to the rappeller. This method provides a faster and more frictional descent than other methods. Gloves can be worn to prevent rope burns.

Figure 7-6. Seat-hip rappel.

(1) An alternate technique is to insert two carabiners opposite and opposed. Then insert a locking carabiner into the two carabiners with opening gate on brake hand side. Then run the rope through the single carabiner. This helps to keep the rappel rope away from the harness.

(2) To hook up for the seat-hip method, stand to one side of the rope. If using a right-hand brake, stand to the left of the rappel rope facing the anchor; if using a left-hand brake, stand to the right of the rappel rope. Place the rappel rope(s) into the locking carabiner; slack is taken between the locking carabiner and anchor point and wrapped around the shaft of the locking carabiner and placed into the gate so that a round turn is made around the shaft of the locking carabiner (Figure 7-7, page 7-12). Any remaining slack is pulled toward the uphill anchor point. If a single rope is used, repeat this process to place two round turns around the shaft of the locking carabiner. Face the anchor point and descend using the upper hand as the guide and the lower hand as the brake. This method has minimal friction, and is fast and safe. However, care is taken that the rope is hooked correctly into the carabiner to avoid the gate being opened by the rope. Loose clothing or equipment around the waist may be accidentally pulled into the locking carabiner and lock (stop) the rappel. For this reason, the rappeller must tuck in his shirt and keep his equipment out of the way during his descent.

Figure 7-7. Proper hookup using carabiner wrap.

d. **Figure-Eight Descender**. The figure-eight descender puts less kinks in the rope, and it can be used with one or two ropes (Figure 7-8).

(1) To use the figure-eight descender, pass a bight through the large eye and then over the small eye onto the neck. Place the small eye into a locking carabiner. To reduce the amount of friction on the figure-eight, place the original bight into the carabiner and not around the neck of the descender. (Less friction requires more braking force from the rappeller.)

(2) The guide hand goes on the rope that is running from the anchor. The brake hand goes on the slack rope. The brake is applied by moving the brake hand to the rear or downward.

Figure 7-8. Figure-eight descender.

d. **Other Devices**. Many different types of devices are similar in design and operation to the basic plate. These include slots or plates and tubers. Most of these devices can accommodate two ropes not greater than 7/16 of an inch in size. Follow manufacturer's directions for using these devices for rappelling.

e. **Extending the Rappel Device**. The rappel device can be extended using either a piece of webbing or cordage to move the device away from the body and the harness, preventing accidental damage (Figure 7-9, page 7-14). It also allows for easier self-belay.

f. **Self-Belay Techniques.** A friction knot can be used as a belay for a rappeller (Figure 7-9, page 7-14). The knot acts as the brake hand when the rappeller must work or negotiate an obstacle requiring the use of both hands. The knot acts as a belay if the rappeller loses control of the rope.

Figure 7-9. Extended hookup with self-belay.

Section III. ONE-ROPE BRIDGE

The one-rope bridge is constructed using a static rope. The rope is anchored with an anchor knot on the far side of the obstacle and is tied off at the near end with a tightening system. A one-rope bridge may be built many ways, depending upon the tactical situation and area to be crossed (crossing a gorge above the tree line may require constructing artificial anchors). However, they all share common elements to safely construct and use the bridge: two suitable anchors; good loading and unloading platforms; a rope about 1-meter (waist) high for loading and unloading; a tightening system; and a rope tight enough for ease of crossing. Which side the tightening system is utilized, or whether an anchor knot or retrievable bowline is used, depends on the technique.

7-10. SITE SELECTION

A suitable crossing site must have "bombproof" anchors on both the near side and far side. These anchors must be extremely strong due to the amount of tension that will be placed upon them. Natural anchors, such as large trees and solid rock formations, are always preferred. The site must also have suitable loading and off-loading platforms to facilitate safe personnel movement.

7-11. INSTALLATION USING TRANSPORT TIGHTENING SYSTEM

The transport tightening system provides a mechanical advantage without requiring additional equipment.

a. The rope must first be anchored on the far side of the obstacle. If crossing a stream, the swimmer must be belayed across. If crossing a ravine or gorge, crossing may involve rappelling and a roped climb. Once across, the swimmer/climber will temporarily anchor the installation rope.

b. One man on the near side ties a fixed-loop knot (for example, wireman's, figure-eight slip knot) approximately 3 feet from the near side anchor and places the carabiner into the loop of the knot. The opening gate must be up and away from the loop. If two carabiners are used, the gates will be opposing. At that time, soldiers route the remainder of the rope around the near side anchor point and hook the rope into the carabiner. This system is known as a transport-tightening system (Figure 7-10). The man on the far side pulls the knot out four to six feet from the near anchor.

c. Once the knot has been pulled out, the far side man anchors the rope using a tensionless anchor. The anchor should be waist high.

Figure 7-10. Transport tightening system.

d. A three-man pull team on the near side pulls the slack out of the installation rope. The knot should be close enough to the near side anchor to allow personnel to easily load the installation.

Note: No more than three personnel should be used to tighten the rope. Using more personnel can over-tighten the rope and bring the rope critically close to failure.

e. The rope the can be secured using one of three methods: transport knot (Figure 7-11), round turn around anchor and two half hitches on a bight (Figure 7-12), or a tensionless anchor knot (Figure 7-13).

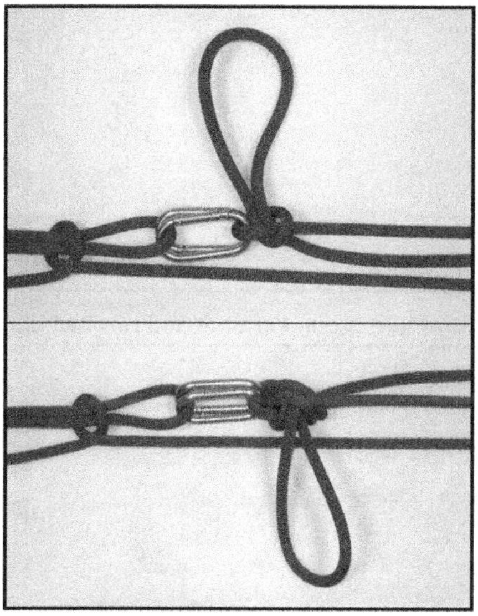

Figure 7-11. Transport knot.

Figure 7-12. Round turn around anchor and two half hitches on a bight.

Figure 7-13. Tensionless anchor knot.

Note: During training, a second static rope may be installed under less tension and alongside the tight rope to increase safety. An individual would clip into both ropes when crossing, thus having a backup in case of failure of the tighter rope.

7-12. INSTALLATION USING Z-PULLEY TIGHTENING SYSTEM

The Z-pulley tightening system (Figure 7-14) is another method for gaining a mechanical advantage.

a. The rope is brought across the obstacle the same way as discussed in paragraph 7-10.

b. Once across, the far side man anchors the rope.

c. One soldier ties a friction knot (autoblock, web wrap, Kleimheist) with a sling rope onto the bridging rope on the near side bank. Two steel carabiners are inserted with opposing gates into the friction knot.

d. The rope is routed around the near side anchor and through the carabiners, from inside to outside, and is run back to the near side anchor.

e. A second sling rope is tied to the bridge rope and then anchored to the near side anchor. This knot will be used as a progress capture device.

Figure 7-14. Z-pulley tightening system.

f. The three-man pull team on the near side then pulls on the rope, creating a pulley effect that tightens the system. As the rope is pulled tight, one man pushes the friction knot back toward the far side.

g. When the rope is tight, it is tied off with a tensionless anchor knot, transport knot, or round turn around anchor and two half hitches on a bight.

7-13. UTILIZATION

The rope bridge can be used to move personnel and equipment over obstacles. There are several methods of accomplishing this.

a. **Method of Crossing**. If dry crossing is impossible, soldiers will use the rope bridge as a hand line. Preferably, all soldiers will tie a safety line and attach it to the rope installation as they cross. If the soldier must cross with his rucksack, he may wear it over both shoulders, although the preferred method is to place another carabiner into the top of the rucksack frame, attach it to the bridge, and pull the rucksack across. Soldiers will always cross on the downstream side of the installation. If a dry crossing is possible soldiers will use one of three methods: commando crawl, monkey crawl, and Tyrolean traverse.

(1) ***Commando Crawl*** (Figure 7-15). The soldier lies on top of the rope with the upstream foot hooked on the rope and the knee bent close to the buttocks; the downstream leg hangs straight to maintain balance. He progresses by pulling with his hands and arms. To recover if he falls over, the soldier hooks one leg and the opposite arm over the rope, and then pushes down with the other hand to regain position.

Figure 7-15. Commando crawl.

Note: Only one man at a time is allowed on the bridge while conducting a commando crawl.

(2) ***Monkey Crawl*** (Figure 7-16, page 7-20). The soldier hangs below the rope suspended by his hands with both heels crossed over the rope. He pulls with his hands and arms, and pushes with his feet to make progress.

Figure 7-16. Monkey crawl.

(3) ***Rappel Seat Method*** (Figure 7-17). The soldier ties a rappel seat (or dons a seat harness) with the carabiner facing up and away from his body. He then faces the rope and clips into the rope bridge. He rotates under the rope and pulls with his hands and arms to make progress. The rappel seat method is the preferred method. If crossing with rucksacks, a carabiner is inserted into the frame and attached to the rope bridge. The soldier the places one or both legs through the shoulder carrying straps and pulls the rucksack across.

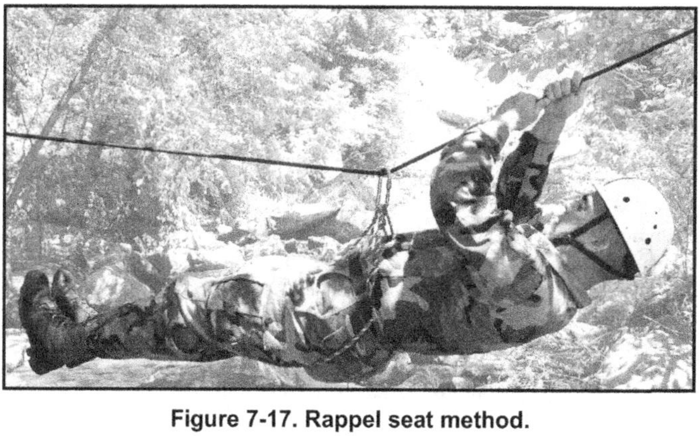

Figure 7-17. Rappel seat method.

b. **Rigging Special Equipment**. Any special equipment, such as crew-served weapons, ammunition, or supplies, must be rigged for movement across the rope bridge. A unit SOP may dictate the rigging of these items, but many expedient methods exist. The rigging should use various items that would be readily available to a deployed unit. Some of these items include tubular nylon webbing, cordage (various sizes), and carabiners.

(1) ***Machine Guns***. To rig machine guns, use a sling rope and tie a rerouted figure-eight around the spine of the front sight post. Then tie two evenly spaced fixed loops. Finally, anchor the sling rope to the buttstock of the machine gun. Additional tie downs may be necessary to prevent accidental disassembly of the weapon.

(2) ***ALICE Packs.*** ALICE packs can be joined together with a sling to facilitate moving more than one rucksack at one time.

7-14. HAULING LINE

A hauling line may be used to move rucksacks or casualties across the rope bridge (Figure 7-18).

a. **Construction**. An additional rope is brought across the rope bridge and anchored to the far side. The other end is anchored on the near side. All the slack is pulled to the near side, and a figure-eight slip knot is tied at the loading platform. A carabiner is inserted into the loop and clipped onto the rope bridge.

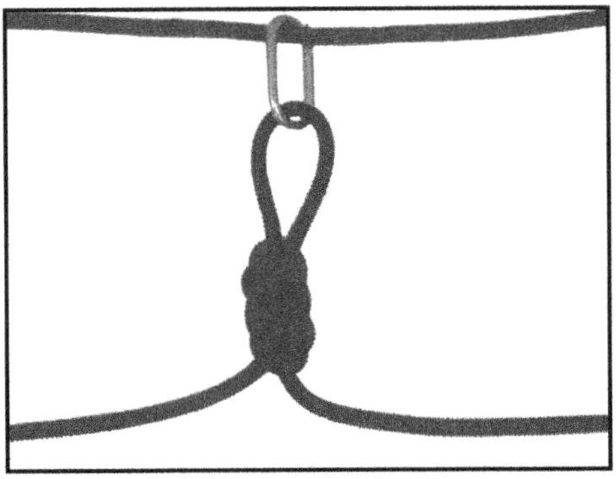

Figure 7-18. Hauling line.

b. **Moving Rucksacks**. Use carabiners to attach the rucksack frames to the rope bridge. Then clip the carabiner of the hauling line into the carabiner of the rucksack closest to the far side. Personnel on the far side pull the rucksacks across using the hauling line while personnel on the near side manages the slack at all times.

c. **Moving Litters**. The carabiner of the hauling line will remain on the rope bridge. On each side of this carabiner, using the hauling line tie a middle-of-the-rope clove hitch around both of the horizontal lift straps of the litter. Remove the slack between the carabiners. Then place the carabiners in each of the lift straps onto the rope bridge. The same technique used for the rucksacks is used to pull the litter across.

7-15. RETRIEVAL

Once all except two troops have crossed the rope bridge, the bridge team commander (BTC) chooses either the wet or dry method to dismantle the rope bridge.

a. If the BTC chooses the dry method, he should have anchored his tightening system with the transport knot.

(1) The BTC back-stacks all of the slack coming out of the transport knot, then ties a fixed loop and places a carabiner into the fixed loop.

(2) The next to last man to cross attaches the carabiner to his rappel seat or harness, and then moves across the bridge using the Tyrolean traverse method.

(3) The BTC then removes all knots from the system. The far side remains anchored. The rope should now only pass around the near side anchor.

(4) A three-man pull team, assembled on the far side, takes the end brought across by the next to last man and pulls the rope tight again and holds it.

(5) The BTC then attaches himself to the rope bridge and moves across.

(6) Once across, the BTC breaks down the far side anchor, removes the knots, and then pulls the rope across.

(b) If the BTC chooses a wet crossing, any method can be used to anchor the tightening system.

(1) All personnel cross except the BTC or the strongest swimmer.

(2) The BTC then removes all knots from the system.

(3) The BTC ties a fixed loop, inserts a carabiner, and attaches it to his rappel seat or harness. He then manages the rope as the slack is pulled to the far side.

(4) The BTC then moves across the obstacle while being belayed from the far side.

Section IV. SUSPENSION TRAVERSE

The suspension traverse is used to move personnel and equipment over rivers, ravines, chasms, and up or down a vertical rock face (Figure 7-19). The system may be established on a plane, varying from horizontal to near vertical.

Figure 7-19. Suspension traverse.

7-16. SITE SELECTION

The crossing site must have bombproof anchors at the near side and the far side, and suitable loading and off-loading platforms. If the anchors do not provide sufficient height to allow clearance, an A-frame must be used.

7-17. INSTALLATION

Installation of a suspension traverse can be time-consuming and equipment-intensive. All personnel must be well trained and well rehearsed in the procedures.

a. **A-frames**. Even in wooded mountainous terrain constructing an A-frame may be necessary due to the lack of height where the installation is needed. Site selection determines whether more height is needed; mission requirements determine site selection. The two main installations that use A-frames are the suspension traverse and vertical hauling line.

b. **Equipment**. Two sturdy poles are needed. The exact size of the poles depends on the type of load and location of the installation. The average size A-frame pole should be at least 3 inches in diameter and 9 to 12 feet long. Three to five 14-foot sling ropes are needed, depending on the size of the poles used for the A-frame.

c. **Construction**. Place two poles with the butt ends flush, and mark the apex on both poles.

(1) Ensure that proper height is attained and that the installation runs in a straight line between the two anchors. An A-frame placed out of proper alignment can cause the system to collapse. Try to find natural pockets in which to place the base of the A-frame poles.

(2) With a sling rope, tie a clove hitch around the left pole (standing at the base of the poles and facing the top) 3 inches above the apex marking, leaving about 18 inches of the sling rope free on top of the clove hitch. Place the locking bar on the outside edge of the pole. Make sure the rope end is pointing down as it is tied. (See Figure 7-20A, page 7-24.)

(3) Place the poles side by side and wrap the sling rope horizontally around both poles six to eight times, wrapping down from the clove hitch (Figure 7-20B, page 7-24). It may be necessary to join another sling rope to the first by using a square knot secured with overhand knots. Position this knot on the outside of one of the poles so as not to interfere with the vertical wraps. Make at least two additional wraps below the joining square knot. (See Figure 7-20C, page 7-24.)

(4) On the last horizontal wrap (ensure there are at least two wraps below the joining knot) to which the clove hitch is not tied, pass the rope between the poles below the wraps, and make four to six tight vertical wraps around the horizontal wraps (Figure 7-20D, page 7-24). Make the wraps as tight as possible. The vertical wraps must be as flat as possible next to each other. When starting the first vertical wrap, ensure it is in the same direction as the 18-inch tail on the top of the clove hitch. Insert a carabiner into the last two vertical wraps (Figure 7-20E, page 7-24).

(5) On the last vertical wrap, pass the rope between the poles above the horizontal wraps. Tie it off with a square knot in the section of rope coming from the clove hitch. Secure with overhand knots tied in the tails. (See Figure 7-20F, page 7-24.)

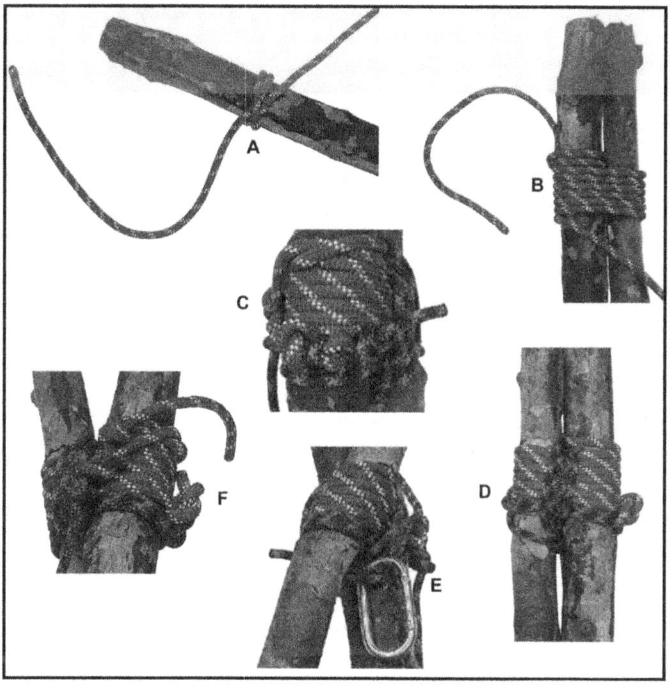

Figure 7-20. A-frame horizontal and vertical wraps.

(6) Use a spreader rope to prevent the A-frame from collapsing from pressure applied at the apex (Figure 7-21). If the ground is soft, dig the legs in about 6 inches. Tie a sling rope between the legs with a round turn with two half hitches around each leg. Remove all slack in the rope between the legs.

(7) If the ground is a hard surface, tie end-of-the-rope clove hitches with the locking portions facing to the rear, the direction of kick. Tie the tails off at a 45-degree angle with a round turn and two half hitches to a secondary anchor point. The spreader rope should be no more than 6 inches above ground level. The use of clove hitches and half hitches permits easy adjustment of the spreader rope. If more than one sling rope is needed, tie the two ropes together with a square knot and secure with half hitches or overhand knots.

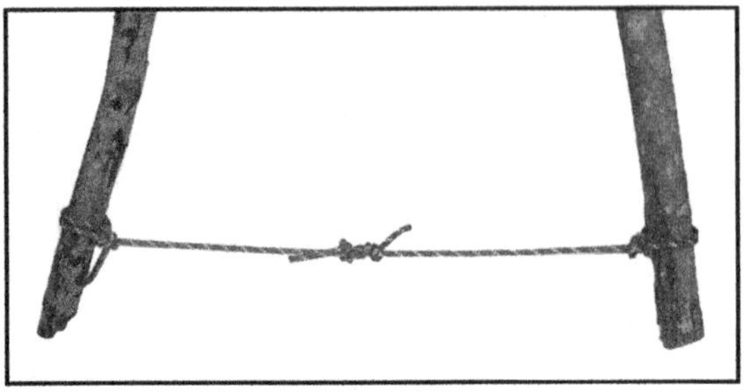

Figure 7-21. A-frame spreader.

d. **Installation Construction**. One man rappels down the pitch and secures two installation (traverse) ropes to the far anchor with an anchor knot. Place a transport tightening system in each installation rope at the near (upper) anchor. Run the installation ropes through or around the anchor in opposite directions and tie off. Anchor the traverse ropes as close together as possible so that the ropes do not cross.

(1) Place the A-frame (if needed) so that both traverse ropes run over the apex and the A-frame splits the angle formed between the near (upper) and far (lower) anchors, with the legs firmly emplaced or anchored with pitons. Ensure that the A-frame is in line with the anchors. Adjust the A-frame under the traverse ropes after tightening to firmly implant the A-frame.

(2) Tighten the installation ropes using either the transport tightening system (paragraph 7-11) or the z-pulley tightening system (paragraph 7-12).

(3) Anchor the A-frame to the traverse rope. Tie a clove hitch at the center of a sling rope. Place it over one of the poles above the apex and move down to the apex so that the locking bar of the clove hitch is to the inside of the A-frame. Secure each end of the sling rope to one of the tightened static lines with two Prusik knots-one forward and one to the rear of the A-frame on the same static line rope (Figure 7-22).

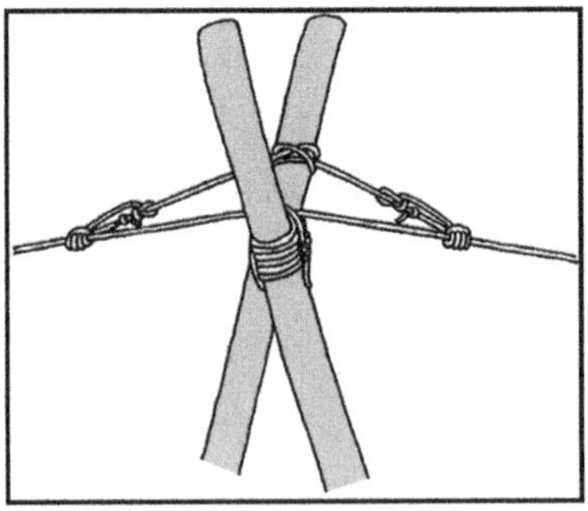

Figure 7-22. Anchoring the A-frame to the traverse rope.

Note: The A-frame should be positioned so that the angles created by the A-frame bisecting the installation rope are approximately equal on both sides. This creates downward pressure holding the A-frame in position, not forcing it in a lateral direction. It must also be placed in a straight line between the upper and lower anchor points.

(4) Use a carrying rope to attach loads to the traverse ropes (Figure 7-23). Join the ends of a 14-foot sling rope with a square knot and two overhand knots. Displace the knot one-third of the distance down the loop and tie an overhand knot both above and below the square knot. This forms two small loops and one large loop that is longer than the two small loops combined.

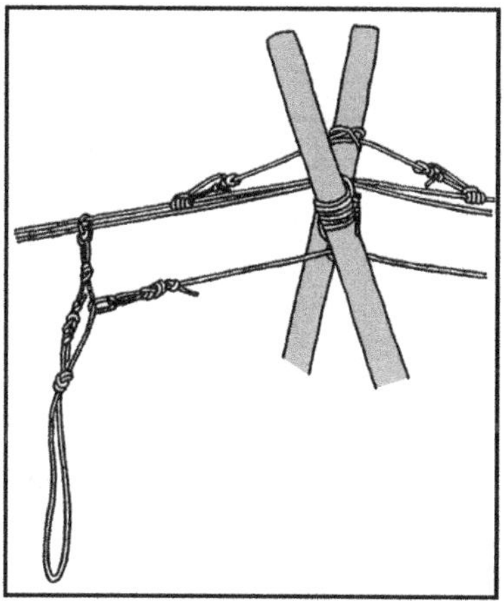

Figure 7-23. Carrying rope for use on a traverse.

(5) Attach the carrying rope to the traverse ropes with carabiners (or a pulley) that have the gates reversed and opening in opposite directions. Attach a belay rope to the center loop of the carrying rope using a fixed loop or locking carabiner on the side opposite the joining knot (Figure 7-23). When the suspension traverse is near horizontal, a second rope may be needed to pull the load across and should be attached to the carrying rope the same as the first.

(6) Insert second carabiner into the one placed into the wraps of the A-frame. This is where a belay rope will be attached

(7) With a sling rope, tie a six wrap middle-of-the-rope Prusik knot to both static ropes near the far side off-loading point. This acts as a stopper knot for the man descending, preventing him from hitting the lower anchor.

(8) Attach the load by running the long loop of the carrying rope through the load or through the soldier's harness and attaching the bottom loop to the traverse rope carabiner.

Descent must be belayed slowly and be controlled. Soldiers descending should hold onto the carrying rope and keep their feet high to avoid contact with the ground. Due to the constant tension maintained on the belayer, use a mechanical belay. If the belayer cannot view the entire descent route, use a relay man.

7-18. RETRIEVAL

The suspension traverse is not as readily retrievable as the one-rope bridge. Therefore, the installing unit should dismantle it after it is no longer needed.

Section V. VERTICAL HAULING LINE

The vertical hauling line is an installation used to move men and equipment up vertical or near-vertical slopes (Figure 7-24). It is often used with a fixed rope for personnel movement. The hauling line is used to move equipment, such as mortars or other crew-served weapons, rucksacks, or supplies

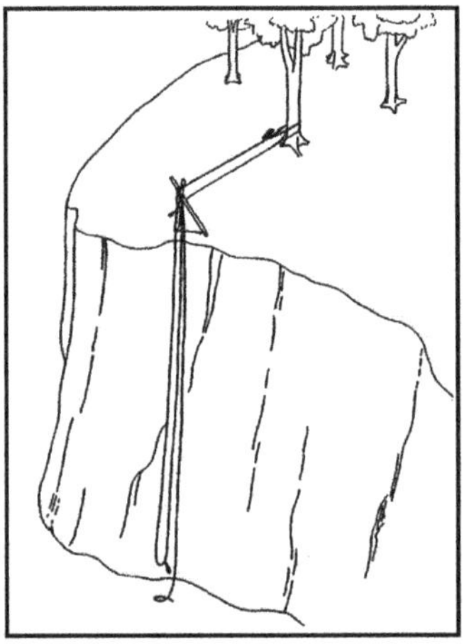

Figure 7-24. Vertical hauling line.

7-19. SITE SELECTION

The first and most important task is to determine where to construct the vertical hauling line. The site must have an appropriate top anchor that is secure enough to hold the system and load. Loading and unloading platforms should be easily accessible natural platforms that provide a safe working area. The ideal platform at the top allows construction of the vertical hauling line without the use of an A-frame. The site should also have sufficient clearance to allow for space between the slope and pulley rope for easy hauling of troops or equipment.

7-20. INSTALLATION

Construct an A-frame, if necessary, and anchor it. Double one installation rope, find the middle, and lay the middle of the installation rope over the apex of the A-frame; a 30-centimeter (12-inch) bight should hang below the apex.

a. To maintain the 12-inch bight, tie clove hitches above the A-frame lashing on each side of the apex with the installation rope, ensuring that the locking bars of the clove hitches are on the inside. Ensure that the portion of the rope that forms the bight comes out of the bottom of the clove hitch. (See Figure 7-25.)

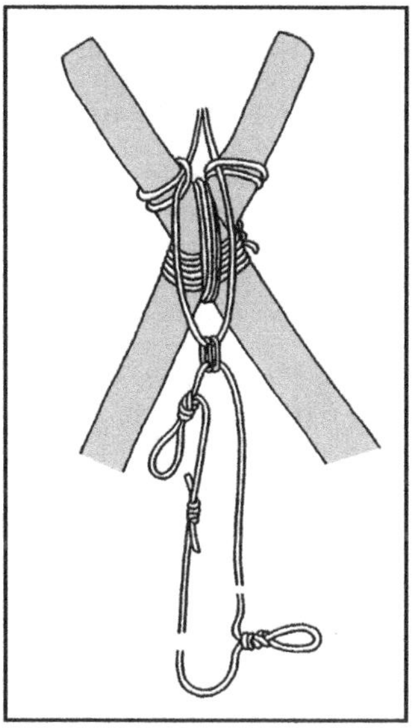

Figure 7-25. Attaching the anchor rope to the A-frame.

b. To anchor the A-frame, use a transport tightening system with the doubled rope, which is tied to the A-frame. Tie this off at an anchor point to the rear of the A-frame installation and adjust the angle of the A-frame so it leans out over the cliff edge. The angle should be 15 to 25 degrees unloaded. The A-frame should not lean outward more than 45 degrees once loaded since the legs can lose their position.

c. Tie the ends of another installation rope together with a joining knot to form the hauling line. Attach the rope to the system by two carabiners with gates up and opposed or one mountain rescue pulley with a locking steel carabiner in the 12-inch bight hanging from the apex of the A-frame. Tie fixed loops (wireman's, directional figure-eight, or single butterfly) on opposite sides of the endless rope at the loading and unloading platforms.

d. Attach equipment to the hauling line 12 inches above the joining knot by a carabiner in the fixed loop.

e. Additional fixed loops may be tied in the hauling line for more control over the object when moving large loads. Attach personnel to the hauling line by use of a rappel seat or seat harness.

Note: Mortar tubes and similar objects are attached to the line by two knots so that the tube stays parallel and as close to the hauling line as possible.

f. When personnel are moved using a vertical hauling line, make a knotted hand line; anchor it in line with, or to, the primary anchor (round turn with a bowline); and place it over the spreader on the legs of the A-frame. Space the overhand knots in the knotted hand line 12 inches apart, with about 20 feet of rope without knots at one end for the anchor. Throw the knotted hand line over the A-frame spreader rope and down the side of the cliff. Personnel ascending the vertical hauling line use this as a simple fixed rope.

g. Use as many men as needed to pull the load to the top by pulling on the rope opposite the load. If equipment and personnel are only being lowered, belay from the top using the hauling line. Station two climbers at the unloading platform to retrieve loads.

h. If only equipment is being hauled up, it is not necessary to use the knotted hand line rope, but it may be necessary to use a belay rope. To move materials or troops up on one side of the hauling line, pull the other side from below.

Note: Personnel using the hauling line for movement must apply all related principles of climbing. Always station two operators at the top of the vertical hauling line to aid men or to retrieve loads when they reach the top. They will always be safetied while working near the edge. When in use, the A-frame should lean slightly over the edge of the cliff to prevent excessive wear on the ropes that pass over sharp rocks. Reduce excessive friction on the system. Remove all obstacles and any loose objects that could be dislodged by personnel and equipment.

7-21. RETRIEVAL

The vertical hauling line is used along a main supply route. When it is no longer needed, the installing unit will return and dismantle the system.

Section VI. SIMPLE RAISING SYSTEMS

Moving heavy objects with limited manpower may be necessary in mountainous terrain. To reduce fatigue of those personnel moving the load, simple rigging techniques can be used to increase the mechanical advantage of the hauling system.

7-22. Z-PULLEY SYSTEM

The Z-pulley system is a simple, easily constructed hauling system (Figure 7-26, page 7-30).

a. **Considerations**. Anchors must be sturdy and able to support the weight of the load. Site selection is governed by different factors: tactical situation, weather, terrain, equipment, load weight, and availability of anchors.

b. **Theory.** Use carabiners as a substitute if pulleys are not available. The mechanical advantage obtained in theory is 3:1. The less friction involved the greater the mechanical advantage. Friction is caused by the rope running through carabiners, the load rubbing against the rock wall, and the rope condition.

c. **Construction**. Use the following procedures to construct a Z-pulley system.

(1) Establish an anchor (anchor pulley system [APS]). Place a carabiner on the runner at the anchor point, place a pulley into the carabiner, and run the hauling rope through the pulley.

(2) With a sling rope (preferably 7 millimeter), tie a middle-of-rope Prusik knot secured with a figure-eight knot on the load side of the pulley. This will be used as a progress capture device (PCD). A mechanical descender may be used in place of the Prusik knot. Take the tails exiting the figure-eight and tie a Munter hitch secured by a mule knot. Ensure the Munter hitch is loaded properly before tying the mule knot.

(3) At an angle away from the APS, establish a moveable pulley system (MPS) to create a "Z" in the hauling rope. Tie another Prusik knot on the load side of the hauling rope. Secure it with a figure-eight knot. Using the tails tie a double-double figure-eight knot. Insert a locking carabiner into the two loops formed, then place the working end into the carabiner. Mechanical ascenders should not be used as an MPS. Move the working end back on a parallel axis with the APS. Provide a pulling team on the working end with extra personnel to monitor the Prusik knots.

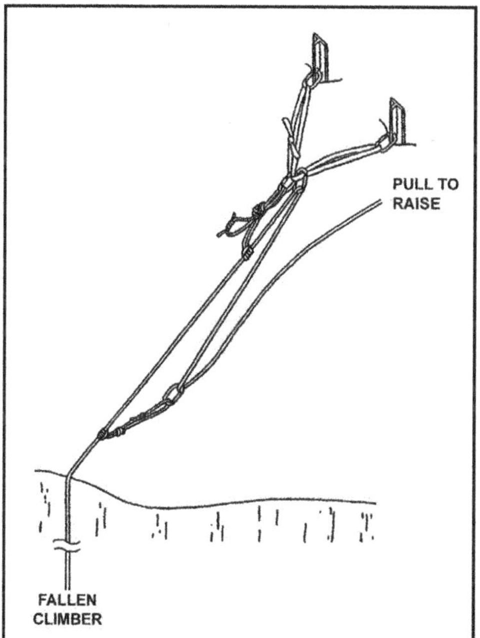

Figure 7-26. Z-pulley system.

d. **Other Factors**. If the two pulleys touch, the "Z" is lost along with the mechanical advantage. For greater efficiency, the main anchor should be well back from the edge and all ropes should pull parallel to the load.

Note: Avoid the possibility of overstressing the anchors. Be aware of reduced sensitivity to the load due to the mechanical advantage. Use belays and backup safeties. Protect the rope from edges and other abrasive parts of the rock.

7-23. U-PULLEY SYSTEM

The U-pulley system is another simple, easily-constructed hauling system (Figure 7-27, page 7-32).

a. **Considerations**. Anchors must be sturdy and able to support the weight of the load. Site selection is governed by different factors: tactical situation, weather, terrain, equipment, load weight, and availability of anchors.

b. **Theory.** Use carabiners as a substitute if pulleys are not available. The mechanical advantage obtained in theory is 2:1. The less friction involved the greater the mechanical advantage. Friction is caused by the rope running through carabiners, the load rubbing against the rock wall, and the rope condition.

c. **Construction.** Use the following procedures construct a U-pulley system.

(1) Anchor the hauling rope.

(2) Prepare the load or casualty for hauling. Place a locking carabiner the on to the harness or the rigged load.

(3) Lower a bight to the casualty or the load.

(4) Place the bight into the carabiner; or place the bight on to a pulley and then place pulley into the carabiner.

(5) Construct a second anchor. Attach a locking carabiner to the anchor.

(6) Tie a middle of the rope Prusik onto the haul rope exiting the pulley. Secure the Prusik with a double-double figure eight. This is the PCD. Place the fixed loops into the locking carabiner of the second anchor.

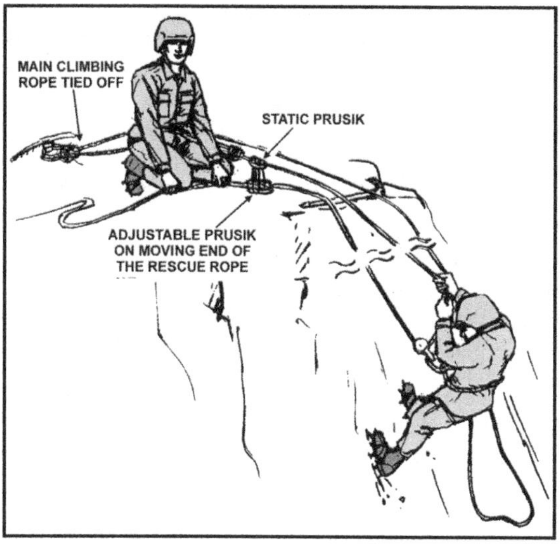

Figure 7-27. U-pulley system.

CHAPTER 8
MOUNTAIN WALKING TECHNIQUES

Mountain travel encompasses the full spectrum of techniques used to negotiate steep, rugged terrain. Mountain walking on rock and snow, technical rock and ice climbing, skiing or snow shoeing, rappelling, and stream crossing are the key travel skills a military mountaineer must possess.

8-1. BASIC PRINCIPLES

Up scree or talus, through boulder fields or steep wooded mountainsides, over snow or grass-covered slopes, the basic principles of mountain walking remain the same.

a. The soldier's weight is centered directly over the feet at all times. He places his foot flat on the ground to obtain as much (boot) sole-ground contact as possible. Then, he places his foot on the uphill side of grass tussocks, small talus and other level spots to avoid twisting the ankle and straining the Achilles tendon. He straightens the knee after each step to allow for rest between steps, and takes moderate steps at a steady pace. An angle of ascent or descent that is too steep is avoided, and any indentations in the slope are used to advantage.

b. In addition to proper technique, pace is adapted to conditions. The mountaineer sets a tempo, or number of steps per minute, according to the pace of the unit in which he is moving. (Physical differences mean that the tempos of two people moving at the same speed will not always be the same.) The soldier maintains tempo and compensates for changes of slope or terrain by adjusting the length of his stride. Tempo, pace, and rhythm are enhanced when an interval of three to five paces is kept between individuals. This interval helps lessen the "accordion" effect of people at the end of the file who must constantly stop and start.

c. The terrain, weather, and light conditions affect the rate of climb. The more adverse the conditions, the slower the pace. Moving too fast, even under ideal conditions, produces early fatigue, requires more rest halts, and results in loss of climbing time. A soldier can only move as fast as his lungs and legs will allow. The trained, conditioned and acclimatized soldier has greater endurance and moves more efficiently. Rest, good nutrition and hydration, conditioning, acclimatization, proper training, and the will to climb are key to successful mountain operations.

d. Breaks are kept to a minimum. When a moderate pace is set, the need for rest halts decreases, the chance of personnel overheating is lessened, and a unit can cover a given distance in a minimal time. If possible, rests should be taken on level ground avoiding steeper inclines.

(1) During the first half-hour of movement an adjustment halt should be taken. Soldiers will loosen or tighten bootlaces as needed, adjust packs and add or remove appropriate layers of clothing.

(2) Following the first halt, a well-conditioned party may take a short rest every 1 to 1.5 hours. If possible, soldiers lean against a tree, rock, or hillside to relieve the shoulders of pack weight, breathe deeply, hydrate, and snack on trail food. These halts are kept short enough to avoid muscles stiffening (one to two minutes).

(3) Later in the march longer halts may be necessary due to fatigue or mission requirements. At these halts soldiers should immediately put on additional clothing to avoid chilling—it is much easier to keep a warm body warm than to warm up a cold one.

(4) After a climb, a good rest is needed to revive tired muscles.

e. The rest step is used for steep slopes, snowfields, and higher elevations. It controls pace and limits fatigue by giving the lungs and legs a moment to recuperate between steps. Pace is kept slow and rhythmic.

(1) After each step forward, the soldier pauses briefly, relaxing the muscles of the forward leg while resting his entire bodyweight on the rear leg. The rear leg is kept straight with the knee locked so that bone, not muscle, supports the weight.

(2) Breathing is synchronized with the rest step. The number of breaths per step will change depending on the difficulty of the climb. Steeper slopes or higher elevations may require several breaths per step. When the air thins at altitude it is especially important to breathe deeply, using the "pressure breathing" technique. The soldier exhales strongly, enabling an easier, deeper inhale.

(3) This slow, steady, halting rest step is more efficient than spurts of speed, which are rapidly exhausting and require longer recovery.

f. Downhill walking uses less energy than uphill but is much harder on the body. Stepping down can hammer the full bodyweight onto the feet and legs. Blisters and blackened toenails, knee damage, and back pain may follow. To avoid these problems the soldier should start by tightening bootlaces to ensure a snug fit (also keep toenails trimmed). A ski pole, ice ax, or walking stick will help take some of the load and give additional stability. (Refer to Chapter 11 for techniques and use of the ice ax.) Keep a moderate pace and walk with knees flexed to absorb shock.

g. Side hill travel on any surface should be avoided whenever possible. Weighted down with a rucksack, the soldier is vulnerable to twisted ankles, back injury, and loss of balance. If side hill travel is necessary, try to switchback periodically, and use any lower angle flat areas such as rocks, animal trails, and the ground above grass or brush clumps to level off the route.

8-2. TECHNIQUES

Mountain walking techniques can be divided according to the general formation, surface, and ground cover such as walking on hard ground, on snow slopes and grassy slopes, through thick brush, and on scree and talus slopes.

a. **Hard Ground.** Hard ground is firmly compacted, rocky soil that does not give way under the weight of a soldier's step. It is most commonly found under mature forest canopy, in low brush or heather, and areas where animals have beaten out multiple trails.

(1) When ascending, employ the rest step to rest the leg muscles. Steep slopes can be traversed rather than climbed straight up. To turn at the end of each traverse, the soldier should step off in the new direction with the uphill foot. This prevents crossing the feet and possible loss of balance. While traversing, the full sole-to-ground principle is accomplished by rolling the ankle downhill on each step. For small stretches the herringbone step may be used—ascending straight up a slope with toes pointed out. A normal progression, as the slope steepens, would be from walking straight up, to a herringbone step, and then to a traverse on the steeper areas.

(2) Descending is best done by walking straight down the slope without traversing. The soldier keeps his back straight and bends at the knees to absorb the shock of each step. Body weight is kept directly over the feet and the full boot sole is placed on the ground with each step. Walking with a slight forward lean and with the feet in a normal position make the descent easier.

b. **Snow Slopes.** Snow-covered terrain can be encountered throughout the year above 1,500 meters in many mountainous areas. Talus and brush may be covered by hardened snowfields, streams made crossable with snowbridges. The techniques for ascending and descending moderate snow slopes are similar to walking on hard ground with some exceptions.

(1) ***Diagonal Traverse Technique.*** The diagonal traverse is the most efficient means to ascend snow. In conjunction with the ice ax it provides balance and safety for the soldier. This technique is a two-step sequence. The soldier performs a basic rest step, placing the leading (uphill) foot above and in front of the trailing (downhill) foot, and weighting the trail leg. This is the in-balance position. The ice ax, held in the uphill hand, is placed in the snow above and to the front. The soldier shifts his weight to the leading (uphill) leg and brings the unweighted trail (downhill) foot ahead of the uphill foot. He shifts weight to the forward (downhill) leg and then moves the uphill foot up and places it out ahead of the trail foot, returning to the in-balance position. At this point the ax is moved forward in preparation for the next step.

(2) ***Step Kicking.*** Step kicking is a basic technique used when crampons are not worn. It is best used on moderate slopes when the snow is soft enough to leave clear footprints. On softer snow the soldier swings his foot into the snow, allowing the leg's weight and momentum to carve the step. Fully laden soldiers will need to kick steps, which take half of the boot. The steps should be angled slightly into the slope for added security. Succeeding climbers will follow directly in the steps of the trailbreaker, each one improving the step as he ascends. Harder snow requires more effort to kick steps, and they will not be as secure. The soldier may need to slice the step with the side of his boot and use the diagonal technique to ascend.

(3) ***Descending Snow.*** If the snow is soft and the slope gentle, simply walk straight down. Harder snow or steeper slopes call for the plunge step, which must be done in a positive, aggressive manner. The soldier faces out, steps off, and plants his foot solidly, driving the heel into the snow while keeping his leg straight. He shifts his weight to the new foot plant and continues down with the other foot. On steeper terrain it may be necessary to squat on the weighted leg when setting the plunge step. The upper body should be kept erect or canted slightly forward.

(4) ***Tips on Snow Travel.*** The following are tips for travelling on snow.

(a) Often the best descent is on a different route than the ascent. When looking for a firmer travel surface, watch for dirty snow—this absorbs more heat and thus hardens faster than clean snow.

(b) In the Northern Hemisphere, slopes with southern and western exposures set up earlier in the season and quicker after storms, but are more prone to avalanches in the spring. These slopes generally provide firm surfaces while northern and eastern exposures remain unconsolidated.

(c) Travel late at night or early in the morning is best if daytime temperatures are above freezing and the sun heats the slopes. The night's cold hardens the snow surface.

(d) Avoid walking on snow next to logs, trees, and rocks as the subsurface snow has melted away creating hidden traps.

c. **Grassy Slopes.** Grassy slopes are usually composed of small tussocks of growth rather than one continuous field.

(1) When ascending, step on the upper side of each hummock or tussock, where the ground is more level.

(2) When descending a grassy slope, the traverse technique should be used because of the uneven nature of the ground. A climber can easily build up too much speed and fall if a direct descent is tried. The hop-skip step can be useful on this type of slope. In this technique, the lower leg takes all of the weight, and the upper leg is used only for balance. When traversing, the climber's uphill foot points in the direction of travel. The downhill foot points about 45 degrees off the direction of travel (downhill). This maintains maximum sole contact and prevents possible downhill ankle roll-out.

Note: Wet grass can be extremely slippery; the soldier must be aware of ground cover conditions.

d. **Thick Brush.** For the military mountaineer, brush is both a help and a hindrance. Brush-filled gullies can provide routes and rally points concealed from observation; on the other hand steep brushy terrain is hazardous to negotiate. Cliffs and steep ravines are hidden traps, and blow downs and thickets can obstruct travel as much as manmade obstacles. When brush must be negotiated take the most direct route across the obstacle; look for downed timber to use as raised paths through the obstacle; or create a tunnel through the obstacle by prying the brush apart, standing on lower branches and using upper limbs for support.

e. **Scree Slopes.** Slopes composed of the smallest rocks are called scree slopes. Scree varies in size from the smallest gravel to about the size of a man's fist.

(1) Ascending scree slopes is difficult and tiring and should be avoided, if possible. All principles of ascending hard ground and snow apply, but each step is carefully chosen so that the foot does not slide down when weighted. This is done by kicking in with the toe of the upper foot (similar to step-kicking in snow) so that a step is formed in the loose scree. After determining that the step is stable, weight is transferred to the upper leg, the soldier then steps up and repeats the process with the lower foot.

(2) The best method for descending scree slopes is to come straight down the slope using a short shuffling step with the knees bent, back straight, feet pointed downhill, and heels dug in. When several climbers descend a scree slope together, they should be as close together as possible (one behind the other at single arm interval) to prevent injury from dislodged rocks. Avoid running down scree as this can cause a loss of control. When the bottom of the slope (or run out zone) cannot be seen, use caution because drop-offs may be encountered.

(3) Scree slopes can be traversed using the ice ax as a third point of contact. Always keep the ax on the uphill side. When the herringbone or diagonal method is used to ascend scree, the ax can be used placing both hands on the top and driving the spike into the scree slope above the climber. The climber uses the ax for balance as he moves up to it, and then repeats the process.

f. **Talus Slopes.** Talus slopes are composed of rocks larger than a man's fist. When walking in talus, ascending or descending, climbers should always step on the uphill side of rocks and stay alert for movement underfoot. Disturbing unstable talus can cause rockslides. Climbers must stay in close columns while walking through talus so that dislodged rocks do not reach dangerous speeds before reaching lower soldiers. To prevent rock fall injuries, avoid traversing below other climbers. All other basics of mountain walking apply.

8-3. SAFETY CONSIDERATIONS

The mountain walking techniques presented here are designed to reduce the hazards of rock fall and loss of control leading to a fall. Carelessness can cause the failure of the best-planned missions.

a. Whenever a rock is kicked loose, the warning, "Rock!" is shouted immediately. Personnel near the bottom of the cliff immediately lean into the cliff to reduce their exposure, and do not look up. Personnel more than 3 meters away from the bottom of the cliff may look up to determine where the rock is heading and seek cover behind an obstacle. Lacking cover, personnel should anticipate which way the rock is falling and move out of its path to the left or right.

b. If a soldier slips or stumbles on sloping terrain (hard ground, grass, snow, or scree) he must immediately self-arrest, digging into the slope with hands, elbows, knees and toes. If he falls backwards and rolls over he must immediately try to turn over onto his stomach with his legs downhill and self-arrest with hands and toes.

c. When traveling through steep terrain, soldiers should be trained in the use of the ice ax for self-arrest. The ax can be used to arrest a fall on solid ground, grass and scree as well as snow. It may also be used as a third point of contact on difficult terrain. If not in use the ice ax is carried in or on the rucksack with its head down and secured.

8-4. NAVIGATION

Navigation is the process of determining one's present position, the location of a target objective, and selecting and following a route between these two points. Navigation consists of three distinct stages: orientation, navigation, and route finding.

- Orientation is simply figuring out exactly where one is. The use of the map, compass and identifiable terrain features, assisted by an altimeter and GPS, is the foundation of good navigation.
- Navigation includes the determination of the objective's location and the direction from the soldier's starting point to the objective. The same skills and equipment used in orientation are essential for good navigation.
- Route finding is picking the best line of travel that matches the equipment and capabilities of the team. Good route finding incorporates a comprehensive awareness of terrain, a solid base of mountaineering experience, good judgement and sound tactical instincts.

a. **Compasses.** The magnetic compass is the simplest and most widely used instrument for measuring directions and angles in the mountains. The lensatic compass is most commonly used in the military and can be employed in a variety of ways for either day or night navigation.

b. **Altimeters.** The altimeter is a vital piece of navigational equipment that can save valuable time in determining position through elevation.

(1) The standard altimeter is a modified barometer. A barometer is an instrument that measures the weight of a column of air above itself and displays the result on a scale marked in units of pressure, usually inches of mercury, millimeters of mercury, or millibars. Since air pressure drops uniformly as elevation is gained, it can be used to read altitude by means of the altimeter's scale, marked in feet or meters of elevation above sea level. By measuring air pressure, the altimeter/barometer gives the navigator new techniques for position finding, route planning, checking progress and terrain identification. It also gives the navigator valuable weather information specific to his immediate location.

(2) Changes in the weather are usually accompanied by air pressure changes, which are reflected in the altimeter. As the air pressure drops due to the approach of inclement weather for instance, the displayed elevation will rise by a corresponding amount. This means that a barometric pressure change of one inch of mercury equals roughly 1,000 feet of elevation. If the altimeter displays an elevation gain of 300 feet, a loss of barometric pressure of .3 inches has occurred, and bad weather should be expected.

(3) Altimeters come in two types: wrist-mounted digital altimeters and analog altimeters, usually attached to a cord.

(4) Because the altimeter is sensitive to changes in air pressure it must be recalibrated whenever a point of known elevation (summits, saddles, stream-trail intersections, survey monuments, and so forth) is reached. This is especially important when weather fronts are moving rapidly through the area.

(5) The altimeter may expand or contract because of changes in temperature. This can result in faulty elevation readings. Although some altimeters are temperature-compensated, rapid ascents or descents sometime overcome the adjustment, causing them to give poor readings.

(6) Keep the altimeter at a constant temperature. This is best accomplished by storing the altimeter (analog) in a pocket or on a cord around the neck, or on the wrist under the parka and hand gear (digital).

(7) Even though altimeters can be precise they are affected by both pressure and temperature changes and should be monitored carefully. The soldier should become familiar with the specific altimeter he employs and understand its capabilities and limitations.

c. **Global Positioning System.** The GPS is a space-based, global, all-weather, continuously available radio positioning navigation system. It is highly accurate in determining position location derived from a satellite constellation system. It can determine the latitude, longitude and elevation of the individual user. Location information is also displayed in military grid coordinates.

(1) The GPS provides precise steering information as well as position locations. The receiver can accept many checkpoints entered in any coordinate system by the user and convert them to the desired coordinate system. The user then calls up the desired checkpoint and the receiver will display direction and distance to the checkpoint. It can also compute travel time to the next checkpoint.

(2) Because the GPS does not need visible landmarks to operate, it can provide position (accurate up to 16 meters) in whiteouts or on featureless terrain. It also does not compound navigational errors as compass use can.

(3) During route planning, after choosing critical checkpoints, start point and objective, enter their coordinates as way points. The best use of the GPS is to verify these as they are reached, as a backup to terrain association and compass navigation.

(4) Since the 21-satellite constellation is not yet complete, coverage may be limited to specific hours of the day in certain areas of the world. The GPS navigational signals are similar to light rays, so anything that blocks light will reduce or block the effectiveness of the signals. The more unobstructed the view of the sky, the better the system performs. Although the GPS can be used in any terrain, it is performs best in more open areas such as the desert.

(5) Because the GPS requires horizon to horizon views for good satellite reception its use can be limited in the mountains. Canyons, deep valleys, saddles, and steep mountainsides are all problematic spots to use for shots. Ridgelines, spurs, summits, open valleys, or plateaus are better.

(6) When using GPS in regions with questionable surveying and mapping products, operational datum of the local maps must be reconciled with the datum used in navigational and targeting systems. Identify the spheroid and datum information on the pertinent map sheets and then check that the GPS receiver has the compatible datum loaded. If not then you must contact the S2 for updated datum or maps. Otherwise, the GPS will show different locations than those on the map.

(7) Extremely cold temperatures (-4 degrees F and below) and high elevations will adversely affect the operation of the GPS, due to the freezing of the batteries and the LCD screen. Battery life and overall performance can be improved by placing the GPS inside the parka or coat.

d. **Navigation Techniques.** The choice of movement technique often determines the route and navigational technique. For navigation, three techniques can be used: dead reckoning, terrain association, or altimeter navigation. The three are not mutually exclusive and are normally used together, with one chosen as the primary technique. The GPS can be used to supplement these techniques, but due to the problems associated with the restricted line of sight in the mountains, it should not be used as the main technique.

(1) ***Dead Reckoning.*** Because of the complex nature of mountainous terrain, dead reckoning is usually of limited value on most movements. The compass is generally employed more to support terrain association and to orient the map, than as a primary navigational aid. The main exception is during periods of limited visibility on featureless terrain. Heavy fog, snowy or whiteout conditions on a snowfield, glacier, large plateau or valley floor all would call for dead reckoning as a primary navigational technique.

(2) ***Terrain Association.*** The standard terrain association techniques all apply. Handrails, checkpoints, catching features, navigational corridors, boxing-in areas, and attack points are all used. When a small objective lies near or on an easily identifiable feature, that feature becomes an expanded objective. This simplifies the navigational problem by giving a large feature to navigate to first. The altimeter may finalize the search for the objective by identification through elevation. Rough compass headings are used to establish a general direction to the next checkpoint; used when the checkpoint headed toward is a linear feature, and not a precise point. The shape, orientation, size,

elevation, slope (SOSES) strategy is especially valuable in mountain terrain association and should be practiced extensively (FM 3-25.26).

(a) After extensive study of the map and all available sources of information it helps to create a mental image of the route. This will enable the navigator to make the terrain work in his favor. Avoid brush for speed and ease of movement; the military crest of spurs and ridgelines generally provides the best route while providing terrain masking effects. When clear cut, burned-over, or large avalanche slide areas are encountered, it may be necessary to box or contour around them as they may be full of slash or brushy second-growth small trees. Old-growth forest provides the easiest travel.

(b) The following situations will result in objects appearing closer than they actually are:

- When most of the object is visible and offers a clear outline.
- When you are looking across a partially cleared depression.
- When looking down a straight, open road or track.
- When looking over a smooth, uniform surface, such as snow, water, or desert.
- When the light is bright and the sun is shining from behind the observer.
- When the object is in sharp contrast to the background.
- When seen in the clear air of high altitude.
- When looking down from high ground to low ground.

(c) The following situations will result in objects appearing farther away than they actually are:

- When only part of the object is seen or it is small in relation to its surroundings.
- When you are looking across an exposed depression.
- When looking up from low ground to high ground.
- When your vision is narrowly confined.
- When the light is poor, such as dawn, dusk, or low visibility weather; or when the sun is in your eyes, but not behind the object being viewed.
- When the object blends into the background.

(2) ***Altimeter Navigation.*** Altimeters provide assistance to the navigator in several ways. They aid in orientation, in computing rates of ascent or descent, in resection, and in weather prediction.

(a) When moving along any linear feature such as a ridge, watercourse, or trail which is shown on the map, check the altimeter. The point where the indicated elevation contour crosses that feature is your location.

(b) The navigator frequently finds it necessary to determine his position through the use of resection. A modified resection can be performed by shooting an azimuth to a known, clearly visible summit or similar feature and then plotting the back azimuth on the map. By determining your present elevation and finding where that particular contour crosses the back azimuth you should locate your position. This can be difficult when in low ground, as mountain summits can rarely be clearly seen from valley floors. In addition, most mountaintops are so large that there is usually no specific point to shoot at. In this case, the soldier should take multiple azimuths to known features. If he is located on a good linear feature he will have a decent idea of where he is. The altimeter can be used to verify elevation and establish a notional linear feature—a contour line. The point where the resecting back azimuths cross the contour line is the navigator's location.

(c) Using the altimeter to calculate rates of ascent can help in sound decision-making. Rates of travel, along with weather conditions, light conditions (time of day), and the physical condition of the team, are all key variables that can influence the success or failure of the mission.

(d) Altimeters can be used as barometers to assist in weather prediction.

e. **Approach Observations.** Watch the mountain during the approach march, studying it for climbing routes. Distant views can reveal large-scale patterns of ridges, cliffs, snowfields and glaciers. General angles of the large rock masses can be seen from afar.

(1) Closer viewing displays these patterns and angles on a smaller scale. Fault lines, gross bedding planes of rock, cliff bands, and crevasse zones become visible. Snowy or vegetated ledge systems appear. Weaknesses in the mountain walls, such as couloirs or gullies, may present themselves.

(2) Most of these features repeat themselves at increasingly finer levels, as they are generally derived from the overall structure of the particular mountain group. A basic knowledge of mountain geology, combined with the specific geological background of the operational area, pays off in more efficient travel.

f. **Natural Indicators of Direction in the Northern Hemisphere.** Southern slopes are sunnier and drier than northern slopes, with sparser or different types of vegetation. Northern slopes can be snowier and, because of more intense glaciation in past ages, are often steeper.

Note: Opposite rules apply in the Southern Hemisphere.

g. **Winter Route Selection.** The following must be considered when selecting a route in the winter.

(1) Conduct a thorough map reconnaissance considering the weather, individual ski abilities, avalanche danger, vegetation, water features, terrain relief, and the size of the unit.

(2) Weather conditions will affect the chosen route. During calm weather, your rate of movement will be significantly faster than during periods of inclement weather.

(3) Individual ski abilities will affect your rate of movement, constrain your choice of terrain, and impact on your route choices.

(4) Avalanche danger zones must be identified by map review and data gathered during route planning. During movement, snow pits, shovel tests, and ski shear tests must be conducted prior to crossing an avalanche danger zone. Bottom line: avoid avalanche danger areas. If you must cross one, cross above the starting zone or below the run-out zone.

(5) Vegetation can work for you or against you. Thickly forested areas usually have a deep snow pack. For weaker skiers, forested areas are full of potentially dangerous obstacles. On slopes with an angle of 30 to 45 degrees that are sparsely vegetated an avalanche danger is still present. If the weather turns bad, forested areas provide welcome relief from wind and blowing snow.

(6) Water features provide valuable navigation aids. Under deep snow pack small creeks and ponds may be hard to locate. Large frozen lakes and rivers can provide excellent means of increasing your rate of march.

(7) During ski movements, efficient use of terrain will greatly improve morale and reduce fatigue. While traveling in mountainous terrain, do not needlessly give up elevation gained. Maintain a steady climb rate and avoid over exertion. Avoid north, east, and south facing slopes when the avalanche danger is high. Avoid cornices and be aware of their probable and improbable fracture lines. Weather and tactical situation permitting, travel on the windward side of ridgelines. If weak skiers are in the group, stay away from restrictive terrain with sheer drop-offs. When touring use climbing skins to maintain control and lessen lost time per hour due to individuals falling.

(8) The following are additional hints for navigation in snowy conditions:

- Keep the compass warm.
- If no terrain features exist for steering marks, use your back azimuth and tracks to maintain course.
- Limit steering marks to shorter distances since visibility can change quickly.
- Never take azimuths near metallic objects. Hold the compass far enough from your weapon, ice ax, and so on to get accurate readings.
- Make frequent compass checks.
- Preset azimuths on your compass.
- Use a steady, unshifting wind to aid you in maintaining course.

h. **Problems.** The following conditions and characteristics of cold weather and mountainous regions make accurate navigation difficult.

(1) In winter, short hours of daylight, fog, snowfall, blizzards, whiteouts, and drifting snow, especially above tree line, drastically limit visibility. At times, an overcast sky and snow-covered terrain create a phenomenon called flat light, which makes recognition of irregularities in the terrain extremely difficult.

(2) Heavy snow may completely cover existing tracks, trails, outlines of small lakes and similar landmarks. Because the appearance of the terrain is quite different in winter from that in summer, particular attention must be paid to identifying landmarks, both on the ground and from aerial photographs.

(3) Magnetic disturbances, caused by large ore deposits, are frequently encountered and make magnetic compass readings difficult and sometimes unreliable.

(4) Handling maps, compasses, and other navigation instruments in low temperatures with bare hands is difficult. Removing hand wear may only be possible for short periods.

(5) Keeping count of pace is extremely difficult in winter and mountain environments. Thick vegetation and rough, steep slopes hamper attempts at accurate pace counts. The most reliable method is the use of a 50-meter long piece of field wire or rope.

8-5. ROUTE PLANNING

Proper route planning can make the difference between success and failure on long mountain movements. Careful map reconnaissance, knowledge of the enemy situation, terrain analysis of the operational area, and an accurate assessment of the unit's capabilities are all key parts of the planning process.

a. **Map Reconnaissance.** Topographic maps provide the primary source of information concerning the area of operations. A 1:25,000 map depicts greater detail than a 1:50,000 map and should be used whenever possible. Because examination of the micro-terrain is so important for mountain operations, even larger scale maps are desirable. Civilian 1:12,000 maps can be used if available. Aerial, oblique angle, photographs give details not always shown on maps (crags and overhangs). Sketch maps supplement other sources of information but should not be relied on for accuracy since they are seldom drawn to scale. Along with sketch maps, verbal descriptions, documented information gathered from units previously in the area, or published sources such as alpine journals or climbing guides may help. Forest service and logging and mining company maps provide additional information, often showing the most recent changes to logging trails and mining access roads. Standard military topographic maps are generally accurate graphic depictions of the operational area.

(1) When conducting a map reconnaissance, pay close attention to the marginal information. Mountain-specific terrain features may be directly addressed in the legend. In addition, such facilities as ski lifts, cable and tramways are often found. Check the datum descriptor (for foreign maps) to ensure compatibility with entered datum in GPS units. Along with the standard topographic map color scheme, there are some commonly seen applications for mountainous terrain. White with blue contours indicates glaciers or permanent snowfields. The outline of the snow or ice is shown by dashed blue lines while their contour lines are solid blue. High ice cliffs which are equal to or exceed the contour interval will be shown. Low ice cliffs and ice caves may be indicated if they provide local landmarks. Brown contour lines on white mean dry areas without significant forest cover. Areas above tree line, clear cuts, rock or avalanche slide paths and meadows are all possible. Study the surrounding terrain and the legend for other clues. An important point to remember is that thick brush in small gullies and streambeds may not be depicted by green, but should still be expected.

(2) Obstacles, such as rivers and gorges, will require technical equipment to cross if bridges are not present. Fords and river crossing sites should be identified. Due to the potential for hazardous weather conditions, potential bivouac sites are noted on the map. Ruins, barns, sheds and terrain-protected hollows are all possible bivouac sites. Danger areas in the mountains; isolated farms and hamlets, bridges, roads, trails, and large open areas above tree line, are factored in, and plans made to avoid them. Use of terrain-masking becomes essential because of the extended visibility offered by enemy observation points on the dominant high ground.

(3) Helicopter support, weather permitting, requires identification of tentative landing zones for insertions, extractions, resupply and medevac. The confined nature of mountain travel means that crucial passes become significant chokepoints and planners should designate overwatches/surveillance positions beforehand. Alternate routes should be chosen with weather imposed obstacles in mind: spring flood or afternoon snowmelt turns small streams into turbulent, impassable torrents. Avalanche danger prohibits travel on certain slopes or valley floors.

b. **Enemy Situation.** Route selection should only be done after reviewing all available information about the friendly and enemy situation.—Is the enemy force on his own ground? Are they accustomed to the terrain and the weather? Are they trained mountain troops with specialized equipment?—Only after answering these and other questions can an effective route plan be completed. If the enemy force is better prepared to maneuver in the mountains, they have a marked advantage, and route selection must be scrutinized.

c. **Analysis of the Operational Area.** Not all mountainous terrain is created equal and not all movement plans have the same expectation of success. Planners must undertake a thorough analysis of the general terrain to be crossed, including the geology, mountain structure and forms, and ground cover.

(1) Heavily glaciated granite mountains pose different problems than does river-carved terrain. The U-shaped valley bulldozed out by a glacier forces maneuver elements down to the valley floor or up to the ridge tops, while the water-cut V-shape of river valleys allows movement throughout the compartment.

(2) Routes through granite rock (long cracks, good friction; use of pitons, chocks and camming units) will call for different equipment and technique than that used for steep limestone (pockets, smooth rock; bolts, camming units).

(3) Operations above tree line in temperate climates or in the brushy zone of arid mountains means that material for suspension traverse A-frames must be packed. The thick brush and krummholtz mats of subalpine zones and temperate forested mountains can create obstacles that must be bypassed.

(4) Heavy spruce/fir tangles slow progress to a crawl, therefore planners should ensure routes do not blindly traverse these zones.

d. **Unit Assessment.** When assessing unit movement capabilities the key indices are training and conditioning levels. Soldiers who have received basic military mountaineer training, who know how to move through rough terrain, and who have been hardened with training hikes through the mountains, will perform better than troops without this background.

e. **Time-Distance Formulas.** Computing march rates in the mountains is extremely difficult, especially when there is snow cover. The following rates are listed as a guide (Table 8-1). Rates are given for movement over flat or gently rolling terrain for individuals carrying a rifle and loaded rucksack.

	UNBROKEN TRAIL	BROKEN TRAIL
On foot (no snow cover)	2 to 3 kph (cross-country)	3 to 4 kph (trail walking)
On foot (less than 1 foot of snow)	1.6 to 3.2 kph	2 to 3.2 kph
On foot (more than 1 foot of snow)	.4 to 1.2 kph	2 to 3.2 kph
Snowshoeing	1.6 to 3.2 kph	3.2 to 4 kph
Skiing	1. to 5.6 kph	4.8 to 5.6 kph
Skijoring	N/A	3 to 24 kph

Table 8-1. Time-distance formulas.

(1) March distances in mountainous terrain are often measured in time rather than distance units. In order to do this, first measure the map distance. This distance plus 1/3 is a good estimate of actual ground distance. Add one hour for each 1,000 feet of ascent or 2,000 feet of descent to the time required for marching a map distance.

(2) As Table 8-1 indicates, snow cover will significantly affect rates of march. Since snow can be expected in the mountains most months of the year, units should have some experience at basic snow travel.

(3) Individual loads also affect march rates. Combined soldier loads that exceed 50 pounds per man can be expected to slow movement significantly in mountainous terrain. Given the increased weight of extra ammunition for crew-served weapons, basic mountaineering gear, and clothing for mountain travel, it becomes obvious that soldiers will be carrying weights well in excess of that 50-pound limit. Units should conduct cross-country movements in the mountains with the expected rucksack and LCE weights in order to obtain accurate, realistic rates of march.

(4) In the harsh environment of the mountains, helicopter support cannot be relied on. The process of transporting extra equipment and sustainment supplies will result in vastly increased movement times. The heavier loads will exhaust soldiers mentally and physically. Tactical movements, such as patrolling or deliberate assaults, should take this into account.

8-6. ROUTE SELECTION

Many variables affect the selection of the proper route. The following guidelines apply to all situations.

a. **Select a Current Map.** Check the date of the map for an indication of the reliability of the map in depicting vegetation, clearings, roads, and trails accurately. The leader should use all the latest topographic data he can find.

b. **Gather Intelligence Information.** The most important consideration in every leader's mind when plotting a movement is "where is the enemy?" The latest intelligence reports are essential. Additionally, weather reports, snow condition reports, avalanche probability, aerial photos, and any past or recent history of operations in the area may be of help.

c. **Select a Route.** Identify the starting point and determine the movement objective. Plot start and end points. Carefully scrutinize the area in between and begin to select the route. Consider the following:

(1) ***Trafficability***. This includes degree of slopes, vegetation, width of trails, snow depth, avalanche probability, and the likelihood of crevasses.

(2) ***Time-Distance Formula.*** Time allotted and distance to be covered must be considered.

(3) ***Required Equipment.*** Carry enough equipment to move along the route and to survive if an extended stay becomes necessary. Do not plan a route beyond the means of your equipment.

(4) ***Location of Enemy.*** Plan a route that allows maximum use of the masking effect of the terrain. Avoid danger areas or areas of recent enemy activity unless required by the mission. Use vegetation to mask your movement if possible (especially coniferous forests). Avoid silhouetting on ridgelines.

(5) ***Communications.*** Communications will be severely limited in the mountains. Dead spaces or communications holes are common. Use all available information and plan accordingly.

(6) ***Conditions/Capabilities of Unit.*** The unit must be able to negotiate the route chosen. Take into consideration their present health, as well as their training level when selecting your intended route.

(7) ***Checkpoints/Control Points.*** When plotting a route on the map, utilize prominent terrain features on either side of the route as checkpoints. Ensure that when you select

your checkpoints they are visually significant (elevation) and that they are easily identifiable. Avoid the use of manmade features as checkpoints due to their unreliability and lack of permanence. Select features that are unique to the area.

CHAPTER 9

MOUNTAIN STREAM CROSSINGS

Operations conducted in mountainous terrain may often require the crossing of swift flowing rivers or streams. Such crossings should not be taken lightly. The force of the flowing water may be extremely great and is most often underestimated. All rivers and streams are obstacles to movement. They should be treated as danger areas and avoided whenever possible. When rivers or streams must be crossed, there are a variety of techniques the small-unit leader may choose from, depending upon the type of stream, its width, speed of the current, and depth of the water.

There are limits on the safe use of these techniques. Not all mountain rivers or streams will be fordable with these techniques. If a water obstacle is too wide, swift, or deep, an alternate route should be used, or the crossing will require major bridging by engineers. It may require the use of rafts or boats. Reconnaissance of questionable crossing sites is essential. This chapter covers the techniques for crossing mountain streams that have a depth generally not exceeding waist deep.

9-1. RECONNAISSANCE

Reconnaissance of the route (map, photo, and or aerial) may not always reveal that a water obstacle exists. In a swamp, for example, unfordable sloughs may not show on the map, and they may be concealed from aerial observation by a canopy of vegetation. Whenever it is possible that a unit will be required to cross a water obstacle, its commander must plan some type of crossing capability.

a. Site selection is extremely important once you determine that you must make a crossing (Figure 9-1). Look for a high place from which you can get a good view of the obstacle and possible crossing sites. A distant view, perhaps from a ridge, is sometimes better than a hundred close views from a riverbank. Site selection must be made before the arrival of the main body.

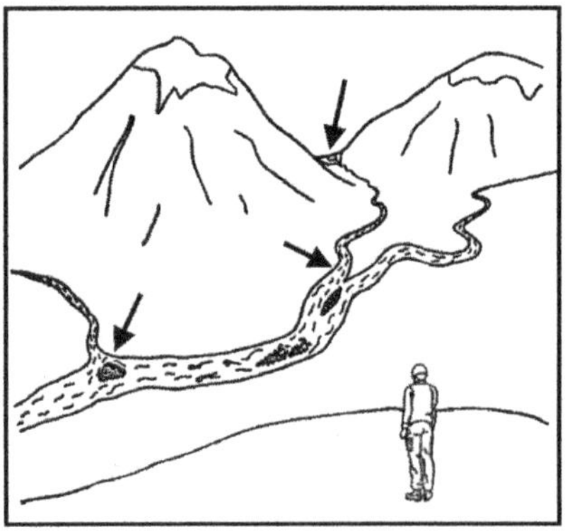

Figure 9-1. Normal locations of shallowest water and safest crossing sites.

b. A dry crossing on fallen timber or log jams is preferable to attempting a wet crossing. Depending upon the time of year, the closer you are to the source, or headwaters, the better your chances are of finding a natural snow or ice bridge for crossing. If a dry crossing is unavailable, the following considerations should be made:

(1) The time of day the crossing can be an important factor. Although early morning is generally best because the water level is normally lower during this period, recent weather is a big factor; there may have been heavy rain in the last eight hours. As glaciers, snow, or ice melt during the day, the rivers rise, reaching their maximum height between mid afternoon and late evening, depending on the distance from the source. Crossings, if made during the early morning, will also allow clothing to dry more quickly during the heat of the day.

(2) A crossing point should normally be chosen at the widest, and thus shallowest, point of the river or stream. Sharp bends in the river should be avoided since the water is likely to be deep and have a strong current on the outside of the bend. Crossings will be easiest on a smooth, firm bottom. Large rocks and boulders provide poor footing and cause a great deal of turbulence in the water.

(3) Many mountain streams, especially those which are fed by glacier run-off, contain sections with numerous channels. It is often easier to select a route through these braided sections rather than trying to cross one main channel. A drawback to crossing these braided channels, however, is the greater distance to the far bank may increase exposure time and often the sand and gravel bars between the channels will offer little cover or concealment, if any.

(4) The crossing site should have low enough banks on the near and far side to allow a man carrying equipment to enter and exit the stream with relative ease. If a handline or rope bridge is to be constructed, the crossing site should have suitable anchors on the near and far bank, along with safe loading and unloading areas. Natural anchors are not a necessity, however the time required to find a site with solid natural anchors will probably be less than the time required to construct artificial anchors. In some areas, above the tree line for example, artificial anchors may have to be constructed. Deadman anchors buried in the ground, or under a large pile of boulders work well.

(5) Log jams and other large obstructions present their own hazards. Logs floating downstream will generally get hung up in shallower sections creating the jam. Once a log jam is formed, however, the water forced to flow around it will erode the stream bottom. Eventually deep drop-offs or holes may develop, especially around the sides and off the downstream end of the log jam. A log jam that totally bridges a section of the stream may be the best way to cross. A wet crossing in the vicinity of a log jam should be performed a good distance below or above it. Some things to consider when crossing near log jams are:

- Cross well to the downstream side when possible.
- Keep a sharp lookout for floating timber that could knock you off your feet.
- If you must cross on the upstream side, stay well upstream from the log jam. If a person is swept off his feet and caught in the debris of the jam, he could easily drown. A handline will greatly increase safety here.

(6) When possible, select a crossing site that has enough natural protection on the near and far banks so that security teams may be placed out and enough cover and concealment is available for the size of the element making the crossing. When cover and

concealment is minimal, as in the higher alpine zones, the crossings must be conducted as efficiently as possible to minimize exposure to enemy observation.

9-2. PREPARATION OF TROOPS AND EQUIPMENT

Prepare men and equipment for a crossing as far in advance as feasible. Final preparation should be completed in a security perimeter on the near side just before crossing. Preparation includes the following.

a. Waterproof water-sensitive items. Wrap radios, binoculars, SOI, papers, maps and any extra clothing in waterproof bags (trash bags also work well), if available. These bags also provide additional buoyancy in case of a fall.

b. Trousers are unbloused and shirts are pulled out of the trousers. All pockets are buttoned. This allows water to escape through the clothing. Otherwise the clothing would fill up and retain water, which would weigh the body down. This is especially critical if an individual must swim to shore. Depending on the circumstances of the crossing (for example, tactical situation, temperature of the air and water), the crossing can be made in minimal clothing so that dry clothing is available after the crossing. Boots should be worn to protect feet from rocks; however, socks and inner soles should be removed. On the far side, the boots can be drained and dry socks replaced.

c. Load-carrying equipment harness and load-bearing vest (LBV) is unbuckled and worn loosely. It is extremely difficult to remove a buckled harness in the water in an emergency.

d. Helmets are normally removed and placed in the rucksack in slow moving streams with sandy or gravel bottoms. If you have to resort to swimming it is easier done without the helmet. However, when crossing swift flowing streams, especially those with large rocks and other debris, the risk of head injury if a person slips is high. In this case the helmet should be worn with the chinstrap fastened.

e. The rucksack should be worn well up on the shoulders and snug enough so it does not flop around and cause you to lose your balance. The waist strap MUST be unbuckled so you can get rid of the pack quickly if you are swept off your feet and have to resort to swimming. If a pack has a chest strap it must also be unbuckled. Secure everything well within your pack. It is easier to find one large pack than to find several smaller items.

f. Individual weapons should be attached to the pack or slung over the shoulder.

9-3. INDIVIDUAL CROSSINGS

Whenever possible, and when the degree of experience permits, streams should be forded individually for a speedier crossing. The average soldier should be able to cross most streams with mild to moderate currents and water depths of not much more than knee deep using proper techniques.

a. The individual should generally face upstream and slightly sideways, leaning slightly into the current to help maintain balance. At times, he may choose to face more sideways as this will reduce the surface area of the body against the current, thus reducing the current's overall force on the individual.

b. The feet should be shuffled along the bottom rather than lifted, with the downstream foot normally in the lead. He should take short, deliberate steps. Lunging steps and crossing the feet result in a momentary loss of balance and greatly increase the chance of a slip.

c. The individual should normally cross at a slight downstream angle so as not to fight the current. There is normally less chance of a slip when stepping off with the current as opposed to stepping off against the current.

d. The individual must constantly feel for obstacles, holes and drop-offs with the lead foot and adjust his route accordingly. If an obstacle is encountered, the feet should be placed on the upstream side of it where the turbulence is less severe and the water normally shallower.

e. To increase balance, and if available, a long ice ax, sturdy tree limb, or other staff can be used to give the individual a third point of contact (Figure 9-2). The staff should be used on the upstream side of the individual and slightly leaned upon for support. The staff should be moved first, then the feet shuffled forward to it. This allows two points of contact to be maintained with the streambed at all times. The individual still moves at a downstream angle with the downstream foot in the lead.

Figure 9-2. Individual crossing with staff.

9-4. TEAM CROSSING

When the water level begins to reach thigh deep, or anytime the current is too swift for personnel to safely perform an individual crossing, a team crossing may be used. For chain crossing, two or more individuals cross arms with each other and lock their hands in front of themselves (Figure 9-3). The line formed faces the far bank. The largest individual should be on the upstream end of the line to break the current for the group. The line formed will then move across the stream using the same principles as for individual crossings, but with the added support of each other. The line should cross parallel to the direction of the current. The team still moves at a slight downstream angle, stepping off with the downstream foot in the lead.

Figure 9-3. Chain method for stream crossing.

9-5. ROPE INSTALLATIONS

When the water level begins to reach waist deep or the current is too swift for even a team crossing, the chosen site must be closely examined. The stream at this point may be impassable. Many times though, a crossing site which may be unsafe for individual or team crossings can be made safe with the installation of a handline or rope bridge. Crossing on a handline will still require each individual to enter the water and get wet. If a one-rope bridge can be constructed, it may require only a couple of individuals to enter the water. Deciding whether to install a handline or a rope bridge will depend on the anchors available, height of the anchors above the water, and the distance from the near and far anchors. The maximum distance a one-rope bridge is capable of spanning is approximately 1/2 to 2/3 the length of the rope in use.

a. **Establishing the Far Anchor.** Whether a handline or rope bridge is to be installed, someone must cross the stream with one end of the rope and anchor it on the far side. This duty should be performed by the most capable and strongest swimmer in the party. The swimmer should be belayed across for his own safety. The belay position should be placed as far above the crossing as possible. In the event that the current is too strong for the individual, he will pendulum back to the near bank. Rescuers should be poised on the near bank at points where the individual will pendulum back, should he fail to reach the far bank. The initial crossing site should be free of obstacles that would snag the rope and prevent the pendulum back to the bank for an easy recovery.

(1) The individual may attach the belay rope to his seat harness or a swami belt with a carabiner. He should NEVER tie directly into the rope when being belayed for a stream crossing. If the swimmer should be swept away and become tangled, he must be able to release himself quickly from the rope and swim to shore as best he can. The individual may also choose to tie a fixed loop into the end of the belay rope and hang on to it, where he can immediately release it in an emergency.

(2) Anytime a crossing site must be used where the swimmer may encounter problems getting to the far bank, he should have on a life vest or other personal flotation

device (PFD). If the swimmer must release the rope at any time, he will have to rely on his own water survival skills and swimming ability to get to shore. A PFD will greatly increase his own personal safety. A PFD may also be used by the last man across, as he will release the rope from the anchor and be belayed across as the first man

b. **Installation of a Handline.** If it is possible to use a rope high enough above the water to enable soldiers to perform a dry crossing, then a rope bridge should be installed as such. If this is impossible, and the rope must be installed to assist in a wet crossing, then it should be installed as a handline (Figure 9-4).

(1) The far anchor should be downstream from the near anchor so that the rope will run at an angle downstream from the near anchor, approximately thirty to forty-five degrees, rather than straight across the stream. Here again, it is easier to move with the current as opposed to directly across or against it.

(2) The rope may be anchored immediately on the far bank, pulled tight, and anchored on the near bank, or it may be installed with a transport tightening system if a tighter rope is required.

(3) Crossing will always be performed on the downstream side of the handline, shuffling the feet with the downstream foot in the lead.

(4) A second climbing rope is used as a belay (Figure 9-5). One end of the belay rope will be on the near bank and the other end on the far bank. It should be sent across with the strong-swimmer. An appropriate knot is tied into the middle of the belay rope to form two fixed loops with each loop being approximately 6 inches long. One loop is connected to the handline with carabiner(s) and the individual crossing connects one loop to himself. The loops are short enough so the individual is always within arms reach of the handline should he slip and let go. The individuals are belayed from both the near and far banks. If a mishap should occur the individual can be retrieved from either shore, whichever appears easiest. The belay on the opposite shore can be released allowing the individual to pendulum to the bank. It is important that the belay rope NOT be anchored or tied to the belayer so that it may be quickly released if necessary.

Figure 9-4. Stream crossing using a handline.

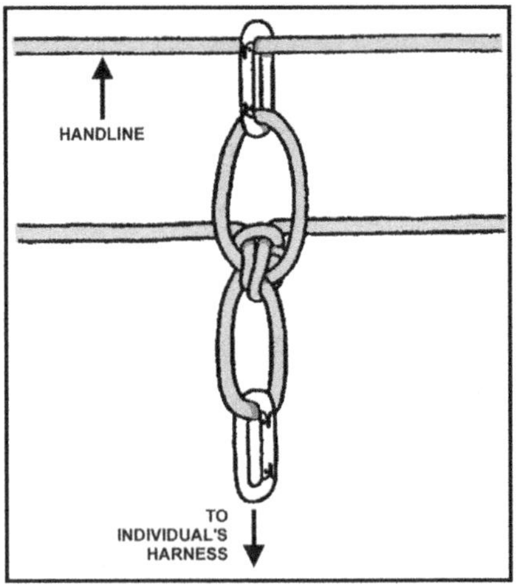

Figure 9-5. Belay rope for crossing using a handline.

(5) Under most circumstances, the handline should be crossed one person at a time. This keeps rope stretch and load on the anchors to a minimum.

(6) Rucksacks can be either carried on the back the same way as for individual crossings, or they can be attached to the handline and pulled along behind the individual.

(7) If a large amount of equipment must be moved across the stream, especially heavier weapons, such as mortars, recoilless rifles, and so on, then a site should be selected to install a rope bridge.

9-6. SAFETY

River and stream crossings present one of the most hazardous situations faced by the military mountaineer. The following safety procedures are minimum guidelines that should be followed when conducting a river or stream crossing.

a. All weak and nonswimmers should be identified before a crossing so that stronger swimmers may give assistance in crossing.

b. Not every river or stream can be crossed safely. It is always possible to cross at a different time or place, use a different technique, or choose another route.

c. The technique used is directly dependent upon water depth, speed of the current, stream bottom configuration, width of the stream, and individual experience.

d. The safest methods of crossing are always with the use of a handline or one-rope bridge.

e. If the installation of a handline or rope bridge becomes too difficult at a given crossing site, then that site should be considered too hazardous and another site selected.

f. A lookout should be posted 50 to 100 meters upstream to watch for any obstacles that may be carried downstream and interfere with the crossing.

g. When conducting individual crossings (those without a handline or rope bridge), lifeguards should be posted downstream with poles or ropes prepared to throw, for assistance or rescue.

h. When the unit knows a rope installation will be required for crossing, at least two life vests or other PFDs should be on hand to provide additional safety for the strong swimmer who must establish the far anchor, and the last man across who retrieves the system.

9-7. SWIMMING

There are times when you might be alone and have no choice but to swim across, or there may be a time that you find yourself suddenly plunged into a swift river or rapids. In either case, the following techniques could save your life.

a. Immediately jettison any equipment or clothing that restricts movement.

b. Do not try to fight the current. Maneuver towards shore in a position with the feet downstream, facing downstream, and fanning the hands alongside the body to add buoyancy and to fend off submerged rocks. Use the feet to protect the rest of the body and to fend off submerged rocks.

c. Keep the head above water to observe for obstacles and attempt to maneuver away from them.

d. Try to avoid backwater eddies and converging currents as they often contain dangerous swirls. Avoid bubbly water under falls as it has little buoyancy. Breath between the wave troughs.

e. If the shore is too difficult to reach, seek out the closest and safest spot, such as a sandbar, to get yourself out of the water as quickly as possible. Hypothermia will set in quickly in colder waters.

CHAPTER 10

MOVEMENT OVER SNOW AND ICE

Movement over snow- and ice-covered slopes presents its own unique problems. Movement on steeper slopes requires an ice ax, crampons, and the necessary training for this equipment. Personnel will also have to learn how to place solid anchors in snow and ice to protect themselves during these movements if roped. Snow-covered glaciers present crevasse fall hazards even when the slope is relatively flat, requiring personnel to learn unique glacier travel and crevasse rescue techniques.

All the principles of rock climbing, anchor placement, belays, and rope usage discussed throughout the previous chapters apply to snow and ice climbing as well. This chapter will focus on the additional skills and techniques required to move safely through snow-covered mountains and over glaciated terrain.

10-1. MOVEMENT OVER SNOW

The military mountaineer must be equally adept on both snow and ice due to route necessity and rapidly changing conditions. On steep slopes in deep snow, the climber may climb straight up facing the slope. The ice ax shaft, driven directly into the snow, provides a quick and effective self-belay in case of a slip—the deeper the shaft penetrates the snow, the better the anchor (Figure 10-1). It is usually best, however, to climb snow-covered slopes in a traversing fashion in order to conserve energy, unless there is significant avalanche danger.

Figure 10-1. Self-belay on snow.

a. The progression from walking on flat terrain to moving on steep terrain is the same as for moving over snow-free terrain. If the snow is packed the sole of the boot will generally hold by kicking steps, even on steep slopes. Where it is difficult to make an effective step with the boot, a cut made with the adze of the ice ax creates an effective step. In these situations crampons should be used for faster and easier movement.

b. When descending on snow, one can usually come straight downhill, even on steep terrain. Movement downhill should be slow and deliberate with the climber using an even pace. The heels should be kicked vigorously into the snow. The body may be kept erect with the aid of an ice ax, which may be jammed into the snow at each step for additional safety. Here again, crampons or step cutting may be necessary. A technique known as glissading may also be used as an easy method of descent and is covered in detail later in this chapter.

10-2. MOVEMENT OVER ICE

Ice is found in many areas of mountains when snow is present, and during the summer months also where perennial snowpack exists. Many times an ice area will be downslope of a snowfield and sometimes the ice pack itself will be lightly covered with snow. Even if using an ice ax and or crampons, movement will still be difficult without proper training.

10-3. USE OF ICE AX AND CRAMPONS

Movement over snow and ice is almost impossible without an ice ax and or crampons.

a. **Ice Ax.** When walking on snow or ice, the ice ax can be used as a third point of contact. When the terrain steepens, there are a number of ways to use the ice ax for snow or ice climbing. Some positions are more effective than others, depending on the intended result. You may find other ways to hold and use the ax, as long the security remains in effect.

(1) ***Cane Position.*** The ice ax can be used on gentle slopes as a walking stick or cane (Figure 10-2). The ax is held by the head with the spike down and the pick facing to the rear in preparation for self-arrest. When moving up or down gentle slopes the ice ax is placed in front as the third point of contact, and the climber moves toward it. When traversing, the ax is held on the uphill side, in preparation for a self-arrest.

Figure 10-2. Using the ice ax in the cane position.

(2) ***Cross Body Position or Port Arms Position.*** On steeper slopes the ax can be used in the port arms position, or cross body position (Figure 10-3). It is carried across the chest, upslope hand on the shaft, spike towards the slope. The head of the ax is held away from the slope with the pick to the rear in preparation for self-arrest. Ensure the leash is connected to the upslope hand, which allows the ax to be used in the hammer position on the upslope side of the climber. The spike, in this case, is used as an aid for maintaining balance.

Figure 10-3. Ice ax in the cross body or port arms position.

(3) ***Anchor Position.*** As the slope continues to steepen, the ax may be used in the anchor position (Figure 10-4). The head is held in the upslope hand and the pick is driven into the slope. The spike is held in the downhill hand and pulled slightly away from the slope to increase the "bite" of the pick into the ice. If the climber is wearing a harness, the pick can be deeply inserted in the ice or hard snow and the ax leash could be connected to the tie-in point on the harness for an anchor (ensure the ax is placed for the intended direction of pull).

Figure 10-4. Ice ax in the anchor position.

(4) ***Push-Hold Position.*** Another variation on steep slopes is the push-hold position (Figure 10-5). The hand is placed on the shaft of the ax just below the head with the pick forward. The pick is driven into the slope at shoulder height. The hand is then placed on the top of the ax head for use as a handhold.

Figure 10-5. Ice ax in the push-hold position.

(5) ***Dagger Position.*** The dagger position is used on steep slopes to place a handhold above shoulder height (Figure 10-6). The hand grasps the head of the ax with the pick forward and the shaft hanging down. The ax is driven into the surface in a stabbing action. The hand is then placed on the ax head for use as a handhold.

Figure 10-6. Ice ax in the dagger position.

(6) ***Hammer Position.*** The hammer position will set the pick deepest in any snow or ice condition (Figure 10-7, page 10-6). The ax is used like a hammer with the pick being driven into the slope. On vertical or near-vertical sections, two axes used in the hammer position will often be required.

Figure 10-7. Ice ax in the hammer position.

b. **Crampons.** Walking in crampons is not complicated but it does present difficulties. When walking in crampons, the same principles are used as in mountain walking, except that when a leg is advanced it is swung in a slight arc around the fixed foot to avoid locking the crampons or catching them in clothing or flesh. The trousers should be bloused to prevent catching on crampons. All straps should be secured to prevent stepping on them and, potentially, causing a fall. The buckles should be located on the outside of each foot when the crampons are secured to prevent snagging. Remember, when the crampon snags on the pants or boots, a tear or cut usually results, and sometimes involves the skin on your leg and or a serious fall.

(1) Two methods of ascent are used on slopes: traversing and straight up.

(a) A traverse on ice or snow looks much like any mountain walking traverse, except that the ankles are rolled so that the crampons are placed flat on the surface (Figure 10-8). On snow the points penetrate easily; on ice the foot must be pressed or stamped firmly to obtain maximum penetration. At the turning points of a traverse, direction is changed with the uphill foot as in mountain walking.

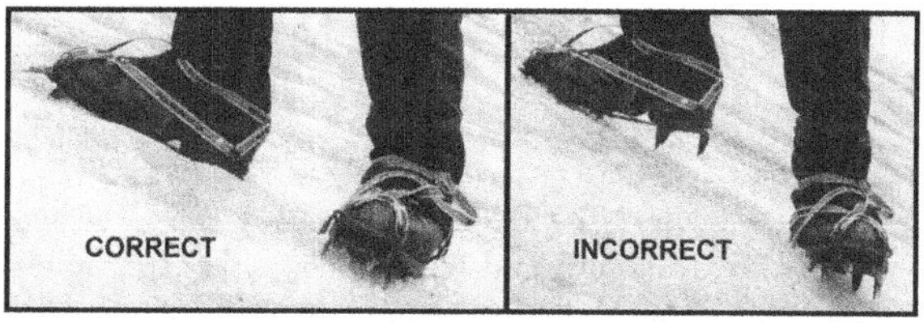

Figure 10-8. Correct and incorrect crampon technique.

(b) A straight up method is for relatively short pitches, since it is more tiring than a traverse. The climber faces directly up the slope and walks straight uphill. As the slope steepens, the herringbone step is used to maintain the flatfoot technique. For short steep pitches, the climber may also face downslope, squatting so the legs almost form a

90-degree angle at the knees, driving the spike of the ice ax into the slope at hip level, and then moving the feet up to the ax. By repeating these steps, the ax and crampon combination can be used to climb short, steep pitches without resorting to step cutting. This method can be tiring. The technique is similar to the crab position used for climbing on slab rock and can also be used for short descents.

(2) A technique known as "front-pointing" may be used for moving straight uphill (Figure 10-9). It is especially useful on steep terrain, in combination with the ice ax in the push-hold, dagger, or hammer position. Front-pointing is easiest with the use of more rigid mountain boots and rigid crampons. The technique is similar to doing calf raises on the tips of the toes and is much more tiring than flat-footing.

(a) The technique starts with the feet approximately shoulder width apart. When a step is taken the climber places the front points of the crampons into the ice with the toe of the boot pointing straight into the slope.

(b) When the front points have bitten into the ice the heel of the boot is lowered slightly so that the first set of vertical points can also bite. The body is kept erect, with the weight centered over the feet as in climbing on rock.

Figure 10-9. Front-pointing with crampons.

c. **Vertical Ice.** When a climb on ice reaches the 60- to 70-degree angle, two ice axes may be helpful, and will become necessary as the angle approaches 90 degrees. The same basic climbing techniques described in Chapter 6 should be applied. If leashes of the correct length and fit are attached to both axes, it may be possible to hang completely from the axes while moving the feet.

d. **Descending with Crampons and Ice Ax.** Whenever possible, descend straight down the fall line. As the slope steepens, gradually turn sideways; on steeper slopes, bend at the waist and knees as if sitting, keeping the feet flat to engage all vertical crampon points and keep the weight over the feet as in descending rock slab (Figures 10-10 and 10-11, page 10-8). On steep terrain, assume a cross body or port arms position with the ax, and traverse. The crab position or front-pointing may also be used for descending. Regardless of the technique used, always ensure the points of the crampons are inserted in the snow or ice and take short, deliberate steps to minimize the chance of tripping and falling down the slope.

Figure 10-10. Flat-footing in the crab position.

Figure 10-11. Use of ice ax in descent.

e. **Normal Progression.** The use of the ice ax and crampons follows a simple, logical progression. The techniques can be used in any combination, dictated by the terrain and skill of the individual. A typical progression could be as follows:

(1) ***Crampons.*** Use crampons in the following situations:

- Walking as on flat ground.

- Herringbone step straight up the slope.
- Traverse with feet flat.
- Backing up the slope (crab position).
- Front-pointing.

(2) ***Ice Ax.*** Use the ice ax in these situations:

- Cane position on flat ground.
- Cane position on uphill side as slope steepens.
- Port arms position with spike on uphill side.
- Anchor position with pick on uphill side.
- Push-hold position using front-pointing technique.
- Dagger position using front-pointing technique.
- Hammer position using front-pointing technique.

e. **Climbing Sequence.** Using most of these positions, a single ax can be "climbed" in steps to move upslope on low-angle to near vertical terrain (Figure 10-12). Begin by positioning the feet in a secure stance and placing the ax in the hammer position as high as possible. Slowly and carefully move the feet to higher positions alternately, and move the hand up the ax shaft. Repeat this until the hand is on top of the head of the ax. Remove the ax and place it at a higher position and begin again.

Figure 10-12. Climbing sequence.

f. **Step Cutting.** Step cutting is an extremely valuable technique that is a required skill for any military mountaineer (Figure 10-13). Using cut steps can save valuable time that would be spent in donning crampons for short stretches of ice and can, in some cases, save the weight of the crampons altogether. Steps may also have to be cut by the lead team to enable a unit without proper equipment to negotiate snow- or ice-covered terrain. As units continue to move up areas where steps have been cut they should continue to improve each step. In ascending, steps may be cut straight up the slope, although a traverse will normally be adopted. In descending, a traverse is also the preferred method. When changing direction, a step large enough for both feet and crampons must be made. Once the step is formed, the adze is best used to further shape and clean the step.

(1) ***Snow.*** On slopes of firm snow and soft ice, steps may be cut by swinging the ax in a near-vertical plane, using the inside corner of the adze for cutting. The step should be fashioned so that it slopes slightly inward and is big enough to admit the entire foot. Steps used for resting or for turning must be larger.

(2) ***Ice.*** Hard ice requires that the pick of the ax be used. Begin by directing a line of blows at right angles to the slope to make a fracture line along the base of the intended step. This technique will reduce the chance of an unwanted fracture in the ice breaking out the entire step. Next, chop above the fracture line to fashion the step. When using the pick it should be given an outward jerk as it is placed to prevent it from sticking in the ice.

(3) ***Step Cutting in a Traverse.*** When cutting steps in a traverse, the preferred cutting sequence is to cut one step at an arm's length from the highest step already cut, then cut one between those two. Cutting ahead one step then cutting an intermediate step keeps all of the steps relatively close to one another and maintains a suitable interval that all personnel can use.

(4) ***Handholds.*** If handholds are cut, they should be smaller than footholds, and angled more.

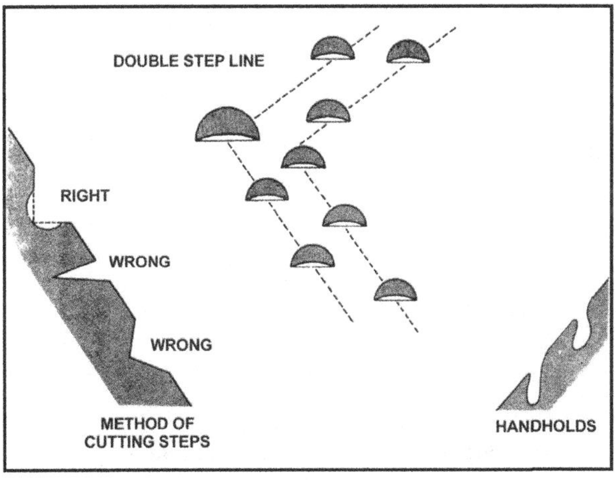

Figure 10-13. Step cutting and handhold cutting.

g. **Self-Arrest.** The large number of climbers injured or killed while climbing on snow and ice can be attributed to two major failings on the part of the climber: climbing unroped, and a lack of knowledge and experience in the techniques necessary to stop, or arrest, a fall (Figure 10-14). A climber should always carry an ice ax when climbing on steep snow or ice; if a fall occurs, he must retain possession and control of his ice ax if he is to successfully arrest the fall. During movement on steep ice, the ax pick will be in the ice solidly before the body is moved, which should prevent a fall of any significance (this is a self belay not a self-arrest).

CAUTION

Self-arrest requires the ax pick to gradually dig in to slow the descent. Self-arrest is difficult on steep ice because the ice ax pick instantly "bites" into the ice, possibly resulting in either arm or shoulder injury, or the ax is deflected immediately upon contact.

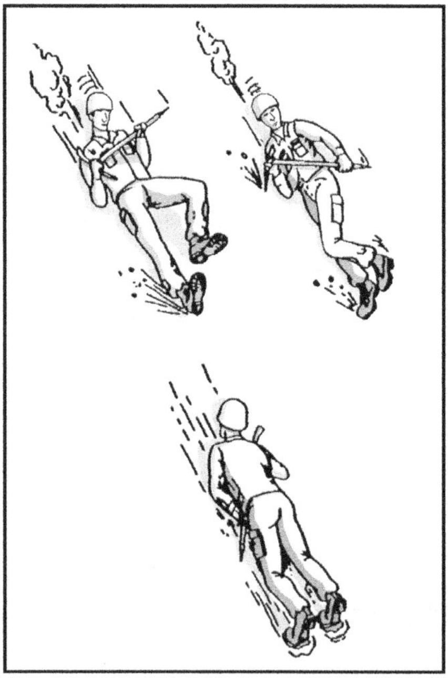

Figure 10-14. Self-arrest technique.

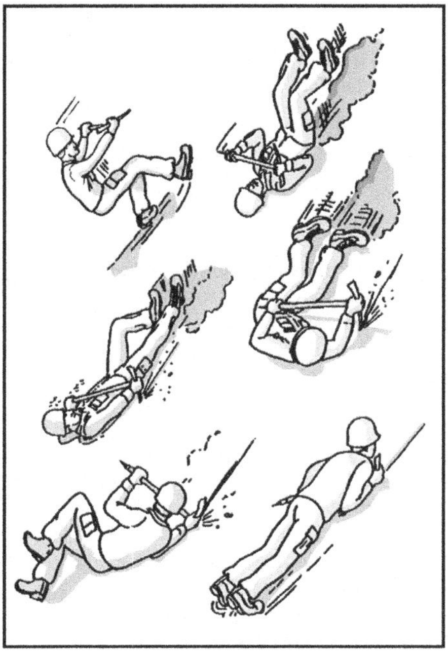

Figure 10-14. Self-arrest technique (continued).

(1) A climber who has fallen may roll or spin; if this happens, the climber must first gain control of his body, whether it is with his ice ax or simply by brute force. Once the roll or spin has been controlled, the climber will find himself in one of four positions.

- Head upslope, stomach on the slope, and feet pointed downslope.
- Head upslope, back to the slope, and feet pointed downslope.
- Head downslope, stomach on the slope, and feet pointed upslope.
- Head downslope, back to the slope, and feet pointed upslope.

(2) To place the body in position to arrest from the four basic fall positions the following must be accomplished.

(a) In the first position, the body is in proper relation to the slope for an arrest.

(b) In the second position, the body must first be rotated from face up to face down on the slope. This is accomplished by rolling the body toward the head of the ax.

(c) In the third position, the pick of the ice ax is placed upslope and used as a pivot to bring the body into proper position.

(d) In the fourth position, the head of the ax must be driven into the snow to the climber's side. This will cause the body to rotate into a head up, stomach down position.

(3) The final position when the arrest of the fall is completed should be with the head upslope, stomach on the slope, with the feet pointed downslope. If crampons are not

worn, the toe of the boots may be dug into the slope to help arrest the fall. The ax is held diagonally across the chest, with the head of the ax by one shoulder and the spike near the opposite hip. One hand grasps the head of the ax, with the pick pointed into the slope, while the other hand is on the shaft near the spike, lifting up on it to prevent the spike from digging into the slope.

Note: If crampons are worn, the feet must be raised to prevent the crampons from digging into the snow or ice too quickly. This could cause the climber to tumble and also, could severely injure his ankles and legs.

(4) When a fall occurs, the climber should immediately grasp the ax with both hands and hold it firmly as described above. Once sufficient control of the body is attained, the climber drives the pick of the ice ax into the slope, increasing the pressure until the fall is arrested. Raising the spike end of the shaft increases the biting action of the pick. It is critical that control of the ice ax be maintained at all times.

10-4. GLISSADING

Glissading is the intentional, controlled, rapid descent, or slide of a mountaineer down a steep slope covered with snow (Figure 10-15, page 10-14). Glissading is similar to skiing, except skis are not used. The same balance and control are necessary, but instead of skis the soles of the feet or the buttocks are used. The only piece of equipment required is the standard ice ax, which serves as the rudder, brake, and guide for the glissade. The two basic methods of glissading are:

a. **Squatting Glissade.** The squatting glissade is accomplished by placing the body in a semi-crouched position with both knees bent and the body weight directly over the feet. The ice ax is grasped with one hand on the head, pick, and adze outboard (away from the body), and the other hand on the shaft. The hand on the shaft grips it firmly in a position that allows control as well as the application of downward pressure on the spike of the ax.

b. **Sitting Glissade.** Using this method the glissader sits on the snow with the legs flat, and the heels and feet raised and pointed downslope. The ice ax is firmly grasped in the same manner as the squatting glissade, with the exception that the hand on the shaft must be locked against the hip for control. The sitting glissade is slower but easier to control than the squatting glissade.

c. **Safety.** A glissade should never be attempted on a slope where the bottom cannot be seen, since drop-offs may exist out of view. Also, a sitting glissade should not be used if the snow cover is thin, as painful injury could result.

Figure 10-15. Glissading techniques.

10-5. SNOW AND ICE ANCHORS

Ice and snow anchors consist of snow pickets, flukes, deadman-type anchors, ice screws, and ice pitons. Deadman anchors can be constructed from snowshoes, skis, backpacks, sleds, or any large items.

a. **Ice Pitons.** The ice piton is used to establish anchor points. The ice piton is not seen in modern ice climbing but may still be available to the military. The standard ice piton is made of tubular steel and is 10 inches in length. Ice pitons installed in pairs are a bombproof anchor; however, ice pitons have no threads for friction to hold them in the ice once placed and are removed easily. Safe use of ice pitons requires placement in pairs. Used singularly, ice pitons are a strong anchor but are easily removed, decreasing the perceived security of the anchor. Follow the instructions below for placing ice pitons in pairs.

(1) Cut a horizontal recess into the ice, and also create a vertical surface (two clean surfaces at right angles to each other).

(2) Drive one piton into the horizontal surface and another into the vertical surface so that the two pitons intersect at the necessary point (Figure 10-16).

(3) Connect the two rings with a single carabiner, ensuring the carabiner is not cross-loaded. Webbing or rope can be used if the rings are turned to the inside of the intersection.

(4) Test the piton pair to ensure it is secure. If it pulls out or appears weak, move to another spot and replace it. The pair of pitons, when placed correctly, are multidirectional.

Figure 10-16. Ice piton pair.

(5) The effective time and or strength for an ice piton placement is limited. The piton will heat from solar radiation, or the ice may crack or soften. Solar radiation can be nearly eliminated by covering the pitons with ice chips once they have been placed. If repeated use is necessary for one installation, such as top roping, the pitons should be inspected frequently and relocated when necessary. When an ice piton is removed, the ice that has accumulated in the tube must be removed before it freezes in position, making further use difficult.

c. **Ice Screws**. The ice screw is the most common type of ice protection and has replaced the ice piton for the most part (Figure 10-17). Some screws have longer "hangers" or handles, which allow them to be easily twisted into position by hand. Place ice screws as follows:

(1) Clear away all rotten ice from the surface and make a small hole with the ax pick to start the ice screw in.

(2) Force the ice screw in until the threads catch.

Figure 10-17. Placement of ice screw using the pick.

(3) Turn the screw until the eye or the hanger of the ice screw is flush with the ice and pointing down. The screw should be placed at an angle 90 to 100 degrees from the

lower surface. Use either your hand or the pick of the ice ax to screw it in. If you have a short ax (70 centimeters or less), you may be able to use the spike in the eye or hanger to ease the turning. (Remember that you may only have use of one hand at this point depending on your stance and the angle of the terrain.)

(4) As with ice pitons, melting of the ice around a screw over a period of time must be considered. The effective time and or strength of an ice screw placement is limited. The screw will heat from solar radiation, or the ice may crack or soften. Solar radiation can be nearly eliminated by covering the screw with ice chips once it has been emplaced. If repeated use is necessary for one installation, such as top roping, the screws should be inspected frequently and relocated when necessary. When an ice screw is removed, the ice that has accumulated in the tube, must be removed before further use.

d. **Horseshoe or Bollard Anchor.** This is an artificial anchor shaped generally like a horseshoe (Figure 10-18). It is formed from either ice or snow and constructed by either cutting with the ice ax or stamping with the boots. When constructed of snow, the width should not be less than 10 feet. In ice, this width may be narrowed to 2 feet, depending on the strength of the ice. The length of the bollard should be at least twice the width. The trench around the horseshoe should be stamped as deeply as possible in the snow and should be cut not less than 6 inches into the ice after all rotten ice is removed. The backside of the anchor must always be undercut to prevent the rope from sliding off and over the anchor.

(1) This type of anchor is usually available and may be used for fixed ropes or rappels. It must be inspected frequently to ensure that the rope is not cutting through the snow or ice more than one-third the length of the anchor; if it is, a new anchor must be constructed in a different location.

(2) A horseshoe anchor constructed in snow is always precarious, its strength depending upon the prevailing texture of the snow. For dry or wind-packed snow, the reliability of the anchor should always be suspect. The backside of the bollard can be reinforced with ice axes, pickets, or other equipment for added strength.

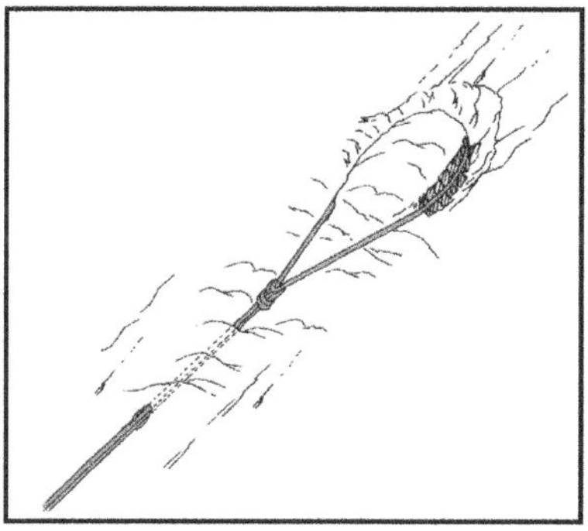

Figure 10-18. Horseshoe or bollard anchor.

e. **Pickets and Ice Axes.** Pickets and ice axes may be used as snow anchors as follows.

(1) The picket should be driven into the snow at 5 to 15 degrees off perpendicular from the lower surface. If the picket cannot be driven in all the way to the top hole, the carabiner should be placed in the hole closest to the snow surface to reduce leverage. The picket may also be tied off with a short loop of webbing or rope as in tying off pitons.

(2) An ice ax can be used in place of a picket. When using an ice ax as a snow anchor, it should be inserted with the widest portion of the ax shaft facing the direction of pull. The simplest connection to the ax is to use a sling or rope directly around the shaft just under the head. If using the leash ensure it is not worn, frayed, or cut from general use; is strong enough; and does not twist the ax when loaded. A carabiner can be clipped through the hole in the head, also.

(3) Whenever the strength of the snow anchor is suspect, especially when a picket or ax cannot be driven in all the way, the anchor may be buried in the snow and used as a "dead man" anchor. Other items suitable for dead man anchor construction are backpacks, skis, snowshoes, ski poles, or any other item large enough or shaped correctly to achieve the design. A similar anchor, sometimes referred to as a "dead guy," can be made with a large sack either stuffed with noncompressible items or filled with snow and buried. Ensure the attaching point is accessible before burying. The direction of pull on long items, such as a picket or ax, should be at a right angle to its length. The construction is identical to that of the dead man anchor used in earth.

f. **Equalized Anchors.** Snow and ice anchors must be constantly checked due to melting and changing snow or ice conditions.

(1) Whenever possible, two or more anchors should be used. While this is not always practical for intermediate anchor points on lead climbs or fixed ropes, it should be mandatory for main anchors at all belay positions, rappel points, or other fixed rope installations. (Figure 10-19, page 10-18, shows an example of three snow pickets configured to an equalized anchor.)

(2) As with multipoint anchors on rock, two or more snow or ice anchors can be joined together with a sling rope or webbing to construct one solid, equalized anchor. A bowline on a bight tied into the climbing rope can also be used instead of sling ropes or webbing.

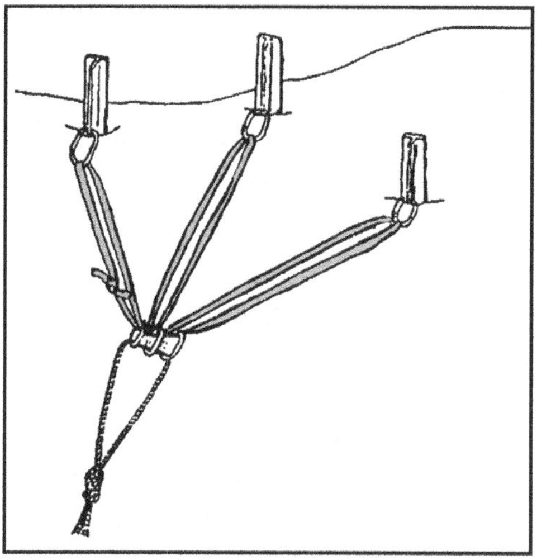

Figure 10-19. Equalized anchor using pickets.

10-6. ROPED CLIMBING ON ICE AND SNOW

When climbing on ice or snow team members tie into a climbing rope the same as when they climb on rock. When crevasses are expected, a three-man rope team is recommended.

a. **Tie-In Method.** For climbing on snow and ice, the tie-in procedure is normally the same as for rock climbing; however, when moving over snow-covered glaciers, the tie-in is modified slightly. (See paragraph 10-7, Movement on Glaciers, for more information).

b. **Movement.** For movement on gentle or moderate slopes where there is little chance of a serious fall, all climbers move simultaneously. Normally the climbers move in single file using the steps created by the lead climber and improving them when necessary. The rope between the climbers should be fully extended and kept reasonably tight. Should any member fall, he immediately yells "FALLING." The other rope team members immediately drop into a self-arrest position. The fallen climber also applies the self-arrest procedure. By using this method, called the "team arrest," the entire team as a whole arrests the fall of one member. On steeper slopes, and when crossing snow-covered crevasses where the snow bridges appear weak, the climbers use belayed climbing techniques as in rock climbing.

c. **Belaying on Snow and Ice.** The principles of belaying on ice and snow are the same as on rock. Generally, the high-force falls found in rock climbing are not present on snow and ice unless the pitch being climbed is extremely steep.

(1) ***Boot-Ax Belay.*** This belay can be useful in areas where the full length of the ice ax can penetrate the snow. The holding strength of the boot-ax belay is directly related to the firmness of the snow and to the strength of the ice ax shaft. The shaft of the ax is

tilted slightly uphill and jammed into the snow. The belayer places his uphill foot against the downhill side of the ax for support. A bight formed in the rope is placed over the boot and around the shaft of the ice ax. The brake is applied by wrapping the rope around the heel of the boot (Figure 10-20).

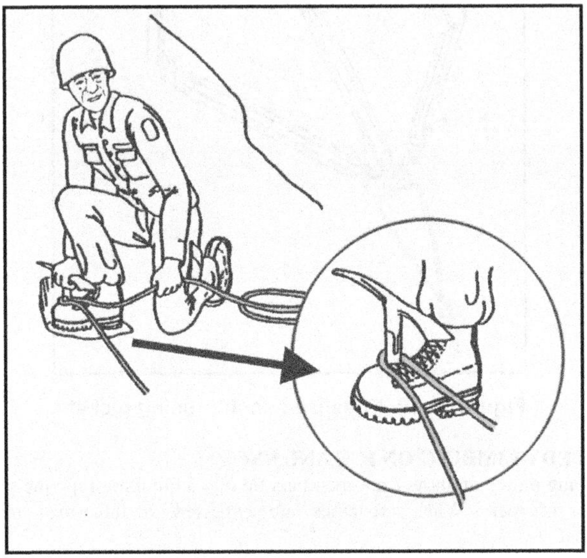

Figure 10-20. Boot-ax belay.

(2) ***Body Belay.*** The body belay can be used on snow and ice, also. The principles are the same as for belays on rock—solid anchors must be used and a well-braced position assumed. The position can be improved by digging depressions into the snow or ice for a seat and footholds. A strong platform should be constructed for the standing body belay.

(3) ***Munter Hitch.*** This belay technique is also used on snow and ice. When using the hitch off of the anchor, a two-point equalized anchor should be constructed as a minimum.

d. **Fixed Ropes.** The use of fixed ropes on ice is recommended for moving units through icefall areas on glaciers or other steep ice conditions. The procedures for emplacing fixed ropes on ice are basically the same as on rock with the exception that anchors need more attention, both in initial placement and in subsequent inspection, and steps may have to be cut to assist personnel.

10-7. MOVEMENT ON GLACIERS

Movement in mountainous terrain may require travel on glaciers. An understanding of glacier formation and characteristics is necessary to plan safe routes. A glacier is formed by the perennial accumulation of snow and other precipitation in a valley or draw. The accumulated snow eventually turns to ice due to metamorphosis. The "flow" or movement of glaciers is caused by gravity. There are a few different types of glaciers identifiable primarily by their location or activity.

- Valley glacier—resides and flows in a valley.
- Cirque glacier—forms and resides in a bowl.
- Hanging glacier—these are a result of valley or cirque glaciers flowing and or deteriorating. As the movement continues, portions separate and are sometimes left hanging on mountains, ridgelines, or cliffs.
- Piedmont glacier—formed by one or more valley glaciers; spreads out into a large area.
- Retreation glacier—a deteriorating glacier; annual melt of entire glacier exceeds the flow of the ice.
- Surging glacier—annual flow of the ice exceeds the melt; the movement is measurable over a period of time.

a. **Characteristics and Definitions.** This paragraph describes the common characteristics of glaciers, and defines common terminology used in reference to glaciers. (Figure 10-21 shows a cross section of a glacier, and Figure 10-22 depicts common glacier features.)

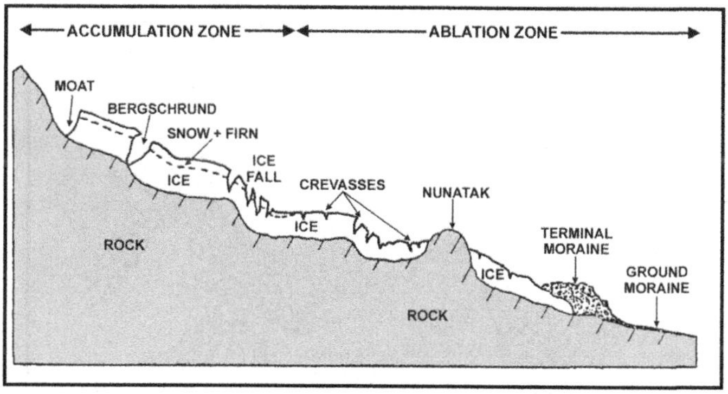

Figure 10-21. Glacier cross section.

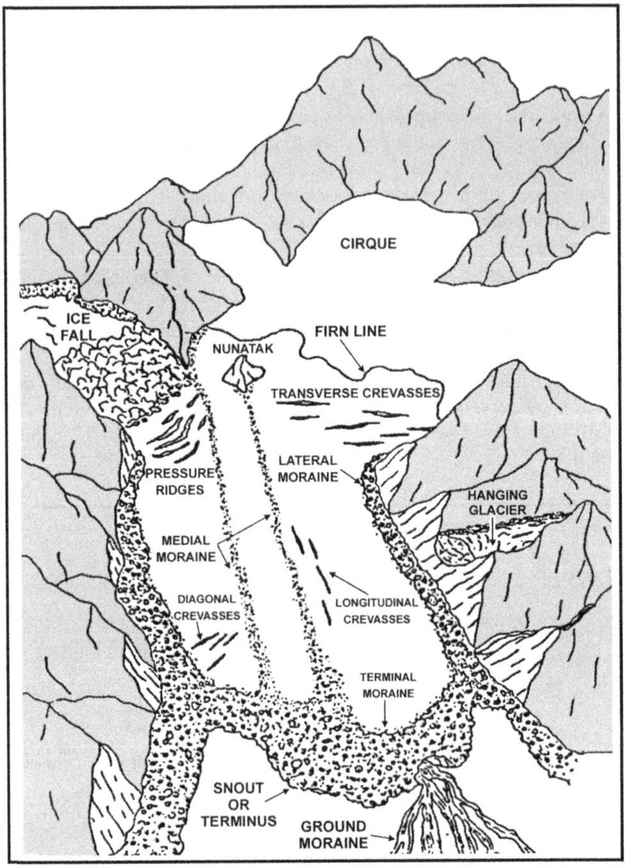

Figure 10-22. Glacier features.

(1) Firn is compacted granular snow that has been on the glacier at least one year. Firn is the building blocks of the ice that makes the glacier.

(2) The accumulation zone is the area that remains snow-covered throughout the year because of year-round snowfall. The snowfall exceeds melt.

(3) The ablation zone is the area where the snow melts off the ice in summer. Melt equals or exceeds snowfall.

(4) The firn line separates the accumulation and ablation zones. As you approach this area, you may see "strips" of snow in the ice. Be cautious, as these could be *snow bridges* remaining over crevasses. Remember that snow bridges will be weakest lower on the glacier as you enter the accumulation zone. The firn line can change annually.

(5) A bergschrund is a large crevasse at the head of a glacier caused by separation of active (flowing) and inactive (stationary) ice. These will usually be seen at the base of a major incline and can make an ascent on that area difficult.

(6) A moat is a wall formed at the head (start) of the glacier. These are formed by heat reflected from valley wall.

(7) A crevasse is a split or crack in the glacier surface. These are formed when the glacier moves over an irregularity in the bed surface.

(8) A transverse crevasse forms perpendicular to the flow of a glacier. These are normally found where a glacier flows over a slope with a gradient change of 30 degrees or more.

(9) Longitudinal crevasses form parallel to the flow of a glacier. These are normally found where a glacier widens.

(10) Diagonal crevasses form at an angle to the flow of a glacier. These are normally found along the edges where a glacier makes a bend.

(11) A snow bridge is a somewhat supportive structure of snow that covers a crevasse. Most of these are formed by the wind. The strength of a snow bridge depends on the snow itself.

(12) Icefalls are a jumble of crisscross crevasses and large ice towers that are normally found where a glacier flows over a slope with a gradient change of 25 degrees or more.

(13) Seracs are large pinnacles or columns of ice that are normally found in icefalls or on hanging glaciers.

(14) Ice avalanches are falling chunks of ice normally occurring near icefalls or hanging glaciers.

(15) The moraine is an accumulation of rock or debris on a glacier caused by rockfall or avalanche of valley walls.

(16) The lateral moraine is formed on sides of glacier.

(17) The medial moraine is in the middle of the glacier. This is also formed as two glaciers come together or as a glacier moves around a central peak.

(18) The terminal moraine is at the base of a glacier and is formed as moraines meet at the snout or terminus of a glacier.

(19) The ground moraine is the rocky debris extending out from the terminus of a glacier. This is formed by the scraping of earth as the glacier grew or surged and exposed as the glacier retreats.

(20) A Nunatak is a rock projection protruding through the glacier as the glacier flows around it.

(21) An ice mill is a hole in the glacier formed by swirling water on the surface. These can be large enough for a human to slip into.

(22) Pressure ridges are wavelike ridges that form on glacier normally after a glacier has flowed over icefalls.

(23) A glacier window is an opening at the snout of the glacier where water runs out of the glacier.

b. **Dangers and Obstacles.** The principle dangers and obstacles to movement in glacial areas are crevasses, icefalls, and ice avalanches. Snow-covered crevasses make movement on a glacier extremely treacherous. In winter, when visibility is poor, the difficulty of recognizing them is increased. Toward the end of the summer, crevasses are widest and covered by the least snow. Crossing snow bridges constitutes the greatest potential danger in movement over glaciers in the summer. On the steep pitch of a glacier, ice flowing over irregularities and cliffs in the underlying valley floor cause the ice to break up into ice blocks and towers, criss-crossed with crevasses. This jumbled cliff of ice is known as an icefall. Icefalls present a major obstacle to safe movement of troops on glaciers.

(1) Moving on glaciers brings about the hazard of falling into a crevasse. Although the crevasses are visible in the ablation zone in the summer (Figure 10-23), the accumulation zone will still have hidden crevasses. The risk of traveling in the accumulation zone can be managed to an acceptable level when ropes are used for connecting the team members (Figure 10-24, page 10-24). Crampons and an ice ax are all that is required to safely travel in the ablation zone in the summer.

Figure 10-23. Ablation zone of glacier in summer.

Figure 10-24. Rope teams moving in the accumulation zone of a glacier.

(2) When conditions warrant, three to four people will tie in to one rope at equal distances from each other. To locate the positions, if three people are on a team, double the rope and one ties into the middle and the other two at the ends. If four people are on a team, form a "z" with the rope and expand the "z" fully, keeping the end and the bight on each "side" of the "z" even. Tie in to the bights and the ends.

(3) Connect to the rope with the appropriate method and attach the Prusik as required. The rope should be kept relatively tight either by Prusik belay or positioning of each person. If the team members need to assemble in one area, use the Prusik to belay each other in.

(4) If a team member falls into a crevasse, the remaining members go into team arrest, assess the situation, and use the necessary technique to remove the person from the crevasse. The simplest and most common method for getting someone out of a crevasse is for the person to climb out while being belayed.

(5) All items should be secured to either the climber or the rope/harness to prevent inadvertent release and loss of necessary items or equipment. Packs should be secured to the rope/harness with webbing or rope. If traveling with a sled in tow, secure it not only to a climber to pull it, but connect it to the rope with webbing or rope also.

(6) If marking the route on the glacier is necessary for backtracking or to prevent disorientation in storms or flat-light conditions, use markers that will be noticeable against the white conditions. The first team member can place a new marker when the last team member reaches the previous marker.

c. **Roped Movement.** The first rule for movement on glaciers is to rope up (Figure 10-25). A roped team of two, while ideal for rock climbing, is at a disadvantage on a snow-covered glacier. The best combination is a three-man rope team. Generally, the rope team members will move at the same time with the rope fully extended and reasonably tight between individuals, their security being the team arrest. If an individual should break through a snow bridge and fall into a crevasse, the other members immediately perform self-arrest, halting the fall. At points of obvious weakness in the snow bridges, the members may decide to belay each other across the crevasse using one of the established belay techniques.

Figure 10-25. Preparation for roped movement.

(1) Even with proper training in crevasse rescue techniques, the probability exists that an individual may remain suspended in a crevasse for a fairly lengthy amount of time while trying to get himself out or while awaiting help from his rope team members. Because of this, it is strongly recommended that all personnel wear a seat/chest combination harness, whether improvised or premanufactured.

(2) Rope team members must be able to quickly remove the climbing rope from the harness(es) during a crevasse rescue. The standard practice for connecting to the rope for glacier travel is with a locking carabiner on a figure-eight loop to the harness. This allows quick detachment of the rope for rescue purposes. The appropriate standing part of the rope is then clipped to the chest harness carabiner.

(3) If a rope team consists of only two people, the rope should be divided into thirds, as for a four-person team. The team members tie into the middle positions on the rope, leaving a third of the rope between each team member and a third on each end of the rope. The remaining "thirds" of the rope should be coiled and either carried in the rucksack, attached to the rucksack, or carried over the head and shoulder. This gives each

climber an additional length of rope that can be used for crevasse rescue, should one of the men fall through and require another rope. If necessary, this excess end rope can be used to connect to another rope team for safer travel.

Note: The self-arrest technique used by one individual will work to halt the fall of his partner on a two-man rope team; however, the chance of it failing is much greater. Crevasse rescue procedures performed by a two-man rope team, by itself, may be extremely difficult. For safety reasons, movement over a snow-covered glacier by a single two-man team should be avoided wherever possible.

d. **Use of Prusik Knots.** Prusik knots are attached to the climbing rope for all glacier travel. The Prusiks are used as a self-belay technique to maintain a tight rope between individuals, to anchor the climbing rope for crevasse rescue, and for self-rescue in a crevasse fall. The Prusik slings are made from the 7-millimeter by 6-foot and 7-millimeter by 12-foot ropes. The ends of the ropes are tied together, forming endless loops or slings, with double fisherman's knots. Form the Prusik knot on the rope in front of the climber. An overhand knot can be tied into the sling just below the Prusik to keep equal tension on all the Prusik wraps. Attach this sling to the locking carabiner at the tie in point on the harness.

Note: An ascender can replace a Prusik sling in most situations. However, the weight of an ascender hanging on the rope during movement will become annoying, and it could be stepped on during movement and or climbing.

e. **Securing the Backpack/Rucksack.** If an individual should fall into a crevasse, it is essential that he be able to rid himself of his backpack. The weight of the average pack will be enough to hinder the climber during crevasse rescue, or possibly force him into an upside down position while suspended in the crevasse. Before movement, the pack should be attached to the climbing rope with a sling rope or webbing and a carabiner. A fallen climber can immediately drop the pack without losing it. The drop cord length should be minimal to allow the fallen individual to reach the pack after releasing it, if warm clothing is needed. When hanging from the drop cord, the pack should be oriented just as when wearing it (ensure the cord pulls from the top of the pack).

f. **Routes.** An individual operating in the mountains must appreciate certain limitations in glacier movement imposed by nature.

(1) Additional obstacles in getting onto a glacier may be swift glacier streams, steep terminal or lateral moraines, and difficult mountain terrain bordering the glacier ice. The same obstacles may also have to be overcome in getting on and off a valley glacier at any place along its course.

(2) Further considerations to movement on a glacier are steep sections, heavily crevassed portions, and icefalls, which may be major obstacles to progress. The use of current aerial photographs in conjunction with aerial reconnaissance is a valuable means of gathering advance information about a particular glacier. However, they only supplement, and do not take the place of, on-the-ground reconnaissance conducted from available vantage points.

g. **Crossing Crevasses.** Open crevasses are obvious, and their presence is an inconvenience rather than a danger to movement. Narrow cracks can be jumped, provided the take off and landing spots are firm and offer good footing. Wider cracks will have to be circumvented unless a solid piece of ice joins into an ice bridge strong enough to support at least the weight of one member of the team. Such ice bridges are often formed in the lower portion of a crevasse, connecting both sides of it.

(1) In the area of the firn line, the zone that divides seasonal melting from permanent falls of snow, large crevasses remain open, though their depths may be clogged with masses of snow. Narrow cracks may be covered. In this zone, the snow, which covers glacier ice, melts more rapidly than that which covers crevasses. The difference between glacier ice and narrow snow-covered cracks is immediately apparent; the covering snow is white, whereas the glacier ice is gray.

(2) Usually the upper part of a glacier is permanently snow covered. The snow surface here will vary in consistency from dry powder to consolidated snow. Below this surface cover are found other snow layers that become more crystalline in texture with depth, and gradually turn into glacier ice. It is in this snow-covered upper part of a glacier that crevasses are most difficult to detect, for even wide crevasses may be completely concealed by snow bridges.

h. **Snow Bridges.** Snow bridges are formed by windblown snow that builds a cornice over the empty interior of the crevasse. As the cornice grows from the windward side, a counter drift is formed on the leeward side. The growth of the leeward portion will be slower than that to the windward so that the juncture of the cornices occurs over the middle of the crevasse only when the contributing winds blow equally from each side. Bridges can also be formed without wind, especially during heavy falls of dry snow. Since cohesion of dry snow depends only on an interlocking of the branches of delicate crystals, such bridges are particularly dangerous during the winter. When warmer weather prevails the snow becomes settled and more compacted, and may form firmer bridges.

(1) Once a crevasse has been completely bridged, its detection is difficult. Bridges are generally slightly concave because of the settling of the snow. This concavity is perceptible in sunshine, but difficult to detect in flat light. If the presence of hidden crevasses is suspected, the leader of a roped team must probe the snow in front of him with the shaft of his ice ax. As long as a firm foundation is encountered, the team may proceed, but should the shaft meet no opposition from an underlying layer of snow, a crevasse is probably present. In such a situation, the prober should probe closer to his position to make sure that he is not standing on the bridge itself. If he is, he should retreat gently from the bridge and determine the width and direction of the crevasse. He should then follow and probe the margin until a more resistant portion of the bridge is reached. When moving parallel to a crevasse, all members of the team should keep well back from the edge and follow parallel but offset courses.

(2) A crevasse should be crossed at right angles to its length. When crossing a bridge that seems sufficiently strong enough to hold a member of the team, the team will generally move at the same time on a tight rope, with each individual prepared to go into self-arrest. If the stability of the snow bridge is under question, they should proceed as follows for a team of three glacier travelers:

(a) The leader and second take up a position at least 10 feet back from the edge. The third goes into a self-belay behind the second and remains on a tight rope.

(b) The second belays the leader across using one of the established belay techniques. The boot-ax belay should be used only if the snow is deep enough for the ax to be inserted up to the head and firm enough to support the possible load. A quick ice ax anchor should be placed for the other belays. Deadman or equalizing anchors should be used when necessary.

(c) The leader should move forward, carefully probing the snow and evaluating the strength of the bridge, until he reaches firm snow on the far side of the crevasse. He then continues as far across as possible so number two will have room to get across without number one having to move.

(d) The third assumes the middle person's belay position. The middle can be belayed across by both the first and last. Once the second is across, he assumes the belay position. Number one moves out on a tight rope and anchors in to a self-belay. Number two belays number three across.

(3) In crossing crevasses, distribute the weight over as wide an area as possible. Do not stamp the snow. Many fragile bridges can be crossed by lying down and crawling to the other side. Skis or snowshoes help distribute the weight nicely.

i. **Arresting and Securing a Fallen Climber.** The simplest and most common method for getting someone out of a crevasse is for the person to climb out while being belayed. Most crevasse falls will be no more than body height into the opening if the rope is kept snug between each person.

(1) To provide a quick means of holding an unexpected breakthrough, the rope is always kept taut. When the leader unexpectedly breaks through, the second and third immediately go into a self-arrest position to arrest the fall. A fall through a snow bridge results either in the person becoming jammed in the surface hole, or in being suspended in the crevasse by the rope. If the leader has fallen only partially through the snow bridge, he is supported by the snow forming the bridge and should not thrash about as this will only enlarge the hole and result in deeper suspension. All movements should be slow and aimed at rolling out of the hole and distributing the weight over the remainder of the bridge. The rope should remain tight at all times and the team arrest positions adjusted to do so. It generally is safer to retain the rucksack, as its bulk often prevents a deeper fall. Should a team member other than the leader experience a partial fall, the rescue procedure will be same as for the leader, only complicated slightly by the position on the rope.

(2) When the person falls into a crevasse, the length of the fall depends upon how quickly the fall is arrested and where in the bridge the break takes place. If the fall occurs close to the near edge of the crevasse, it usually can be checked before the climber has fallen more than 6 feet. However, if the person was almost across, the fall will cause the rope to cut through the bridge, and then even an instantaneous check by the other members will not prevent a deeper fall. The following scenario is an example of the sequence of events that take place after a fall by the leader in a three-person team. (This scenario is for a team of three, each person referred to by position; the leader is number 1.)

(a) Once the fall has been halted by the team arrest, the entire load must be placed on number 2 to allow number 3 to move forward and anchor the rope. Number 3 slowly releases his portion of the load onto number 2, always being prepared to go back into self-arrest should number 2's position begin to fail.

(b) Once number 2 is confident that he can hold the load, number 3 will proceed to number 2's position, using the Prusik as a self belay, to anchor the rope. In this way the rope remains reasonably tight between number 2 and number 3. Number 3 must always be prepared to go back into self-arrest should number 2's position begin to fail.

(c) When number 3 reaches number 2's position he will establish a bombproof anchor 3 to 10 feet in front of number 2 (on the load side), depending on how close number 2 is to the lip of the crevasse. This could be either a deadman or a two-point equalized anchor, as a minimum.

(d) Number 3 connects the rope to the anchor by tying a Prusik with his long Prusik sling onto the rope leading to number 1. An overhand knot should be tied into the long Prusik sling to shorten the distance to the anchor, and attached to the anchor with a carabiner. The Prusik knot is adjusted toward the load.

(e) Number 2 can then release the load of number 1 onto the anchor. Number 2 remains connected to the anchor and monitors the anchor.

(f) A fixed loop can be tied into the slack part of the rope, close to number 2, and attached to the anchor (to back up the Prusik knot).

(g) Number 3 remains tied in, but continues forward using a short Prusik as a self-belay. He must now quickly check on the condition of number 1 and decide which rescue technique will be required to retrieve him.

(3) These preliminary procedures must be performed before retrieving the fallen climber. If number 3 should fall through a crevasse, the procedure is the same except that number 1 assumes the role of number 3. Normally, if the middle person should fall through, number 1 would anchor the rope by himself. Number 3 would place the load on number 1's anchor, then anchor his rope and move forward with a Prusik self-belay to determine the condition of number 2.

j. **Crevasse Rescue Techniques.** Snow bridges are usually strongest at the edge of the crevasse, and a fall is most likely to occur some distance away from the edge. In some situations, a crevasse fall will occur at the edge of the snow bridge, on the edge of the ice. If a fall occurs away from the edge, the rope usually cuts deeply into the snow, thus greatly increasing friction for those pulling from above. In order to reduce friction, place padding, such as an ice ax, ski, ski pole, or backpack/rucksack, under the rope and at right angles to the stress. Push the padding forward as far as possible toward the edge of the crevasse, thus relieving the strain on the snow. Ensure the padding is anchored from falling into the crevasse for safety of the fallen climber.

(1) ***Use of Additional Rope Teams.*** Another rope team can move forward and assist in pulling the victim out of a crevasse. The assisting rope team should move to a point between the fallen climber and the remaining rope team members. The assisting team can attach to the arresting team's rope with a Prusik or ascender and both rope teams' members can all pull simultaneously. If necessary, a belay can be initiated by the fallen climber's team while the assisting team pulls. The arresting team member closest to the fallen climber should attach the long Prusik to themselves and the rope leading to the fallen climber, and the assisting team can attach their Prusik or ascender between this long Prusik and the arresting team member. As the assisting team pulls, the Prusik belay will be managed by the arresting team member at the long Prusik.

Note: Safety in numbers is obvious for efficient crevasse rescue techniques. Additional rope teams have the necessary equipment to improve the main anchor or establish new ones and the strength to pull a person out even if he is deep in the crevasse. Strength of other rope teams should always be used before establishing more time-consuming and elaborate rescue techniques.

(2) ***Fixed Rope.*** If the fallen climber is not injured, he may be able to climb out on a fixed rope. Number 1 clips number 3's rope to himself. He then climbs out using number 3's rope as a simple fixed line while number 2 takes up the slack in number 1's rope through the anchor Prusik for a belay.

(3) ***Prusik Ascending Technique.*** There may be times when the remaining members of a rope team can render little assistance to the person in the crevasse. If poor snow conditions make it impossible to construct a strong anchor, the rope team members on top may have to remain in self-arrest. Other times, it may just be easier for the fallen climber to perform a self-rescue. (Figure 10-26 shows the proper rope configuration.) The technique is performed as follows:

(a) The fallen climber removes his pack and lets it hang below from the drop cord.

(b) The individual slides their short Prusik up the climbing rope as far as possible.

(c) The long Prusik is attached to the rope just below the short Prusik. The double fisherman's knot is spread apart to create a loop large enough for one or both feet. The fallen climber inserts his foot/feet into the loop formed allowing the knot to cinch itself down.

(d) The individual stands in the foot loop, or "stirrup," of the long sling.

(e) With his weight removed from the short Prusik, it is slid up the rope as far as it will go. The individual then hangs from the short Prusik while he moves the long Prusik up underneath the short Prusik again.

(f) The procedure is repeated, alternately moving the Prusiks up the rope, to ascend the rope. Once the crevasse lip is reached, the individual can simply grasp the rope and pull himself over the edge and out of the hole.

(g) Besides being one of the simplest rope ascending techniques, the short Prusik acts as a self-belay and allows the climber to take as long a rest as he wants when sitting in the harness. The rope should be detached from the chest harness carabiner to make the movements less cumbersome. However, it is sometimes desirable to keep the chest harness connected to the rope for additional support. In this case the Prusik knots must be "on top" of the chest harness carabiner so they can be easily slid up the rope without interference from the carabiner. The long Prusik sling can be routed through the chest harness carabiner for additional support when standing up in the stirrup.

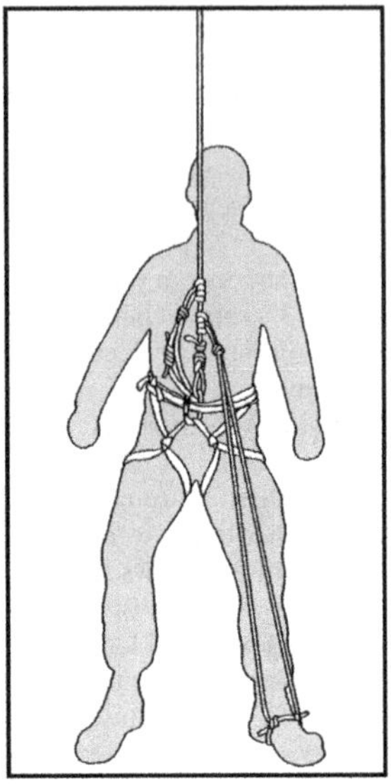

Figure 10-26. Prusik ascending technique.

(4) ***Z-Pulley Hauling System.*** If a fallen climber is injured or unconscious, he will not be able to offer any assistance in the rescue. If additional rope teams are not immediately available, a simple raising system can be rigged to haul the victim out of the crevasse. The Z-pulley hauling system is one of the simplest methods and the one most commonly used in crevasse rescue (Figure 10-27, page 10-32). The basic Z rig is a "3-to-1" system, providing mechanical advantage to reduce the workload on the individuals operating the haul line. In theory, it would only take about 33 pounds of pull on the haul rope to raise a 100-pound load with this system. In actual field use, some of this mechanical advantage is lost to friction as the rope bends sharply around carabiners and over the crevasse lip. The use of mechanical rescue pulleys can help reduce this friction in the system. The following describes rigging of the system. (This scenario is for a team of three, each person referred to by position; the leader is number 1.)

(a) After the rope team members have arrested and secured number 1 to the anchor, and they have decided to install the Z rig, number 2 will attach himself to the anchor without using the rope and clear the connecting knot used. Number 3 remains connected to the rope.

(b) The slack rope exiting the anchor Prusik is clipped into a separate carabiner attached to the anchor. A pulley can be used here if available.

(c) Number 3 will use number 2's short Prusik to rig the haul Prusik. He moves toward the crevasse lip (still on his own self-belay) and ties number 2's short Prusik onto number 1's rope (load rope) as close to the edge as possible.

(d) Another carabiner (and pulley if available) is clipped into the loop of the haul Prusik and the rope between number 3's belay Prusik and the anchor is clipped (or attached through the pulley). Number 3's rope becomes the haul rope.

(e) Number 3 then moves towards the anchor and number 2. Number 2 could help pull if necessary but first would connect to the haul rope with a Prusik just as number 3. If the haul Prusik reaches the anchor before the victim reaches the top, the load is simply placed back on the anchor Prusik and number 3 moves the haul Prusik back toward the edge. The system is now ready for another haul.

CAUTION

The force applied to the fallen climber through use of the Z-pulley system can be enough to destroy the harness-to-rope connection or injure the fallen climber if excess force is applied to the pulling rope.

Notes: 1. The Z-pulley adds more load on the anchor due to the mechanical advantage. The anchor should be monitored for the duration of the rescue.

2. With the "3-to-1" system, the load (fallen climber) will be raised 1 foot for every 3 feet of rope taken up during the haul.

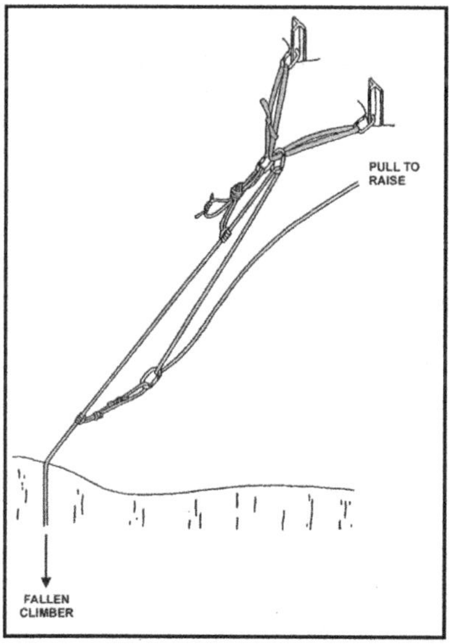

Figure 10-27. Z-pulley hauling system.

10-8. GLACIER BIVOUAC PROCEDURES

When locating a bivouac site or a gathering area where the team might need or want to unrope, at least one person will need to "probe" the area for hidden crevasses. The best type of probe will be the manufactured collapsing probe pole, at least eight feet in length. Other items could be used but the length and strength of the probe is most important. Other rope team members will belay the probers. The prober is "feeling" for a solid platform to place the tent by pushing the probe as hard and deep as possible into the surface. Probing should be in 2-foot intervals in all directions within the site.

a. If the probe suddenly has no resistance while pushing down, a crevasse is present. Attempts to outline the crevasse can be futile if the crevasse is large. Normally, the best decision is to relocate the proposed bivouac area far enough away to avoid that crevasse. (Sometimes only a few feet one way or the other is all that's needed to reach a good platform.) Probe the tent site again after digging to the desired surface. Mark boundaries with wands or other items such as skis, poles, and so on.

b. Occasionally while probing, increased pressure will be noticed without reaching a solid platform. The amount of snowfall may be such that even after digging into the snow, the probe still doesn't contact a hard surface. Try to find a solid platform.

c. There should be no unroped movement outside the probed/marked areas. If a latrine area is needed, probe a route away from the bivouac site and probe the latrine area also. If a dugout latrine is necessary, probe again after digging.

d. Multiple tent sites can be connected, which keeps tents closer together. Probe all areas between the tents if you plan to move in those areas. Closer tents will make communicating between tent groups and rope teams easier.

e. If there is a chance for severe storms with high winds, snow walls may be constructed to protect the tent site from wind. The walls can be constructed from loose snow piled on the perimeter, or blocks can be cut from consolidated snow layers. In deep soft snow, digging three or four feet to find a consolidated layer will result in enough snow moved to build up decent walls around the tent site.

(1) For block construction, move the soft snow from the surface into the wall foundation areas (down to a consolidated layer of snow).

(2) Cut blocks approximately 1 by 1 by 2 feet, and construct the walls by interlocking the blocks with overlapping placements. The walls should be slightly higher than the tent. At a minimum, build walls on the windward side of the tent site.

(3) Snow walls can also provide shelter from wind for food preparation.

CHAPTER 11
MOUNTAIN RESCUE AND EVACUATION

Steep terrain and adverse weather are common in mountainous environments. Under these conditions, relatively minor injuries may require evacuation. The evacuation technique chosen is determined by the type of injury, distance to be moved, terrain, and existing installations. Air evacuation is preferred; however, the weather, tactical situation, or operational ceiling of the aircraft may make this impossible. It is, therefore, imperative that all personnel are trained in mountain evacuation techniques and are self-sufficient. Casualties should be triaged before evacuation. Triage is performed by the most experienced medical personnel available (physician, physician's assistant, medic).

Performing a rescue operation can be a significant emotional event. Rescue scenarios must be practiced and rehearsed until rescue party members are proficient in the many tasks required to execute a rescue. To perform most of the high-angle rescues, Level I and Level II mountaineers are required with a Level III supervising.

11-1. CONSIDERATIONS

The techniques of evacuation are proven techniques. They are, however, all subject to improvement and should be discarded or modified as better methods of handling victims are developed.

a. When evacuating a victim from mountainous areas keep in mind that the purpose of a rescue operation is to save a life, and physical risk to the rescuers must be weighed against this purpose. However, there is no excuse for failing to make the maximum effort within this limitation. Work and expense should be no deterrent when a life is at stake.

b. Rescues will be unplanned (improvised) or planned rescue operations. For a planned rescue, equipment that is especially suited and designed for rescue should be used. For training missions always have a medical plan developed before an emergency arises (plan for the worst and hope for the best). Ensure that the MEDEVAC plan is a comprehensive plan and must be thought out and understood by all that may be involved in a potential rescue.

c. The following actions will be done immediately at the rescue scene.

(1) Assume command. One person, and one person only, is overall in charge at all times.

(2) Prevent further injuries to the victim and to others. Use reasonable care in reaching the victim.

(3) Immediately ensure the victim has an open airway, resume victim's breathing, control serious bleeding, and maintain moderate body warmth. If the victim is unconscious, continually monitor pulse. Protect the patient from environmental hazards.

(4) Do not move the victim until you have ascertained the extent of injuries, unless it is necessary to prevent further injuries or the victim is located in a dangerous location (for example, avalanche run-out zone, hanging glacier, possibility of falling rocks).

(5) Do nothing more until you have thoroughly considered the situation. Resist the urge for action. Speed is less important than correct action.

(6) Decide whether to evacuate with available facilities or to send for help. Speed in getting to a hospital must be balanced against the probability of further injury if working with inexperienced people, lack of equipment or wrong equipment, and terrain at hand.

(7) When the evacuation route is long and arduous, a series of litter relay points or stations should be established. These stations must be staffed with the minimum medical personnel to provide proper emergency treatment. When a victim develops signs of shock or worsens while being evacuated, he should be treated and retained at one of these stations until his condition allows evacuation.

(8) Helicopters or heated vehicles, if available, should be used for evacuation. While the use of aircraft or vehicles is preferred and can expedite a rescue operation, evacuation of a seriously wounded soldier should never be delayed to await aircraft, vehicle, or a change in weather.

11-2. PLANNING RESCUE OPERATIONS

Every commander should have a medical evacuation plan before undertaking an operation. This plan should have contingencies included so as not to rely on a single asset.

a. When rescuing a casualty (victim) threatened by hostile action, environmental hazard, or any other immediate hazard, the rescuer should not take action without first determining the extent of the hazard and his ability to handle the situation. THE RESCUER MUST NOT BECOME A CASUALTY.

b. The rescue team leader must evaluate the situation and analyze the factors involved. This evaluation can be divided into three major steps:

- Identify the task.
- Evaluate the circumstances of the rescue.
- Plan the action.

c. The task must be identified. In planning a rescue, the rescuer tries to obtain the following information:

- Who, what, where, when, why, and how the situation happened.
- Number of casualties by precedence (urgent, priority, routine, tactical immediate),
- number of casualties by type (litter or ambulatory), and the nature of their injuries.
- Terrain features and location of the casualties.
- Tactical situation.
- If adequate assistance is available to aid in security, rescue, treatment, and evacuation.
- If treatment can be provided at the scene; if the victims require movement to a safer location.
- Equipment required for the rescue operation.

d. Circumstances of the rescue are as follows:

(1) After identifying the task, relate it to the circumstances of the situation.

- Are additional personnel, security, medical, or special rescue equipment needed?
- Are there circumstances, such as aircraft accidents (mass casualties), that may require specialized skills?

- What is the weather condition?
- Is the terrain hazardous?
- How much time is available?

(2) The time element may cause a rescuer to compromise planning stages or treatment (beyond first aid). Make a realistic estimate of time available as quickly as possible to determine the action time remaining. The key elements are the casualty's condition and environment.

(3) Mass casualties are to be expected on the modern battlefield. All problems or complexities of rescue are now multiplied by the number of casualties. Time becomes the critical element.

(4) Considerations for the main rescue group for a planned rescue are as follows:

(a) Carry all needed equipment, hot food and drinks, stove, sleeping bags, tents, bivouac sacks, warm clothes, ropes, and stretchers.

(b) Prepare the evacuation route (ground transport to hospital, walking trails, fixed lines, lowering lines, anchor points, and rescue belay points). If the victim is airlifted out, attach a paper with the medical actions that were performed on the ground (for example, blood pressure, pulse rate, drugs started, and so on).

(c) When performing all rescues, the rescuers are always tied in for safety. With all rescue techniques, remember to think things through logically for safety and to prevent the rescuer from accidentally untying himself or the fallen climber.

(d) Constantly inform the casualty (if they are conscious) as to what you are doing and what he must do.

e. The rescue plan should proceed as follows:

(1) In estimating time available, the casualties' ability to endure is of primary importance. Age and physical condition may vary. Time available is a balance of the endurance time of the casualty, the situation, and the personnel and equipment available.

(2) Consider altitude and visibility. Maximum use of secure, reliable trails or roads is essential.

(3) Ensure that blankets and rain gear are available. Even a mild rain can complicate a normally simple rescue. In high altitudes, extreme cold, or gusting winds, available time is drastically reduced.

(4) High altitudes and gusting winds reduce the ability of fixed-wing or rotary-wing aircraft to assist in operations. Rotary-wing aircraft may be available to remove casualties from cliffs or inaccessible sites, and to quickly transport casualties to a medical treatment facility. Relying on aircraft or specialized equipment is a poor substitute for careful planning.

11-3. MASS CASUALTIES

When there are mass casualties, an orderly rescue may involve further planning.

a. To manage a mass casualty rescue or evacuation, separate stages are taken.

- FIRST STAGE: Remove personnel who are not trapped among debris or who can be easily evacuated.
- SECOND STAGE: Remove personnel who may be trapped by debris, but whose extraction only requires the equipment on hand and little time.

- THIRD STAGE: Remove the remaining personnel who are trapped in extremely difficult or time-consuming situations, such as moving large amounts of debris or cutting through a wall.
- FOURTH STAGE: Remove dead personnel.

b. Evacuation of wounded personnel is based on the victim's condition and is prioritized as follows:

- PRIORITY ONE: Personnel with life-threatening injuries that require immediate emergency care to survive; first aid and stabilization are accomplished before evacuation.
- PRIORITY TWO: Personnel with injuries that require medical care but speed of evacuation is not essential.
- PRIORITY THREE: Injured personnel who can evacuate themselves with minimal assistance.
- PRIORITY FOUR: The logistics removal of dead personnel.

11-4. SPECIAL TRAINING

Before receiving training in basic mountain evacuation, litter teams should receive instruction in military mountaineering and basic first aid. Litter bearers and medics must know the use and care of rope as an item of equipment. The members of litter teams must be proficient in the techniques of belaying and choosing belay points. Proper support and protection must be given to victims and litter bearers when evacuating over steep, difficult terrain.

11-5. PREPARATION FOR EVACUATION

Although the wounded soldier's life may have been saved by applying first aid, it can be lost through carelessness, rough handling, or inadequate protection from the elements. Therefore, before trying to move the wounded soldier, the type and extent of his injury must be evaluated. Dressings over wounds must be reinforced, and fractured bones must be properly immobilized and supported. Based upon the evaluation of the type and extent of the soldier's injury, the best method of manual transportation is selected.

11-6. MANUAL CARRIES

Personnel who are not seriously injured but cannot evacuate themselves may be assisted by fellow soldiers. Personnel who are injured and require prompt evacuation should not be forced to wait for mobile evacuation or special equipment.

a. **One-Man Carries.** The basic carries taught in the Soldier's Manual of Common Tasks (fireman's carry, two-hand, four-hand, saddleback, piggyback, pistol belt, and poncho litter) are viable means of transporting injured personnel; however, the mountainous terrain lends itself to several other techniques. One-man carries include the sling-rope carry and the rope coil carry.

(1) ***Sling-Rope Carry***. The sling-rope carry (Figure 11-1) requires a 4.5-meter sling rope and two men—one as the bearer and the other as an assistant to help secure the casualty to the bearer's back. Conscious or unconscious casualties may be transported this way.

(a) The bearer kneels on all fours.

(b) The assistant places the casualty face down on the bearer's back ensuring the casualty's armpits are even with the bearer's shoulders.

(c) The assistant then finds the middle of the sling rope and places it between the casualty's shoulders.

(d) The assistant runs the ends of the sling rope under the casualty's armpits, crosses the ends, and runs the ends over the bearer's shoulders and back under the bearer's arms.

(e) The assistant runs the ends of the rope between the casualty's legs, around the casualty's thighs, and back around to the front of the bearer. The rope is tied with a square knot with two overhand knots just above the bearer's belt buckle.

(f) The rope must be tight. Padding, when available, should be placed where the rope passes over the bearer's shoulders and under the casualty's thighs.

Figure 11-1. Sling-rope carry.

(2) ***Rope Coil Carry.*** The rope coil carry requires a bearer and a 36 1/2-meter coiled rope. It can be used to transport a conscious or unconscious victim.

(a) Place the casualty on his back.

(b) Separate the loops on one end of the coil, forming two almost equal groups.

(c) Slide one group of loops over the casualty's left leg and the other group over the right leg. The wraps holding the coil should be in the casualty's crotch with the loops on the other end of the coil extending upward toward the armpits.

(d) The bearer lies on his back between the casualty's legs and slides his arms through the loops. He then moves forward until the coil is extended.

(e) Grasping the casualty's arm, the bearer rolls over (toward the casualty's uninjured side), pulling the casualty onto his back.

(f) Holding the casualty's wrists, the bearer carefully stands, using his legs to lift up and keeping his back as straight as possible.

(g) A sling rope around both the casualty and bearer, tied with a joining knot at chest level, aids in keeping an unconscious victim upright. This also prevents the coils from slipping off the carrier's chest.

Note: The length of the coils on the rope coil and the height of the bearer must be considered. If the coils are too long and the bearer is shorter, the rope must be uncoiled and recoiled with smaller coils. If this is not done, the casualty will hang too low on the bearer's back and make it a cumbersome evacuation. A sling-rope harness can be used around the victim's back and bearer's chest, which frees the bearer's hands.

b. **Buddy Rappel.** The carrier can also conduct a seat-hip rappel with a victim secured to his back. In this case, the rappeller faces the cliff and assumes a modified L-shape body position to compensate for the weight of the victim on his back. The victim is top-rope belayed from above, which provides the victim with a point of attachment to a secured rope. The methods for securing a victim to a rappeller's back are described below.

(1) To secure the victim to the carrier's back with a rope, the carrier ties a standard rappel seat (brake hand of choice, depending on the injury) and rests his hands on his knees while the victim straddles his back.

(2) A 4.2-meter sling rope is used. A 45-centimeter tail of the sling is placed on the victim's left hip. (This method describes the procedure for a seat-hip rappel with right-hand brake.)

(3) The remaining long end of the sling rope is routed under the victim's buttocks, and passed over the victim's and carrier's right hip. The rope is run diagonally, from right to left, across the carrier's chest, over his left shoulder, and back under the victim's left armpit.

(4) The rope is then run horizontally, from left to right, across the victim's back. The rope is passed under the victim's right armpit and over the carrier's right shoulder.

(5) The rope is run diagonally, from right to left, across the carrier's chest and back across the carrier's and victim's left hip.

(6) The two rope ends should now meet. The two ends are tied together with a square knot and overhand knots.

(7) The knot is positioned on the victim's left hip. The carrier's shoulders may need to be padded to prevent cutting by the rope.

(8) An alternate method is to use two pistol belts hooked together and draped over the carrier's shoulders. The victim straddles the carrier, and the belay man secures the loose ends of the pistol belts under the victim's buttocks. Slack in the pistol belt sling should be avoided, since the carrier is most comfortable when the victim rests high on his back (see FM 8-35).

(9) A large rucksack can be slit on the sides near the bottom so that the victim can step into it. The victim is belayed from the top with the carrier conducting a standard rappel. The carrier wears the rucksack with the victim inside.

(10) A casualty secured to a carrier, as described above, can be rappelled down a steep cliff using a seat-shoulder or seat-hip rappel. The casualty's and rappeller's shoulders should be padded where the sling rope and rappel lines cross if a seat-shoulder

rappel is used. The buddy team should be belayed from above with a bowline tied around the victim's chest under his armpits. The belay rope must run over the rappeller's guide hand shoulder.

11-7. LITTERS

Many types of litters are available for evacuating casualties in rough mountain terrain. Casualties may be secured to litters in many different ways, depending on the terrain, nature of injuries, and equipment available. **All casualties must be secured.** This should be done under medical supervision after stabilization. It is also important to render psychological support to any victim awaiting evacuation.

If the litter must be carried, belayed, and then carried again, a sling rope should be wound around the litter end and tied off in a l-meter-long loop. This enables the carriers to hook and unhook the litter from the belay. Slings are available to aid the soldiers with litter carrying. Utility rope or webbing 6 meters long may be used. The rope is folded in half, and the loose ends are tied together with an overhand knot. These slings are attached to the litter rails (two or three to a side, depending on the number of litter bearers) by a girth hitch, and then routed up along the handling arm, over the shoulder, behind the neck, and then down along the other arm. The knot can be adjusted to help the outside arm grip the webbing. These slings help distribute the load more evenly, which is important if a great distance must be traveled.

a. **Manufactured Litters.** The following litters are readily available to mountaineering units.

(1) The poleless, nonrigid litter (NSN 6530-00-783-7510) is best issued for company medics since it is lightweight, easy to carry, and readily available. Casualties should be secured with the chest strap and pelvic straps, which are sewn on one side. This litter may be used when rappelling, on traverse lines, and on hauling lines in the vertical or horizontal position. It can be improvised with poles.

(2) The poleless semi-rigid litter (NSN 6530-00-783-7600) may be used the same as the nonrigid litter. It offers more victim protection and back support because of the wooden slats sewn into it.

(3) The mountain basket-type rigid litter (NSN 6530-00-181-7767) is best suited for areas where several casualties are to be transported. All other litters may be placed inside this litter basket and transported across traverse lines. This litter is rectangular and has no vertical leg divider so that it will accommodate other litters. It is also known as a modified Stokes litter.

(4) The Stokes metal litter (NSN 6530-00-042-8131) is suited for situations as above; however, the casualty must be moved in and out of the litter since no other litter will fit inside it. Some Stokes litter frames have a central weld on the frame end, which is a potential breaking point. Winding the rope around the frame end will distribute the force over a wider area and stabilize the system. (See FM 8-10-6 or USAF TO 00-75-5 for additional information on the Stokes litter.)

(5) The standard collapsible litter (NSN 6530-00-783-7905) (rigid pole folding litter) is most readily available in all units and, although heavy and unsuited to forward deployment, may be rigged for movement over rough or mountainous terrain. The folding aluminum litter (NSN 6530-00-783-7205) is a compact version of the pole litter and is better suited for forward deployment.

(6) The UT 2000 is manufactured in Austria and is specifically designed for mountaineering operations. The litter consists of two parts that join together to form a rigid litter. Each part has shoulder and waist straps that can be used to man-pack the litter making it extremely light and portable. When joined together the shoulder and waist straps are used to secure the casualty to the litter. Strapping is also provided to make a secure hoist point for aircraft extraction and high-angle rescues. Wheel sets are another accessory to the UT 2000 litter (either two wheels or one); they attach to the litter for use during a low-angle rescue.

(7) The patient rescue and recovery system (NSN 6530-01-260-1222) provides excellent patient support and protection (Figure 11-2). However, it is not a spinal immobilization device. A backboard must be used with this system for patients who have injuries to the shoulder area. This system will accommodate long and short backboards, scoop stretchers, and most other immobilization equipment.

Figure 11-2. Rescue and recovery system (NSN 6530-01-260-1222).

b. **Field-Expedient Litters.** A litter can be improvised from many different things. Most flat-surface objects of suitable size can be used as litters. Such objects include boards, doors, window shutters, benches, ladders, cots, and poles. Some may need to be tied together to obtain the required size. If possible, these objects should be padded.

(1) Litters can also be made by securing poles inside blankets, ponchos, shelter halves, tarpaulin, jackets, shirts, sacks, bags, or mattress covers. Poles can be improvised from strong branches, tent supports, skis, and other similar items.

(2) If poles cannot be found, a large item, such as a blanket, can be rolled from both sides toward the center. Then the rolls can be used to obtain a firm grip to carry the victim. If a poncho is used, the hood must be up and under the victim, not dragging on the ground.

(3) A rope litter is prepared using one rope (Figure 11-3). It requires 20 to 30 minutes to prepare and should be used only when other materials are not available. Four to six bearers are required to carry the litter. The rope litter is the most commonly used field-expedient litter.

Note: Above the tree line, little material exists to construct litters.

Figure 11-3. Rope litter.

(a) Make 24 bights about 45 to 61 centimeters long, starting in the middle of the rope so that two people can work on the litter at one time.

(b) With the remainder of the rope, make a clove hitch over each bight. Each clove hitch should be about 15 centimeters from the closed end of the bight when the litter is complete.

(c) Pass the remainder of the rope through the bights outside of the clove hitches. Dress the clove hitches down toward the closed end of the bight to secure the litter and tie off the ends of the rope with clove hitches.

(d) Line the litter with padding such as clothing, sleeping bags, empty boxes.

(e) Make the rope litter more stable by making it about 6 inches wider. After placing the clove hitches over the bights, slide them in (away from the closed end) about 15 centimeters. Take two 3- to 4-meter poles, 8 centimeters in diameter at the butt ends, and slide each pole down through the bights on each side. Dress down the clove hitches against the poles. Take two 1-meter poles, and tie them off across the head and foot of the litter with the remaining tails of the climbing rope.

Note: The above measurements may have to be altered to suit the overall length of rope available.

11-8. RESCUE SYSTEMS

Rescue systems are indispensable when conducting rescue operations. A large number of soldiers will not always be available to help with a rescue. Using a mechanical advantage rescue system allows a minimal amount of rescuers to perform tasks that would take a larger number of people without it.

a. **Belay Assist.** This system is used to bring a climber over a section that he is unable to climb, but will continue climbing once he is past the difficult section.

(1) First, tie off the following climber at the belay with a mule knot.

(2) Tie a Prusik knot with short Prusik cord about 12 inches below the mule knot, and place a carabiner into the loop. Place the tail from the mule knot into the carabiner in the Prusik cord.

(3) Untie the mule knot without letting the following climber descend any more than necessary. Do not remove the belay.

(4) Maintain control of the brake side of the rope and pull all of the slack through the carabiner in the Prusik cord.

(5) Pull up on the rope. The rope will automatically feed through the belay.

(6) If the leader has to pull for a distance he can tie another mule knot at the belay to secure the second climber before sliding the Prusik down to get more pulling distance.

(7) After the climber can continue climbing, the leader secures the belay with a mule knot.

(8) Remove the Prusik cord and carabiner, then untie the mule knot and continue belaying normally.

Notes: With all rescue techniques make sure that you always think everything through, and double check all systems to ensure that you don't accidentally untie the fallen climber or find yourself without back-up safety. Do not compound the problem! When doing any rescue work the rescuers will always be tied in for safety.

b. **Belay Escape.** The belay escape is used when a climber has taken a serious fall and cannot continue. The belayer is anchored and is performing an indirect belay, and must assist the injured climber or go for assistance. To accomplish this he must escape the belay system. The belayer will remain secured to an anchor at all times.

(1) After a climber has been injured, tie off the belay device on your body using a mule knot. To improve this system, clip a nonlocking carabiner through the loop in the overhand knot and clip it over the rope.

(2) Attach a short Prusik cord to the load rope and secure it to the anchor with a releasable knot.

(3) Using a guard knot or Munter mule, attach the climbing rope from the belay device.

(4) Untie the mule knot in the belay device attached to the harness and slowly lower the climber, transferring the weight of the climber onto the Prusik.

(5) Remove the climbing rope from the belay device attached to the harness.

(6) Release the mule knot securing the Prusik, transferring the weight to the anchor.

(7) At this point the climber is secured by the rope to the anchor system and the belayer can now assist the injured climber.

11-9. LOW-ANGLE EVACUATION

Cliffs and ridges, which must be surmounted, are often encountered along the evacuation path. Raising operations place a greater load on all elements of the system than do lowering operations. Since all means of raising a victim (pulley systems, hand winches, and power winches) depend on mechanical advantage, it becomes easy to overstress and break anchors and hand ropes. Using mechanical raising systems tends to reduce the soldier's sensitivity to the size of the load. It becomes important to monitor the system and to understand the forces involved.

a. **Raising Systems, Belays, and Backup Safeties.** Raising systems, belays, and backup safeties are of special importance in any raising operation. The primary raising system used is the Z-pulley system, which theoretically gives three pounds of lift for each pound of force expended. In practice, these numbers decrease due to rope-pulley friction, rope-edge friction, and other variables. A separate belay rope is attached to the litter and belayed from a separate anchor. Backup Prusik safeties should be installed in case any part of the pulley system fails.

(1) ***Raising System***. When considering a raising system for evacuations, the Z-pulley system is the most adaptable. It can be rigged with the equipment on hand, and can be modified and augmented to handle heavier loads. Although the vertical or horizontal hauling lines can also be used, the Z-pulley system offers a mechanical advantage that requires less exertion by the transport team.

(2) ***Belays***. Whenever ropes are used for an evacuation operation, the overriding safety concern is damage to the ropes. This is the main reason for two-rope raising systems (raising rope and belay rope).

(3) ***Backup Safeties.*** Because the stresses generated by the Z-pulley system can cause anchors to fail, backup safety Prusiks are used to safeguard the system. These should be attached to alternate anchor points, if possible.

b. **Raising the Litter.** The litter is prepared as already described.

(1) The raising ropes and belay ropes are secured to top anchors and are thrown down to the litter crew.

(2) Padding is placed at the cliff edge and over any protrusions to protect the ropes from abrasion.

(3) The litter attendants secure the ropes to the litter.

(4) The raising crew sets up the Z-pulley system.

(5) One member of the crew secures himself to an anchor and moves to the edge of the cliff to transmit signals and directions. (This is the signalman or relay man.)

Note: If the load is too heavy at this time, another pulley is added to the system to increase the mechanical advantage.

(6) Attendants guide the litter around obstacles while the crew continues to raise the system.

(7) As the litter nears the cliff edge, the signalman assists the attendant in moving the litter over the edge and onto the loading platform, taking care not to jar the casualty.

c. **Descending Slopes.** When descending a moderately steep slope that can be down-climbed, the litter and victim are prepared as described earlier (Figure 11-4, page 11-12).

(1) One man serves as the belay man and another takes his position on the rope in front of the belay man, assisting him in lowering the litter. The litter bearers take their positions and move the litter down with the speed of descent controlled by the belay man.

Figure 11-4. Low-angle evacuation—descending.

(2) The extra man may assist with the litter or precede the team to select a trail, clearing away shrubs and vines. He reconnoiters so that the team need not retrace its steps if a cliff is encountered.

(3) The most direct, practical passage should be taken utilizing available natural anchors as belay positions.

11-10. HIGH-ANGLE EVACUATION

Evacuation down cliffs should be used only when absolutely necessary and only by experienced personnel. The cliffs with the smoothest faces are chosen for the route. Site selection should have the following features: suitable anchor points, good loading and unloading platforms, clearance for the casualty along the route, and anchor points for the A-frame, if used. There are many ways to lower a casualty down a steep slope. As long as safety principals are followed, many different techniques can be used. One of the easiest and safest techniques is as follows (Figure 11-5):

a. Use multiple anchors for the litter and litter tenders.

b. Secure the litter to the lowering rope with a minimum of four tie-in points (one at each corner of the litter). Lengths of sling rope or 7-millimeter cordage work best. Make the attached ropes adjustable with Prussik knots so that each corner of the litter can be raised or lowered to keep the litter stable during descent. Tie the top of the ropes with loops and attach to the lowering rope with a pear shaped locking carabiner.

c. Two litter tenders will descend with the litter to control the descent and to monitor the casualty. They can be attached to separate anchors and either self-belay themselves or be lowered by belayers.

d. Once the steep slope has been negotiated, continue the rescue with a low-angle evacuation.

Figure 11-5. Cliff evacuation descent.

APPENDIX A

LEVELS OF MILITARY MOUNTAINEERING

Military mountaineering training provides tactical mobility in mountainous terrain that would otherwise be inaccessible. Soldiers with specialized training who are skilled in using mountain climbing equipment and techniques can overcome the difficulties of obstructing terrain. Highly motivated soldiers who are in superior physical condition should be selected for move advanced military mountaineering training (Levels 2 and 3) conducted at appropriate facilities. Soldiers who have completed advanced mountaineering training should be used as trainers, guides, and lead climbers during collective training. They may also serve as supervisors of installation teams and evacuation teams. Properly used these soldiers can drastically improve mobility and have a positive impact disproportionate to their numbers. Units anticipating mountain operations should strive to have approximately ten percent of their force achieve advanced mountaineering skills.

A-1. LEVEL 1: BASIC MOUNTAINEER

The basic mountaineer should be a graduate of a basic mountaineering course and have the fundamental travel and climbing skills necessary to move safely and efficiently in mountainous terrain. These soldiers should be comfortable functioning in this environment and, under the supervision of qualified mountain leaders or assault climbers, can assist in the rigging and use of all basic rope installations.

a. On technically difficult terrain, the basic mountaineer should be capable of performing duties as the "follower" or "second" on a roped climbing team, and should be well trained in using all basic rope systems. These soldiers may provide limited assistance to soldiers unskilled in mountaineering techniques.

b. Particularly adept soldiers may be selected as members of special purpose teams led and supervised by advanced mountaineers. Figure A-1 lists the minimum knowledge and skills required of basic mountaineers.

c. In a unit training program, Level 1 qualified soldiers should be identified and prepared to serve as assistant instructors to train unqualified soldiers in basic mountaineering skills. All high-risk training, however, must be conducted under the supervision of qualified Level 2 or 3 personnel.

• Characteristics of the mountain environment (summer and winter). • Mountaineering safety. • Use, care, and packing of individual cold weather clothing and equipment. • Care and use of basic mountaineering equipment. • Mountain bivouac techniques. • Mountain communications. • Mountain travel and walking techniques. • Hazard recognition and route selection. • Mountain navigation. • Basic medical evacuation.	• Rope management and knots. • Natural anchors. • Familiarization with artificial anchors. • Belay and rappel techniques. • Use of fixed ropes (lines). • Rock climbing fundamentals. • Rope bridges and lowering systems. • Individual movement on snow and ice. • Mountain stream crossings (to include water survival techniques). • First aid for mountain illnesses and injuries.

Figure A-1. Level 1, basic mountaineering tasks.

A-2. LEVEL 2: ASSAULT CLIMBER

Assault climbers are responsible for the rigging, inspection, use, and operation of all basic rope systems. They are trained in additional rope management skills, knot tying, and belay and rappel techniques, as well as using specialized mountaineering equipment. Assault climbers are capable of rigging complex, multipoint anchors, and high-angle raising/lowering systems. Level 2 qualification is required to supervise all high-risk training associated with Level 1. At a minimum, assault climbers should possess the additional knowledge and skills shown in Figure A-2.

• Use specialized mountaineering equipment.
• Perform multipitch climbing:
  – Free climbing and aid climbing.
  – Leading on class 4 and 5 terrain.
• Conduct multipitch rappelling.
• Establish and operate hauling systems.
• Establish fixed ropes with intermediate anchors.
	• Move on moderate angle snow and ice.
	• Establish evacuation systems and perform high-angle rescue.
	• Perform avalanche hazard evaluation and rescue techniques.
	• Be familiar with movement on glaciers.

Figure A-2. Level 2, assault climber tasks.

A-3. LEVEL 3: MOUNTAIN LEADER

Mountain leaders possess all the skills of the assault climber and have extensive practical experience in a variety of mountain environments in both winter and summer conditions. Level 3 mountaineers should have well-developed hazard evaluation and safe route finding skills over all types of mountainous terrain. Mountain leaders are best qualified to advise commanders on all aspects of mountain operations, particularly the preparation and leadership required to move units over technically difficult, hazardous, or exposed terrain. The mountain leader is the highest level of qualification and is the principle trainer for conducting mountain operations. Instructor experience at a military mountaineering center or as a member of a special operations forces (SOF) mountain team is critical to acquiring Level 3 qualification. Figure A-3 outlines the additional knowledge and skills expected of mountain leaders. Depending on the specific AO, mountain leaders may need additional skills such as snowshoeing and all-terrain skiing.

• Recognizing and evaluating peculiar terrain, weather, and hazards. • Preparing route, movement, bivouac, and risk management plans for all conditions and elevation levels. • Using roped movement techniques on steep snow and ice. • Performing multipitch climbing on mixed terrain (rock, snow, and ice).	• Performing glacier travel and crevice rescue. • Establishing and operating technical high-angle, multipitch rescue and evacuation systems. • Usiing winter shelters and survival techniques. • Leading units over technically difficult, hazardous, or exposed terrain in both winter and summer conditions.

Figure A-3. Level 3, mountaineer leader tasks.

APPENDIX B

MEASUREMENT CONVERSION FACTORS

MULTIPLY	BY	TO OBTAIN
Millimeters	.03937	Inches
Centimeters	.3937	Inches
Centimeters	.03281	Feet
Meters	39.37	Inches
Meters	3.281	Feet
Meters	1.0936	Yards
Kilometers	.62137	Miles
Knots	1.1516	MPH

APPENDIX C

AVALANCHE SEARCH AND RESCUE TECHNIQUES

The effect of an avalanche can be disastrous. Chances of survival after burial by an avalanche are approximately 90 percent if the victim is located within the first 15 minutes. Probability of survival drops rapidly and, after two hours, chances of survival are remote. Suffocation accounts for 65 percent of avalanche fatalities, collision with obstacles such as rocks and trees accounts for 25 percent, and hypothermia and shock accounts for 10 percent.

The best chance of survival in snow country is to avoid an avalanche; but, if a member of your group ***is*** *in an avalanche, they are depending on you for rescue!*

C-1. IMMEDIATE ACTION

Survivors at the avalanche site organize into the first rescue team and immediately start rescue operations. If any indication of the location of the victim is found, random probing starts in that vicinity. The tip and edges of the slide are also likely areas to search. A human body is bulky and is apt to be thrown toward the surface or the sides.

C-2. GENERAL PROCEDURES

Establish from witnesses where the victim was located just before the avalanche to determine the point where the victim disappeared—the "last seen" point. Using this and any other information, establish a probable victim trajectory line leading to high priority search areas. Make a quick but systematic check of the slide area and the deposition area, and mark all clues. Look for skis, poles, ice axes, packs, gloves, hats, goggles, boots, or any other article the person may have been carrying—it might still be attached to the victim.

a. Organize initial searchers and probers. If using avalanche beacons, immediately select personnel to begin a beacon search. Ensure all other beacons are turned off or to receive to eliminate erroneous signals. All personnel should have a shovel or other tool for digging or, if enough personnel are available, a digger can be standing by to assist when needed. If the initial search reveals items from the victim, make an initial probe search in that area. This probing should take only a few seconds.

b. Make a coarse probe of all likely areas of burial, and repeat it as long as a live rescue remains possible. Resort to the fine probe only when the possibility of a live rescue is highly improbable. Unless otherwise indicated, start the coarse probe at the deposition area.

C-3. ESTABLISHING THE VICTIM'S MOST PROBABLE LOCATION

In many respects, a moving avalanche resembles a liquid. A human body, with a higher density than the flowing snow, would be expected to sink deeper and deeper into the avalanche; however, several factors influence the body's location. Turbulence, terrain, and the victim's own efforts to extricate himself all interact to determine the final burial position. Study of a large number of case histories leads to the following conclusions.

- The majority of buried victims are carried to the place of greatest deposition, usually the toe of the slide.
- If two points of the victim's trajectory can be established, a high probability exists that the victim will be near the downhill flow line passing through these two points.
- Any terrain features that catch and hold avalanche debris are also apt to catch a victim.
- If an avalanche follows a wandering gully, all debris deposit areas are likely burial spots. The likelihood of a victim being buried in a particular bend is proportional to the amount of debris deposited there.
- Vegetation, rocks, and other obstacles act as snares. The victim tends to be retained above the obstacle. An obstacle may simply delay the victim's motion, leading to final burial down flow from the obstacle.
- Maximum speed of the flowing snow occurs at the avalanche center. Friction reduces flow velocity along the edges. The closer the victim's trajectory is to the center of the slide, the greater will be his burial depth.
- Efforts of the victim to extricate himself by vigorous motion and "swimming" definitely minimize burial depth. Conversely, the limp body of an unconscious victim is likely to be buried deeply.
- An occasional exception to the above is emphasized. The victim may not be buried but may have been hurled away from the avalanche by wind blast. In the case of large and violent avalanches, a search of the surrounding terrain is advisable. Victims have been located in tree tops outside the slide area.

Use of avalanche transceivers is the most efficient method of searching for an avalanche victim, but only if the victim is wearing an active transceiver. Many models of transceivers are available, each with its own manufacturer's instructions for proper use and care. All currently available transceivers are compatible, although they may operate differently in the search mode.

C-4. PROBING FOR AVALANCHE VICTIMS

Probing offers the advantage of requiring simple equipment that can be operated by personnel with no previous training. Although the probers do not need previous training the search leader must be familiar with the technique to ensure proper execution of the probe line.

a. **Probe Poles.** Rigid steel tubing approximately 3/4-inch in diameter and approximately 10 feet long is recommended for the primary probe pole. Longer poles are difficult to manage, especially in a high wind. Although this type of pole performs best, it is difficult to transport to the avalanche site because of its length and weight.

(1) Each person operating in avalanche areas should carry folding sectional poles. These poles are similar to folding tent poles, but are stronger and are connected with cable instead of bungee cord. These poles should be carried on the outside of the pack for immediate access.

(2) If no probing poles are available, initial probing attempts can be started using ski poles in one of two ways: the ski pole can be reversed, probing with the wrist strap down; or the basket can be removed so that the point is down (the preferred method), which allows the ski pole to penetrate the snow more easily.

b. **Probing Lines.** For the probing operation to be effective, probing lines must be orderly and properly spaced. To ensure systematic and orderly probing, the number of personnel per line should be limited. Twenty per line is satisfactory, while thirty is normally the upper limit. The number of probers in the line will be dictated by not only the width of the area to be probed but the number of personnel available. A string may be used to keep the probe lines aligned, but will require added time to maintain.

(1) The probe line maintains a steady advance upslope. Advancing uphill automatically helps set the pace and permits easy probing to the full length of the probe. Probing does not come to a halt when a possible contact is made. The probe is left in contact and the line continues. A shovel crew follows up on the strike by digging down along the pole. Extra probes are carried by the shovel crew to replace those left in contact. Such a plan of operation is especially important when more than one victim is buried.

(2) Striking a body gives a distinct feel to the probe, which is easily recognizable in soft snow but less recognizable in hard compacted snow. A common problem is encountering debris within the snow that can be mistaken for the victim. The only sure check is by digging.

c. **Probing Techniques.** Two distinct probing methods are recognized: coarse probe and fine probe. As evidenced by their names, coarse probing implies a wider spacing of probe pole insertions with emphasis on speed. Fine probing involves close-spaced probing with emphasis on thoroughness. Coarse probing is used during initial phases of the search when live recovery is anticipated. Fine probing is the concluding measure, which almost guarantees finding the body. The coarse probe technique has a 70-percent chance of locating the victim on a given pass, while the fine probe has, essentially, a 100-percent chance of locating the body.

(1) The coarse probe functions as follows:

(a) Probers are spaced along a line 30 inches center to center, with feet about 15 inches apart.

(b) A single probe pole insertion is made at the center of the straddle span.

(c) On command of the probe line commander, the group advances 20 inches and repeats the single probe.

(d) Three commands are used for the complete sequence:

- "DOWN PROBE."
- "UP PROBE."
- "STEP FORWARD."

By using these commands, the leader can maintain closer control of the advancing probe line. It is important that the commands be adjusted to a rhythm that enforces the maximum reasonable pace. A string may also be used along the probe line to keep the probers dressed, although this requires the use of two soldiers to control the string. Strict discipline and firm, clear commands are essential for efficient probing. The probers themselves work silently.

(2) The fine probe functions as follows:

(a) Probers are spaced the same as for the coarse probe. Each man probes in front of his left foot, then in the center of his straddled position, and finally in front of his right foot.

(b) On command, the line advances 1 foot and repeats the probing sequence. Each probe is made 10 inches from the adjacent one.

(c) The commands for the fine probe are:

- "LEFT PROBE."
- "UP PROBE."
- "CENTER PROBE."
- "UP PROBE."
- "RIGHT PROBE."
- "UP PROBE."
- "STEP FORWARD."

(d) Good discipline and coordinated probing is even more important in fine probing than with the coarse probe. Careless or irregular probing can negate the advantages of fine probing. Use of a string to align the probers is especially important with the fine probe. The three insertions are made along the line established by the string, which is then moved ahead 1 foot.

GLOSSARY

AI	alpine ice
ALICE	all-purpose, lightweight, individual carrying equipment
AMS	acute mountain sickness
APS	anchor pulley system
ATC	air traffic controller
BDU	battle dress uniform
BTC	bridge team commander
C	Celsius
CE	community Europe
CEN	combined European norm
cm	centimeter
CMS	chronic mountain sickness
CPR	cardio-pulmonary resuscitation
CTA	common table of allowance
DA	Department of the Army
ECWCS	extreme cold weather clothing system
EN	European norm
F	Fahrenheit
FM	field manual
ft	foot; feet
GPS	Global Positioning System
GTA	graphic training aid
HACE	high-altitude cerebral edema
HAPE	high-altitude pulmonary edema
IAW	in accordance with
LBV	load-bearing vest
LCD	liquid crystal diode
LCE	load-carrying equipment
MEDEVAC	medical evacuation
MPH	miles per hour
MPS	moveable pulley system
MRE	meal ready-to-eat

NSN	national stock number
OP	observation post
PABA	para-amino benzoic acid
PCD	progress capture device
PFD	personal flotation device
RURP	realized ultimate reality piton
SE	southeast
SLCD	spring-loaded camming device
SOF	special operations forces
SOP	standard operating procedure
SOSES	shape, orientation, size, elevation, slope
SPF	sun protection factor
TOE	table of organization and equipment
UIAA	Union des International Alpine Association
U.S.	United States
USAF	United States Air Force
USN	United States Navy
UV	ultraviolet
UVA	ultraviolet A (radiation wavelengths between 320 and 400 nanometers)
UVB	ultraviolet B (radiation wavelengths between 295 and 320 nanometers)
WI	waterfall ice
YDS	Yosemite Decimal System

INDEX

acute mountain sickness, 2-6, 2-8, 2-21
anchors, 5-1 to 5-16
  artificial, 5-8 to 5-15
    bolts, 5-14, 5-16 (illus)
    chocks, 5-11
      placement, 5-12, 5-13 (illus)
    deadman, 5-8
    pitons, 5-9
      hero loop, 5-10 (illus)
      placement, 5-9, 5-10 (illus)
      removal, 5-11 (illus)
      reusing, 5-11
    spring-loaded camming device, 5-13, 5-14 (illus)
  natural, 5-1
    boulders, 5-2 (illus)
    bushes and shrubs, 5-4 (illus)
    chockstones, 5-2, 5-3 (illus)
    rock projections, 5-3 (illus)
    slinging techniques, 5-4 thru 5-6
    trees, 5-1 (illus)
    tunnels and arches, 5-4
  pre-equalized, 5-16 (illus)
  rope, 5-7 (illus)
    tensionless, 5-7, 5-8 (illus)
  self-equalizing, 5-15 (illus)
ascenders, 3-20 (illus)
avalanche, 1-27
  hazards, 1-29
    checklist, 1-32 (table)
    slope stability, 1-29
    snow pits, 1-30
    triggers, 1-29
  rescue equipment, 3-27 (illus)
    avalanche probe, 3-27, C-2
    avalanche transceivers, C-27
    search and rescue techniques, C-1

belay devices, 3-19 (illus)
bolts, 3-18, 3-19 (illus), 5-14, 6-54

carabiners, 3-11, 3-12 (illus)
chock picks, 3-18 (illus)
chocks, 3-15 (illus), 6-55
climbing, 6-1
  aid, 6-54
  combination techniques, 6-19
    change step, 6-19
    chimney sequence, 6-24, 6-25 (illus)
    down climbing, 6-25
    lieback, 6-21, 6-22 (illus)
    mantling, 6-19, 6-20 (illus)
    stemming, 6-22 (illus)
    undercling, 6-20, 6-21 (illus)
  commands, 6-43
    rope tug, 6-44
    verbal, 6-43, 6-44 (table)
  lead, 6-45
  roped, 6-27
    belay techniques, 6-32
      body belay, 6-34
        sitting, 6-36 (illus)
        standing, 6-36 (illus)
      mechanical, 6-37
    tying-in, 6-27, 6-29 (illus)
  safety, 6-8 thru 6-9
  sequence, 6-4 thru 6-8 (illus)
  shoes, 3-2 (illus)
  stance, 6-3
    correct, 6-3 (illus)
    incorrect, 6-4 (illus)
  top-roped, 6-45
  using the hands, 6-13
    jam hold, 6-17, 6-18 (illus)
    pinch hold, 6-16, 6-17 (illus)
    pull hold, 6-15, 6-16 (illus)
    push hold, 6-14, 6-15 (illus)
  using the feet, 6-10
    edging, 6-11, 6-12 (illus)
    jamming, 6-12, 6-13 (illus)
    smearing, 6-12 (illus)
    sole contact, 6-10, 6-11 (illus)
clothing, 3-2
  eyewear, 3-7 (illus)
  gaiters, 3-4, 3-5 (illus)
  hand wear, 3-5, 3-6 (illus)
  headwear, 3-6 (illus)
  helmets, 3-7 (illus)

insulating layers, 3-3, 3-4 (illus)
outer layers, 3-3
socks, 3-2
underwear, 3-2
clouds, 1-15 to 1-23
cover, 1-25
formation, 1-15
interpretation, 1-23
types, 1-15
altocumulus, 1-18 (illus)
altostratus, 1-18 (illus)
cirrostratus, 1-19, 1-20 (illus)
cirrus, 1-19 (illus)
contrails, 1-22 (illus)
cumulonimbus, 1-20, f1-21 (illus)
cumulus, 1-15, 1-16 (illus), 1-20
lenticular, 1-21, 1-22 (illus)
nimbostratus, 1-15, 1-17 (illus)
orographic, 1-21
stratocumulus, 1-15, 1-17 (illus)
stratus, 1-15, 1-16 (illus)
cord, 3-9
crampons, 3-22, 3-23 (illus), 10-2, 10-8
walking in, 10-6 (illus)
use of, 10-2
crevasse, 10-27
bivouac procedures, 10-33
rescue techniques, 10-29
snow bridge, 10-27

dehydration, 2-14
descenders, 3-20 (illus)

frostbite, 2-11, 2-18 (illus), 2-19 (illus)

gear, 3-30 thru 3-33
personal, 3-30
specialized, 3-31
standard, 3-30
ten essentials, 3-32, 3-33
glacier, 10-19 (illus), 10-22 (illus)
ablation zone, 10-23 (illus)

harness, 3-10
chest, 3-10, 6-28
full-body, 3-10, 6-28
improvised, 6-28, 6-29, 6-30 (illus)
bowline-on-a-coil, 6-28
swami belt, 6-28
seat, 3-10, 3-11 (illus), 6-27
headlamps, 3-30
heat cramps, 2-20
heat exhaustion, 2-21
heat stroke, 2-21
Hibler Pack, 2-17
high-altitude cerebral edema, 2-6, 2-22, 2-24, 2-25
high-altitude pulmonary edema, 2-6, 2-23, 2-25
hypothermia, 2-15

ice ax, 3-21, 3-22 (illus), 10-2, 10-7, 10-9
use of, 10-2
anchor position, 10-4 (illus)
cane position, 10-2, 10-3 (illus)
cross body position, 10-3 (illus)
dagger position, 10-5 (illus)
hammer position, 10-5, 10-6 (illus)
push-hold position, 10-4, 10-5 (illus)
ice hammer, 3-22 (illus)
ice screws, 3-24 (illus), 10-15 (illus)

Kernmantle, 3-8 (illus)
knots, 4-8 to 4-33
Bachman, 4-23, 4-24 (illus)
bowline, 4-13, 4-14 (illus)
bowline-on-a-bight, 4-19, 4-20 (illus)
bowline-on-a-coil, 4-24, 4-25 (illus)
classes of, 4-8
clove hitch, 4-16, 4-17 (illus)
directional figure-eight, 4-18, 4-19 (illus)
double fisherman's, 4-10, 4-11 (illus)
end-of-the-rope Prusik, 4-22, 4-23 (illus)
figure-eight bend, 4-11, 4-12 (illus)
figure-eight loop, 4-21 (illus)
figure-eight retrace, 4-15, 4-16 (illus)
figure-eight slip, 4-26, 4-27 (illus)
fisherman's, 4-9, 4-10 (illus)
frost, 4-29 (illus)

girth hitch, 4-30 (illus)
guarde, 4-32, 4-33 (illus)
Kleimheist, 4-28 (illus)
middle-of-the-rope Prusik, 4-22 (illus)
munter hitch, 4-30, 4-31 (illus)
rappel seat, 4-31, 4-32 (illus)
round turn and two half hitches, 4-14, 4-15 (illus)
square, 4-9 (illus)
transport, 4-27 (illus)
three-loop bowline, 4-25, 4-26 (illus)
two-loop figure-eight, 4-20, 4-21 (illus)
water, 4-12, 4-13 (illus)
wireman's, 4-17, 4-18 (illus)

litters, 11-7
basket type, 11-7
field-expedient, 11-8
patient rescue and recovery system, 11-8 (illus)
poleless, nonrigid, 11-7
poleless, semi-rigid, 11-7
standard collapsible, 11-7
Stokes metal, 11-7
UT 2000, 11-8

mountain, 1-1
building, 1-4
composition, 1-1
structure, 1-4
complex, 1-5
dome, 1-4
fault-block, 1-4
fold, 1-5
volcanic, 1-6
rescue and evacuation, 11-1
stream crossing, 9-1
individual, 9-3, 9-4 (illus)
site selection, 9-1 (illus)
team, 9-4, 9-5 (illus)
walking techniques, 8-1
mountaineering boots, 3-2 (illus)

nutrition, 2-2
carbohydrates, 2-3
fats, 2-3
minerals, 2-4
protein, 2-3
vitamins, 2-3

one-rope bridge, 7-14
commando crawl, 7-19 (illus)
hauling line, 7-21 (illus)
monkey crawl, 7-19, 7-20 (illus)
transport knot, 7-16 (illus)
transport tightening system, 7-15
Z-pulley tightening system, 7-18 (illus)

packs, 3-27
day, 3-31
safety, 3-31
personal hygiene, 2-5
pitons, 3-13, 3-14 (illus), 6-54
ice, 3-24, 3-25 (illus), 10-14
piton hammer, 3-14 (illus)
pulleys, 3-21 (illus)

rappelling, 7-5
commands, 7-7 (table)
types, 7-9
body, 7-10 (illus)
hasty, 7-9 (illus)
seat-hip, 7-10, 7-11 (illus)
rating system, 1-8 (table)
rocks, 1-1
classifications, 1-2
igneous, 1-2
metamorphic, 1-3
sedimentary, 1-3
types, 1-1
chalk and limestone, 1-2
granite, 1-1
sandstone, 1-2
slate and gneiss, 1-2
scree, 1-2
talus, 1-2
rope coil carry, 11-5
ropes, 3-8, 4-1
butterfly coil, 4-6 (illus), 4-7 (illus)
dynamic, 3-8

fixed, 7-1
installation, 7-1
retrieval, 7-2
utilization, 7-1, 7-2 (illus)
mountain coil, 4-4, 4-5 (illus)
mountaineering, 4-1 to 4-3 (illus)
static, 3-8
terminology, 4-3, 4-4 (illus)
route rating, 1-6 to 1-8

sanitation, 2-6
self-arrest, 10-11 (illus)
shock, 2-14
sleds, 3-29
sling-rope carry, 11-4, 11-5 (illus)
ski poles, 3-29 (illus)
skis, 3-28
snowblindness, 2-10
snow picket, 3-25, 3-26 (illus)
snow saw, 3-26
snowshoes, 3-29
snow shovel, 3-26
spring-loaded camming device, 3-17 (illus), 5-13, 6-55
stoves, 3-28
sunburn, 2-10
suspension traverse, 7-22 (illus)
construction, 7-23, 7-24 (illus)
equipment, 7-23

tents, 3-28
three-point camming device, 3-16 (illus)
trench foot, 2-17

Union des International Alpine Association, 1-7, 3-8

vertical hauling line, 7-27 (illus)

wands, 3-27
water, 2-1, 2-2
consumption, 2-1, 2-2
potable, 2-2
weather, 1-11
characteristics, 1-12
high pressure, 1-12
low pressure, 1-13
forecasting methods, 1-24, 1-25
analog, 1-25
climatology, 1-25
numerical weather prediction, 1-25
persistence, 1-24
trends, 1-25
fronts, 1-23
cold, 1-23
occluded, 1-23
stationary, 1-23
warm, 1-23
hazards, 1-28
humidity, 1-14
temperature, 1-23, 1-26
webbing, 3-9
tied runners, 3-9 (illus)
sewn runners, 3-9 (illus)
wind, 1-13
air masses, 1-14
anabatic, 1-14
chill, 2-11, 2-12 (illus)
direction, 1-25
katabatic, 1-14
northeast tradewinds, 1-14
prevailing westerlies, 1-13
speed, 1-25
wired snow anchors, 3-25, 3-26 (illus)

Yosemite Decimal System, 1-6

ROPE LOG *(USAGE AND HISTORY)*

For use of this form, seeFM 3-97.61; the proponent agency is TRADOC

			UNIT ID MARKING
NSN	DOCUMENT NUMBER	SERIAL NUMBER	MFR LOT NUMBER
DATE OF MFR	ISSUE DATE	DATE IN SERVICE	LENGTH
DIAMETER	FIBER	COLOR	CONSTRUCTION

INSPECT ROPE FOR DAMAGE OR EXCESSIVE WEAR EACH TIME IT IS DEPLOYED AND AGAIN AFTER EACH USE. IMMEDIATELY RETIRE ALL SUSPECT ROPES.

DATE USED	LOCATION	TYPE OF USE	ROPE EXPOSURE	INSPECTOR'S INITIAL/DATE	ROPE CONDITION AND COMMENTS

DA FORM 5752-R, MAY 89

DATE USED	LOCATION	TYPE OF USE	ROPE EXPOSURE	INSPECTOR'S INITIAL/DATE	ROPE CONDITION AND COMMENTS

MANUFACTURER ______________________

____

REVERSE OF DA FORM 5752-R, MAY 89

FIELD MANUAL
No. 7-85

HEADQUARTERS
DEPARTMENT OF THE ARMY
Washington, DC, 9 June 1987

FM 7-85
RANGER UNIT OPERATIONS

IMPORTANT

**U.S. Army Infantry School Statement on
U.S. NATIONAL POLICY CONCERNING ANTIPERSONNEL LAND MINES**

TABLE OF CONTENTS

Preface

CHAPTER 1. Introduction

1-1.	Mission	1-1
1-2.	Purpose and Function	1-2
1-3.	Fundamentals of Ranger Operations	1-2
1-4.	Capabilities and Limitations	1-4
1-5.	Employment Considerations	1-5
1-6.	Command and Control	1-6
1-7.	Communications	1-12

CHAPTER 2. Organization and Equipment

2-1.	Headquarters and Headquarters Company	2-1
2-2.	Regimental Headquarters	2-1

i DISTRIBUTION RESTRICTION: Approved for public release; distribution is unlimited.

2-3. Regimental Headquarters Company 2-2
2-4. Ranger Battalion 2-3
2-5. Special Teams and Elements 2-5

CHAPTER 3. Deployment

3-1. Readiness 3-1
3-2. Ranger Ready Force 3-1
3-3. N-Hour Sequence 3-2
3-4. Intelligence and Security 3-2
3-5. Operations 3-3
3-6. Logistics 3-4
3-7. Communications-Electronics 3-7
3-8. Remote Marshalling Base and Intermediate Staging Base 3-7

CHAPTER 4. Insertion, Extraction, Escape, and Evasion

4-1. Intelligence 4-1
4-2. Deception 4-1
4-3. Speed and Mobility 4-2
4-4. Stealth 4-3
4-5. Suppression 4-3
4-6. Security 4-3
4-7. Night Vision and Electronics Devices 4-3
4-8. Rehearsals 4-4
4-9. Sand Tables and Terrain Models 4-4
4-10. Communications 4-4
4-11. Planning Insertions 4-4
4-12. Air Insertion 4-5
4-13. Airborne Insertion 4-7
4-14. Air-Land Insertion (Fixed-Wing) 4-12
4-15. Air Assault Insertion 4-13
4-16. Amphibious Insertion 4-16
4-17. Land Infiltration 4-18
4-18. Stay-Behind Techniques 4-21
4-19. Extraction 4-21
4-20. Escape and Evasion 4-22

CHAPTER 5. Strike Operations

5-1. Application 5-1
5-2. Raid Operations 5-1
5-3. Interdiction Operations 5-7

5-4.	Personnel and Equipment Recovery Operations	5-10
CHAPTER 6.	**Special Light Infantry Operations**	
6-1.	Movement	6-1
6-2.	Ambush Operations	6-4
6-3.	Antiarmor Operations	6-14
6-4.	Military Operations on Urbanized Terrain	6-14
6-5.	Perimeter Defense	6-15
6-6.	Breakout from Encirclement	6-17
6-7.	Linkup Operations	6-18
6-8.	Passage of Lines	6-19
6-9.	Reconnaissance Platoon Operations	6-22
6-10.	Reconnaissance Planning	6-23
6-11.	Execution of Insertion	6-27
CHAPTER 7.	**Combat Support**	
7-1.	Intelligence Support	7-1
7-2.	Electronic Warfare Support	7-3
7-3.	Communications Support	7-4
7-4.	Fire Support	7-5
7-5.	Aviation Support	7-14
7-6.	Engineer Support	7-18
CHAPTER 8.	**Combat Service Support**	
8-1.	Ranger Support Element	8-1
8-2.	Logistics and Personnel	8-3
8-3.	Resupply	8-8
8-4.	Military Police	8-17
8-5.	Counterintelligence	8-18
8-6.	Personnel Service Support	8-18
APPENDIX A.	**Organization and Equipment of the Ranger Regiment**	A-1
APPENDIX B.	**Composition and Equipment of Ranger Regiment Liaison Teams**	B-1
APPENDIX C.	**Equipment Density**	C-1

C-1.	Communication Equipment	C-1
C-2.	Weapons and Equipment Each Battalion	C-2
APPENDIX D.	**Standing Orders, Rogers' Rangers**	D-1
APPENDIX E.	**The Ranger Creed**	E-1
APPENDIX F.	**Ranger History**	F-1
APPENDIX G.	**Training**	G-1
G-1.	Application and Guidance	G-1
G-2.	Philosophy	G-2
G-3.	Mission Analysis	G-4
G-4.	Unit Training	G-4
G-5.	Individual Training	G-4
G-6.	Collective Training	G-5
G-7.	Performance-Oriented Training	G-6
G-8.	Evaluation	G-6
G-9.	Field Training	G-7
G-10.	Professional Development	G-9
G-11.	Physical Training	G-9
G-12.	Environmental and Specialized Training	G-10
G-13.	Ranger Indoctrination Program	G-10
Glossary		Glossary-1
References		References-1
Authorization		Authorization Letter

PREFACE

This publication contains doctrinal guidelines for the employment and training of ranger units. The intended audience is intermediate and higher level commanders, staffs of those headquarters, and newly assigned officers and NCOs of the ranger regiment. It is to be used with all other published US doctrine concerning NBC operations and the climatic or geographic influences on combat. The fundamentals apply to the employment of ranger units on worldwide operations and across the spectrum of conflict from low to high intensity warfare. This includes all aspects of the integrated battlefield: nuclear, biological, and chemical. It is not intended to apply to the employment of long-range reconnaissance patrol (LRRP) companies, even though such units carry the designation "ranger." These LRRP companies, as well as the long-range surveillance company (LRSC) (corps level) and the long-range surveillance detachment (LRSD) (division level), are organized, trained, and equipped for long-range reconnaissance in enemy territory.

The scope of this publication addresses the mission, organization, equipment, capabilities, limitations, planning, training, operations, and logistical support of ranger units. It describes how the ranger regiment and its subordinate battalions train, receive taskings, plan missions, task-organize, deploy, and integrate supporting elements to perform the mission. Once in combat, the ranger regiment maneuvers to accomplish the mission in the manner of other light infantry units. Ranger units make full use of published light infantry doctrinal guidance. This manual will repeat procedures or techniques contained in other publications only if necessary for understanding or continuity.

This publication does not contain everything the ranger needs to know to successfully execute a unit mission. Rather, it supplements unit training and the ranger unit commander's guidance; it does not supersede it.

The proponent of this publication is HQ, TRADOC. Submit changes for improving this publication on DA Form 2028 (Recommended Changes to Publications and Blank Forms) and forward it to Commandant, US Army Infantry School, ATTN: ATSH-B-ID, Fort Benning, GA 31905.

Unless otherwise stated, whenever the masculine gender is used, both men and women are included.

CHAPTER 1

Introduction

The ranger regiment is a major component of the US Army's special operations forces. It is a unique light infantry unit tasked to conduct special military operations in support of national policies and objectives. These operations require highly trained, well-disciplined units capable of employment in any environment, either alone or in concert with other military forces. Within this publication, the term *ranger force* describes any size force consisting mainly of members of the ranger regiment and led by a member of the ranger regiment's chain of command. A ranger force may be a TOE unit or it may be a specially organized task force for a specific mission.

1-1. Mission.

a. The mission of the ranger regiment is to plan and conduct special military operations. These operations are conducted by specially trained, equipped, and organized forces against strategic or tactical targets in pursuit of national military, political, economic, or psychological objectives. They may support conventional military operations or they may be performed independently when conventional forces cannot be used. (See Figure 1-1.)

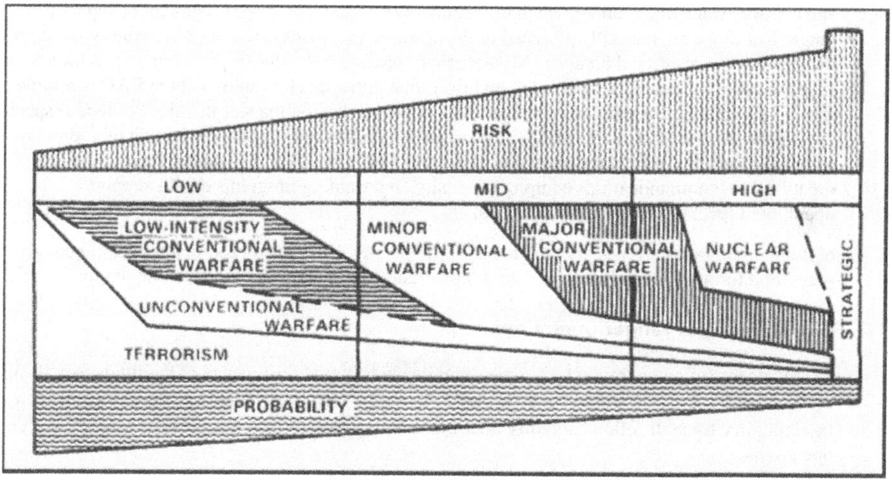

Figure 1-1. Spectrum of conflict.

b. Special military operations conducted by the ranger regiment include strike operations, usually deep penetration, and special light infantry operations. Strike operations include raids, interdiction, and recovery operations. Special light infantry operations include many of the light infantry missions assigned to airborne, air assault, or light infantry battalions and brigades. These operations are conducted in support of the AirLand Battle at all levels of intensity.

1-2. Purpose and function.

The ranger regiment provides the national command authority (NCA) the ability to move a credible military force quickly to any region in the world. The regiment uses the entire spectrum of intelligence support, from national systems to organic assets. Ranger units maintain a readiness posture that supports their immediate commitment to battle once deployed. They are often tailored for specific missions and may require augmentation from external sources. Tactical mobility may be augmented by USAF or Army special operations aviation (SOA) aircraft.

a. The ranger regiment conducts light infantry operations on the integrated battlefield as well as in low intensity conflicts. Special light infantry operations are conducted to accomplish certain contingency missions during peacetime or as part of the echelons above corps (EAC) battle.

b. The regiment conducts both strategic and operational missions. Its efforts are combined into an overall plan to destroy, delay, and disorganize the enemy, or to cause him to divert his attention and combat forces to rear area security. The limited number of ranger units and the diverse targets dictate a careful assignment of missions.

c. The mission, enemy, terrain, troops and time available (METT-T) factors govern the command and control arrangement under which the regiment operates. The ranger regiment or separate ranger battalions are normally assigned to the headquarters whose area of responsibility includes the regiment's operational area(s). The strategic or operational value of potential targets means that ranger units are normally employed at no lower than corps level. Employment at EAC is routine. The EAC headquarters employing ranger units could be the commander in chief's (CINC's) special operations command or field army. The controlling headquarters could also be that of a joint task force (JTF) or the Army component command (ARFOR) of a JTF. The ranger regiment's organization, communications equipment density, and training programs do not support operational level reconnaissance missions.

d. Ranger battalions are not oriented to a specific theater. Current force structure, contingency plans, and training needs (see Appendix G) preclude committing battalions to one region.

1-3. Fundamentals of ranger operations.

a. The success of an operation by a ranger unit depends on the observance of the fundamentals of the US Army's AirLand Battle doctrine (see FM 100-5). The ranger regiment fights the enemy as a light infantry force. It follows infantry doctrine and observes all the basic rules of infantry operations.

b. Along with initiative, depth, agility, and synchronization, ranger operations require--

(1) Detailed planning and coordination that allow the ranger unit to discern and exploit the enemy's weaknesses while avoiding his strength. Both permit the evaluation and use of information gathered from all sources. Detailed coordination integrates all supporting units and services. During planning, the ranger unit conducts pre-mission training, briefings, and rehearsals for all personnel. The use of special mission equipment is practiced and perfected.

(2) Decentralized execution IAW mission orders and the commander's intent. Special operations forces must use individual and unit initiative. Mission-type orders give the ranger force commander the flexibility to take advantage of opportunities on the battlefield.

(3) Surprise, achieved through the ranger unit's ability to move by uncommon means, along unexpected routes, and over rough terrain. Ranger units normally conduct operations during poor weather and reduced visibility, aided by night vision devices. This adds to the attainment of surprise.

(4) Survivability, achieved by using the classic infantry combat techniques of stealth and concealment. The ranger unit engages the enemy at the time and place of its own choosing. It takes full advantage of terrain, and destroys or suppresses enemy weapons. The ranger unit seeks to destroy or neutralize the enemy's command and control systems, his surveillance assets, and his mobility assets. Survivability is enhanced by rapid mission accomplishment and a prompt departure from the objective area.

(5) Mobility, speed, and violence of execution that allow the ranger unit it to close quickly on the objective area and complete the mission before the enemy can react. The speed at which events take place confuses and deceives the enemy as to the intent of the ranger unit. This forces the enemy to react rather than to take the initiative. Tactical mobility lets the ranger unit break contact and withdraw from the objective area.

(6) Shock effect, which is a psychological advantage achieved by the combining of speed and violence with the precision of the ranger attack. The ranger unit strives to apply its full combat power at the decisive time and place, and at the point of the greatest enemy weakness. The ranger unit strives to achieve maximum physical and psychological effect on the enemy by exhibiting aggressiveness and reasoned audacity.

(7) Multiple methods of insertion and attack, trying not to repeat operations thus decreasing the chance the enemy will detect a pattern. This is achieved through imaginative training and planning.

(8) Deception, practiced by the ranger unit during all phases of its operations, from deployment through the insertion phase, to the actions in the objective area and extraction. The enemy is kept unaware of the ranger unit's intentions, is fed conflicting and wrong information, and is kept from knowing the size or mission of the ranger force. The ranger force commander makes full use of ruses, feints, false insertions, electronic countermeasures, and dummy transmissions. Concurrent military operations by US or allied forces may be used as cover for ranger operations. Ranger units are careful to hide every possible aspect of an operation and to disguise those that cannot be hidden. While in the operational area, or when departing any location, ranger units ensure that no material is left that could provide the enemy with information concerning the ranger force.

(9) Audacity, achieved by a willingness to accept risk. The ranger force commander considers what the enemy expects the unit to do and the actual ranger abilities, He then chooses a course of action that may confuse the enemy, while remaining within the capability of the ranger unit. This fundamental is combined with deception and surprise to disrupt the enemy command and control.

c. Detailed planning and coordination finds the enemy. Surprise, mobility and speed, variety, deception, and audacity combine to shock and disorient the enemy, fixing him in place. The violence and precision of the ranger attack finishes the enemy while ensuring the ranger force survives.

1-4. Capabilities and limitations.

a. The ranger regiment has the following capabilities:

(1) Deploying quickly to conduct operations on all types of terrain and in all kinds of weather.

(2) Establishing a credible American presence in any part of the world to show US interest or resolve.

(3) Infiltrating and exfiltrating an area of operations, and assaulting an objective by land, sea, or air.

(4) Conducting strike operations to include raids, personnel and equipment recovery operations, and interdiction of key areas.

(5) Conducting special light infantry operations to include seizing and securing airfields, communications centers, command and control facilities, and key bridges; and other special light infantry operations.

(6) Performing short-duration reconnaissance of assigned ranger objectives for the ranger force commander.

(7) Operating for up to three days without resupply, and for longer periods when provided with accompanying or airdropped supplies.

(8) Providing liaison, communication, and coordination personnel and equipment to integrate the deployed ranger force into the logistical, intelligence, and operational system of the theater or joint task force (JTF) commander.

(9) Assuming operational control, for a limited time, of other US military forces such as engineers or infantry, airborne, or air assault battalions.

(10) Providing the focal point for all-source intelligence support to attached and assigned units of the regimental task force.

(11) Conducting limited combat operations under conditions of chemical, nuclear, or biological contamination.

b. The ranger regiment has the following limitations:

(1) Limited capability against armored or motorized units in open terrain.

(2) No organic transportation.

(3) Limited sustained combat capability due to the shortage of organic combat support and combat service support elements.

(4) Limited organic air defense weapons.

(5) Limited organic indirect fire support.

(6) No casualty evacuation capability.

(7) Reconstitution and retraining needed to replace combat losses.

1-5. Employment considerations.

a. Ranger units are characterized by the quality, motivation, training, and individual skill of their members. This produces units with superb collective abilities, able to adapt well to changing, complex situations.

b. Ranger units can conduct either deliberate or quick-response operations.

(1) Deliberate operations rely on careful planning, reconnaissance and surveillance of the target area, deception, secrecy, thorough preparation and rehearsals, and violent execution. A deliberate operation aims to complete the mission even though the enemy may have heavy forces on or near the objective area. Deliberate operations allow for detailed planning, evaluation, rehearsal, and coordination before insertion. A deliberate operation is likely to succeed against targets that the enemy has protected in depth, that have strong natural defenses, or that need a detailed and long insertion process.

(2) Quick-response operations rely on the high level of training and readiness of the ranger regiment to execute a mission before the enemy can react. These operations are conducted when there is little time for long, detailed planning. They rely on set procedures set forth by the ranger regiment and its supporting elements. A quick-response operation aims to complete the mission before an enemy can react. This type of operation may be chosen due to the time-sensitive nature of the target, political or military goals, the time frame of other operations, or the increased chance of enemy detection.

(3) Whether an operation is to be deliberate or quick response is often a difficult and time-sensitive decision by a high-level command authority. Decision-makers must consider the enemy's strength in the area, his intentions, his ability either to reinforce or to alter the target area, and the consequences of success or failure of diplomatic or military initiatives in related areas. The ranger unit commander tries to use the existing time, manpower, and resources to complete a detailed and coordinated plan. He refines that plan up to the insertion into the objective area.

c. Ranger units train to operate in any environment or weather condition. They regularly perform operations during periods of limited visibility. Ranger units maintain a high state of physical fitness and often train in close quarters combatives. In addition to completing advanced marksmanship training with standard US weapons, each member of a ranger unit trains with many foreign weapons. Ranger units are trained to operate on urbanized terrain, becoming specialists in entry and clearing techniques and quick-fire methods, especially during periods of limited visibility.

d. Because ranger units have limited vehicles, logistics operations capability, indirect fire support, and heavy weapons systems, they are not designed for continuous operations. During all phases of operations and training, ranger units need responsive external support.

e. Ranger units are normally employed against targets and under conditions that need their unique skills. Although targeting priority is set by the overall commander, ranger units are not normally assigned missions that can be done by conventional aerial bombardment or by other units.

f. Ranger units are oriented toward offensive operations. They are not normally employed as a rear area protection force. Although the ranger regiment has a small reconnaissance unit, ranger units do not normally conduct long-range reconnaissance missions. The structure, communications, and training of the ranger unit do not prepare it for LRRP missions. Ranger units engaged in strike and special light infantry operations have a secondary mission to collect and report combat information.

g. Ranger units can be deployed worldwide when US military presence or participation with a host national military activity would serve US interests. This deployment shows a readiness to commit forces into a threatened area or proves US national resolve. After the deployment, other activities include staging operations, rehearsing combat operations, securing base areas for use and deployment of other forces, and so on that provide a clear signal of US intent. Ranger units are not trained or organized to provide mobile training teams (MTT) to train indigenous forces. The US Special Forces or other special operation forces are trained to conduct such security assistance operations.

h. Ranger units can serve as an example to a host country and provide limited military advice and training. They will normally be augmented with linguists and technicians to increase their abilities. The regiment would normally still function as a unit. Most of its assistance would be through short-term, high-impact, unit-oriented operations. They would not be long-term individual efforts associated with advisory-type activities.

i. Ranger units may be deployed to engage in combined training exercises with allies. This enhances US national image by demonstrating the outstanding abilities of the American ranger. These activities may include ranger, light infantry, airborne, air assault, or amphibious operations.

j. Ranger units use standard US nuclear, biological, chemical (NBC) warning; detection; protection; and decontamination equipment and doctrine when operating on the integrated battlefield.

1-6. Command and control.

a. **External.**

(1) The ranger regiment is a key component of the US Army's special operations forces. The other elements of special operations forces are special forces (SF), psychological operations (PSYOP), civil affairs (CA), and special operations aviation.

(2) The 1st Special Operations Command (1st SOCOM) is responsible for command of all active component special operations forces elements, to include the ranger regiment, in peacetime. The 1st SOCOM has command and control of all assigned special operations forces, less operational command (OPCOM) of special operations forces units forward deployed. (See Figure 1-2.) The ranger regiment, while based in the continental United States (CONUS), is under the control of 1st SOCOM. During peacetime, the OPCOM of deployed ranger units is through channels chosen by the NCA through the Joint Chiefs of Staff (JCS) to the unified command.

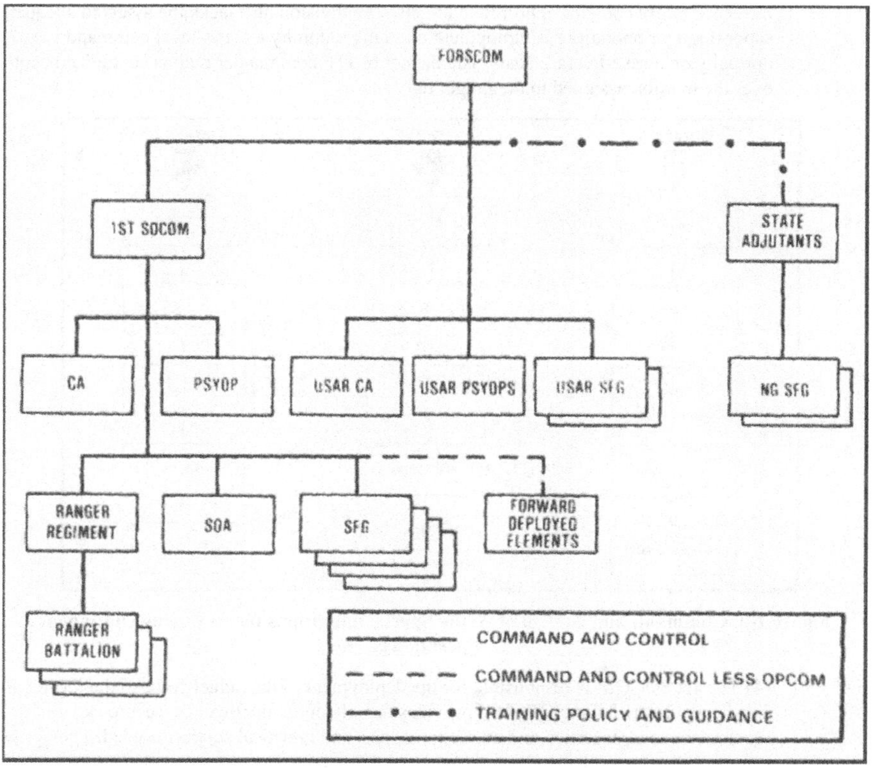

Figure 1-2. Command and control of Army Special operations forces during peacetime.

(3) In support of national policy, the NCA may direct contingency operations involving US

forces. These are politically sensitive military operations normally characterized by the short-term rapid projection or employment of forces in conditions short of conventional war. One aim of the NCA is to conclude them without going to war. The flexibility and effectiveness of the ranger regiment make it possible to use military force quickly to complete the mission and limit the spread of conflict. Command and control of special operations forces in a contingency operation is usually by a special operations task force (SOTF). (See Figure 1-3.) The SOTF and designated special operations forces are under the OPCOM of the highest level of command responsible for the operation. This command may be the unified command responsible for the contingency area (Option A), or a JTF designated by the NCA through the JCS (Option B). If no SOTF is formed, the ranger regiment could be OPCOM to the Army component commander (Option C). The ranger regiment or its battalions are not normally placed OPCOM to a division. A division's area of influence rarely contains appropriate targets. The division also lacks the assets to adequately support ranger operations. Employment of a ranger unit by a corps-level command is normally on a case-by-case basis. The theater or JTF commander retains centralized control over the mission assigned to the ranger unit.

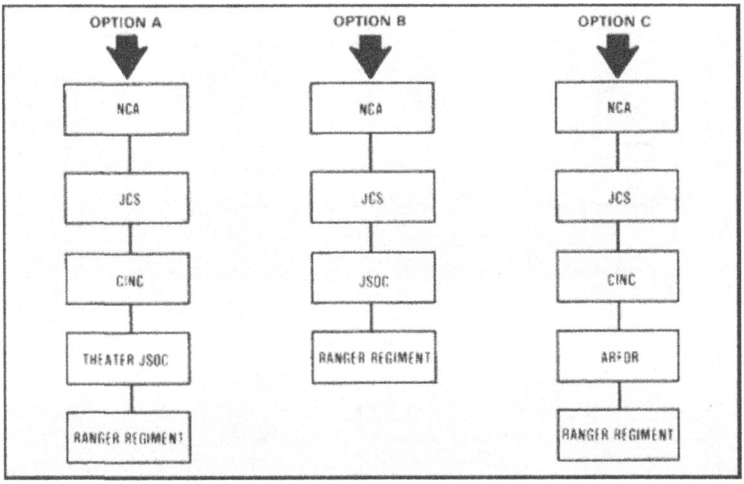

Figure 1-3. Command and control of Army Special operations forces during contingency operations.

(4) The 1st SOCOM is responsible for the deployment of the ranger force to the theater area. The headquarters having OPCOM provides the all-source intelligence, secure communications, insertion and extraction assets, and logistical support needed. The responsible commander in chief, through the Army component commander, provides support during the contingency operation.

(5) During wartime, ranger units are deployed as strategic assets to the theater of operations.

Strategic mission support is provided to the unified commander (CINC) from the NCA through the Joint Chiefs of Staff. It may involve the execution of sensitive operations. The employment of ranger units is controlled at the highest level of command directly responsible for the mission. Ranger units are assigned to a REMAB based on mission, force protection, and OPSEC considerations. The employing command and other elements provide timely and accurate intelligence, communications, logistical, and administrative support. Command and control of these units conforms to the provisions of JCS Pub 2. (See Figure 1-4.)

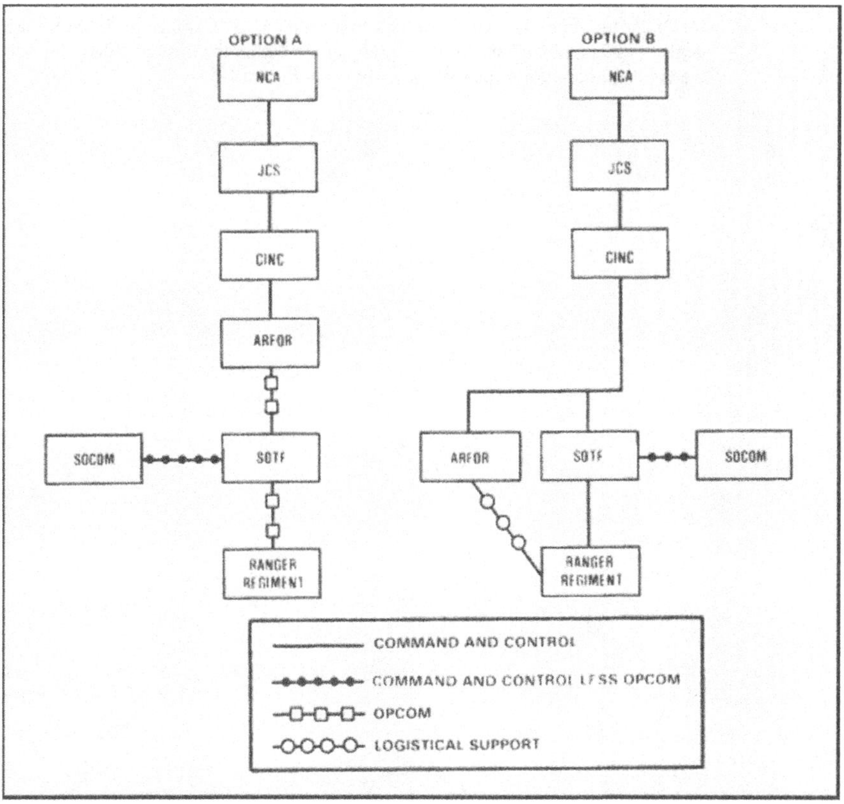

Figure 1-4. Command and control of Army Special operations forces during wartime.

(a) A theater joint special operations command (JSOC) is subordinate to the unified command to supervise the command and control, employment, and support of all US and allied special operations forces. The JSOC is a key element in the command and control of ranger units performing strike missions or special light infantry operations (Option A).

(b) The JSOC may be placed on a lateral line with other in-theater service components (Option B). This can occur when ranger units are -employed far from conventional forces and against strategic targets. It can also occur when the JCS or a CINC forms a JTF to conduct a limited-duration operation in an area where ranger units and conventional forces may operate near each other, or where operations must be coordinated on a regional level.

(c) Under the Army special operations command (ARSOC) concept, 1st SOCOM deploys as a command and control headquarters within the theater. When this occurs, the command relationships will be as shown in Figure 1-5.

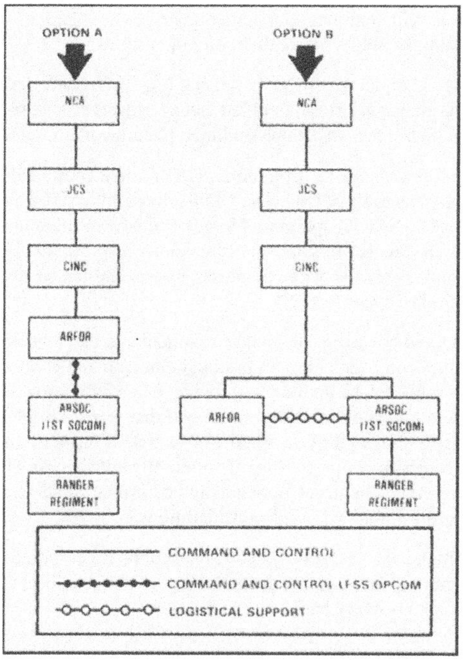

Figure 1-5. Command and control of Army Special operations forces under the ARSOC concept.

(d) Recently passed legislation will modify existing high level SOF command and control arrangements. All US Army, USN, and USAF special operations forces will be consolidated into joint commands at both national and theater level. The ranger regiment will continue to operate under the command of the 1st SOCOM, while increasing its ability to conduct joint operations.

(6) Ranger units conducting strike or special light infantry missions are not normally offered for transfer of authority (TOA) to allied forces. However, this will not preclude allied commands and US elements from combined planning and targeting for ranger missions.

(7) Requests for ranger units, from either US or allied commands, go through normal command channels to the CINC. The CINC, through the JSOC, is responsible for the employment of all special operations forces, including ranger units. This provides flexible response to JCS-directed missions, ensures quick response to the CINC's needs, enhances the joint employment of special operations forces, and lessens the chance of overcommitting and degrading the ranger force.

(8) Command and control of the ranger regiment and its battalions is IAW AirLand Battle doctrine. Before commitment to a mission, command and control is kept at a level where the unit's unique skills can be applied worldwide. Once committed, the ranger unit is placed under the operational control of the command that is responsible for the mission. This normally corresponds to the command whose area of influence includes the target area. Ranger task forces may operate directly under the authority of the senior American representative in the country. Command and control arrangements may vary according to the theater involved and METT-T considerations.

(9) The command and control of special operations is centralized and begins at the highest level of the military and civilian decision-making process. The specific command relationships are normally based on the mission.

b. **Internal.**

(1) The ranger regimental headquarters is organized to operate like that of a brigade headquarters. In addition to commanding and controlling all three ranger battalions, the regimental headquarters can assume operational control (OPCON) of conventional combat and combat support units, and other special operations forces for limited periods. It provides the regimental commander with a well-balanced responsive staff. It is structured as an operational headquarters that can deploy to the mission area and act as the ground tactical headquarters.

(a) The regimental headquarters prepares for combat by forming two small command and control groups. Each of these can deploy to an objective area and control combat operations. Other elements of the staff operate the tactical operations center (TOC) and the logistical operations center (LOC). These may be collocated depending on the situation. The TOC and LOC are normally located well behind the FEBA, at the REMAB or at the intermediate staging base (ISB). The ranger regiment also has two small liaison teams consisting of LNOS, staff representatives, and communications elements. (See Appendix B.)

(b) When a single ranger battalion is committed to an operation, the regimental commander normally exercises control from the TOC. A liaison and communications team is detached from the regimental headquarters to augment the ranger battalion. This team is attached to the deployed battalion until the end of the mission.

(c) If two or more ranger battalions are used in a single operation, the regimental commander normally deploys both command and control groups. He commands from the objective area, The use of two small, mobile control elements makes up for the possible loss of the regimental commander and the primary control element.

(d) Whenever the regimental commander is controlling an operation from the REMAB or the objective area, he normally provides a liaison team with a communications element to the higher headquarters exercising operational control. This cell can provide ranger representation to the controlling staff, operations interface, logistics interface, fire support planning and coordination, intelligence analysis and dissemination, and communications between the higher headquarters and the ranger battalion.

(2) The ranger regimental headquarters can serve as a ranger or special operations force task force headquarters. With augmentation, it can function as the Army special operations force command element. The regimental headquarters can also function as the Army component command of a JTF.

1-7. Communications.

a. Ranger missions need unique and dedicated communication support assets. Along with the ranger regiment's organic communications means, communications support assets are organized within the 1st SOCOM that-can deploy in support of the ranger regiment. This ensures that there are reliable communications to all elements of the SOTF. A special operations communications battalion (SOCB), deployed by the 1st SOCOM, provides more communications systems to support the special operations plan. The SOCB does not replace or duplicate existing in-theater communications. It augments the CINC's communications with the US Army's portion of the critical special operations force command, control, and communications link unique to the SOTF.

b. The ranger regiment's communication platoon can provide secure, long-range communication links from the CINC to a deployed ranger force. It can also enter the SOCB and the joint communications support element (JCSE) nets to provide a direct link with the NCA. Its equipment consists of four light vehicles equipped with various multiple communications systems. The two regimental liaison teams use one of these vehicles each. The TOC and LOC each use one.

c. Secure satellite communications (SATCOM) and radios with AM or FM capability are the main means of communication within the ranger regiment. These are also used for communications within the ranger battalions. The regimental commander has two light vehicles equipped with multiple communications systems. One of these is used to support each of the two mobile command and control elements of the regimental headquarters. Some of the radio systems can be dismounted and man-packed.

d. The ranger regimental headquarters has both man-packed and vehicular tactical satellite communication systems. Each of the ranger battalions has one base station and three man-packed systems.

(1) The man-packed radio is battery-operated. It provides two-way communications in both line-of-sight (mobile) and satellite (at-halt) modes. The digital message device group (DMDG) is the input and output device for data transmission. Secure voice communications is achieved using various encryption devices.

(2) The vehicular net control station uses the basic receiver/ transmitter unit installed on a vehicle. Each can serve up to 15 terminals in a communications net.

e. The reconnaissance platoon is equipped with multiple, man-packed communications systems. These let its teams operate in most of the communications nets within the regiment. Each five-man reconnaissance team has this communications ability, as does the platoon headquarters.

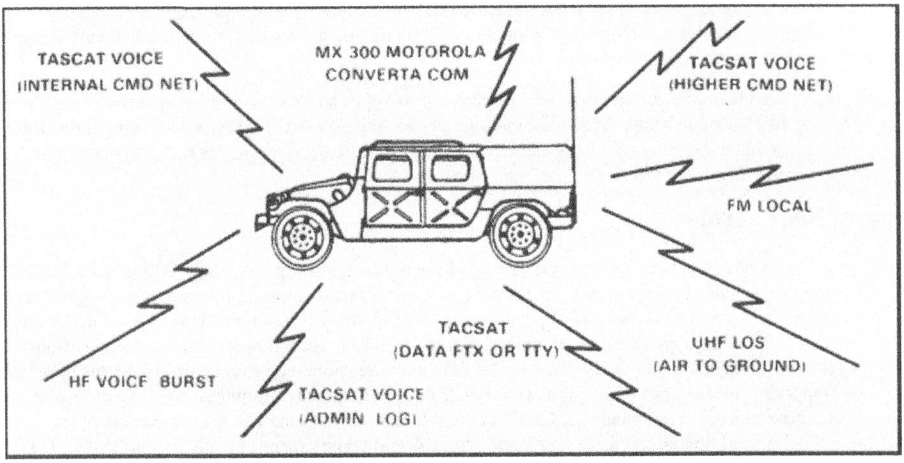

Figure 1-6. Communications capabilities within the regimental headquarters.

CHAPTER 2
Organization and Equipment

The ranger regiment is a specially organized and trained light infantry force consisting of about 2,000 personnel. It includes a regimental headquarters and headquarters company and three ranger battalions. (See Appendix A.) The regiment can operate as a single unit, separate battalions, or separate companies on specific missions. Independent missions are not normally assigned to ranger platoons.

2-1. Headquarters and headquarters company.

The HHC of the ranger regiment consists of 130 personnel, the majority being senior noncommissioned officers, commissioned officers, or soldiers with special skills to support the operations of the ranger battalions. (See Appendix A.) The HHC consists of two elements: the regimental headquarters and the headquarters company (minus).

2-2. Regimental headquarters.

a. The regimental headquarters consists of the regimental commander and his staff. The primary staff is composed of five sections: personnel (S1), intelligence (S2), operations and training (S3), supply (S4), and civil affairs (S5). These staff sections perform the standard functions and are supervised by the executive officer (XO). The regimental commander has a deputy commander and a special staff consisting of a communications officer, fire support officer, surgeon, and staff judge advocate. The S2 officer has a large intelligence analysis and counterintelligence section. The S3 section has extra liaison officers and assistant operations officers to coordinate the widespread and complex ranger missions. A United States Air Force (USAF) staff weather officer and a tactical air control party (TACP) are permanently attached.

b. The regimental headquarters:

- Analyzes assigned missions.
- Plans, coordinates, and controls training and operations.
- Coordinates the infiltration, resupply, exfiltration, and debriefing of the reconnaissance teams.
- Provides liaison support.
- Provides administrative, medical, NBC, and logistical planning and coordination support to the committed ranger battalions.
- Coordinates special support requirements (rigger, transportation, communications security [COMSEC]) for the ranger battalions.

2-3. Regimental headquarters company.

The regimental headquarters company (minus) consists of the company commander and his administrative personnel, plus the limited combat support (CS) and combat service support (CSS) elements assigned to the regiment. These are the fire support element, the communications platoon, the reconnaissance platoon, and the medical treatment team.

a. The company headquarters:

(1) Provides command and control of assigned elements.

(2) Provides administrative, personnel, logistical, and maintenance support to the assigned elements.

b. The fire support element:

(1) Plans fires to support the commander's ground tactical plan.

(2) Coordinates the integration of all available fire support means (mortars, artillery, armed helicopters, close air support, naval gunfire) to support mission execution.

c. The communications platoon (see Appendix A):

(1) Provides secure SATCOM teletype (TTY) and facsimile transmissions (FAX) abilities at the remote marshalling base or intermediate staging base (REMAB or ISB).

(2) Provides secure SATCOM, TTY, and FAX to the regimental headquarters and ranger battalions in the objective area.

(3) Provides cryptographic support to the regimental headquarters.

(4) Provides limited repair of radio and cryptographic equipment.

d. The reconnaissance platoon (see Appendix A):

(1) Trains, equips, and controls the employment of three reconnaissance teams.

(2) Infiltrates the objective area by parachute (high-altitude, low-opening [HALO]; high-altitude, high-opening [HAHO]; static line), helicopter, fixed-wing aircraft, self-contained underwater breathing apparatus (SCUBA), small boat, or other means.

(3) Remains undetected in the objective area for up to five days.

(4) Performs reconnaissance.

(5) Performs demolition target analysis.

(6) Uses small boats and inflatable rafts.

(7) Emplaces and monitors unattended ground sensors.

(8) Uses night image enhancement devices.

(9) Uses photographic equipment.

(10) Emplaces electronic target acquisition and designation devices.

(11) Collects specific combat information to satisfy priority intelligence requirements (PIR).

(12) Reports information using secure, long-range communication equipment.

Figure 2-1. Reconnaissance platoon communications.

(13) Performs drop zone (DZ) and landing zone (LZ) selection, marking, and reception duties.

(14) Escapes and evades enemy areas.

(15) Performs limited attacks or ambushes when tasked by the regimental commander.

(16) Reports weather in the objective area.

e. The medical treatment team:

(1) Provides medical support to the regimental headquarters.

(2) Coordinates required medical support and evacuation.

2-4. Ranger battalion.

a. The ranger battalion is organized and equipped to conduct both special and conventional combat operations. It contains organic combat, CS, and limited CSS elements. It consists of a battalion headquarters, a headquarters company, and three ranger rifle companies. (See Appendix A.)

b. The ranger battalion is the combat element of the ranger regiment. It can conduct independent combat operations or subdivide into its separate ranger rifle companies and their subordinate elements who can conduct semi-independent operations.

c. The ranger battalion headquarters and headquarters company (see Appendix A) includes:

(1) **Battalion headquarters.** The battalion headquarters consists of the commander; the executive officer; the command sergeant major; the S1, S2, S3, S4, and S5 officers; the chaplain; the communications officer; the fire support officer; and two radio operators.

(2) **Battalion headquarters section.** The battalion headquarters section consists of the S3 air and assistant S8, an assistant S2, an intelligence sergeant and a senior intelligence analyst, two operations sergeants and an operations assistant, a personnel staff noncommissioned officer (PSNCO), the chaplain's assistant, two clerk typists, two radio operators, a chemical officer and NCO, a personnel administration center (PAC) NCO, three PAC clerks, and a legal specialist. There is a USAF tactical air control party permanently attached.

(3) **Communications section.** The communications section consists of one telecommunications technician, one tactical communications chief, two combat signal radio team chiefs, one tactical communications systems supervisor, three radio operators, one switchboard operator, one single channel radio operator, and two equipment mechanics. The section provides communications support and associated maintenance.

(4) **Battalion medical section.** The battalion medical section consists of one general medical officer, one physician's assistant, two medical specialists, and twelve aidmen. The section provides unit-level medical support and trains selected members of the battalion in advanced medical procedures.

(5) **Battalion support platoon.** The battalion support platoon consists of a battalion support platoon leader, an ammunition NCO, and a food service section. The platoon provides limited CSS for assigned and attached units. This platoon does not have organic transportation. The battalion food service section consists of a food service sergeant (E7), a first cook (E6), and eight cooks.

(6) **Fire support team headquarters.** The fire support team (FIST) headquarters is supervised by the battalion fire support officer. It consists of the fire support sergeant and two fire support specialists. The headquarters controls three 9-man fire support teams. These teams are normally with the ranger rifle companies in a habitual support relationship. When operations are conducted near water, the fire support team headquarters may be augmented by spotter teams from the naval shore fire control party.

(7) **Company headquarters.** The HHC company headquarters consists of a commanding officer, an executive officer, a first sergeant, a supply sergeant, and an armorer.

d. The ranger rifle company (see Appendix A) includes a company headquarters, three rifle platoons, and a weapons platoon. Each company can conduct semi-independent operations for a short time when augmented by assets from the ranger battalion headquarters company.

(1) **Ranger rifle company headquarters.** The ranger rifle company headquarters consists of a commanding officer, an executive officer, a first sergeant, a supply sergeant, a tactical communications chief, two radiotelephone operators, an armorer, and an NBC NCO. It also has a two-man sniper team equipped with a sniper weapon. It normally has one of the 9-man FISTs attached from the ranger battalion headquarters company. Elements of this FIST may be further attached to the ranger rifle platoons.

(2) **Ranger rifle platoon.** The ranger rifle platoon consists of a headquarters section, three rifle squads, and a machine gun squad. The headquarters section personnel include a platoon leader, a platoon sergeant, and a radiotelephone operator. Each rifle squad consists of a squad leader and two balanced fire teams of four men each. The machine gun squad consists of a squad leader and three M60 machine gun crews of three men each. The machine gun squad is normally employed as a single element to provide quick, accurate, long-range supporting fire. One or more crews may be attached to other squads within the platoon.

(3) **Ranger weapons platoon.** The ranger weapons platoon consists of a headquarters section, a mortar section, and an antiarmor section. The headquarters section consists of a platoon leader, a platoon sergeant, and a radiotelephone operator. The mortar section has two 3-man 60-mm mortar squads. The mortar squads are normally employed as a section. They may be attached by section or squad to other platoons. The antitank section normally consists of a section sergeant and three 3-man 90-mm recoilless rifle or Dragon missile teams. They are normally employed as a section. The antiarmor section may also be attached by section or team to other platoons.

2-5. Special teams and elements.

The ranger regiment forms special teams to do critical tasks. These teams are made up of members of the unit and are used when needed. Examples of these special teams and elements are--

a. SCUBA teams whose members are graduates of the Special Forces Underwater Operations School. They conduct special insertion and reconnaissance duties.

b. HALO teams whose members are graduates of the Special Forces Freefall School and who conduct high-altitude, low-opening parachute operations.

c. HAHO teams whose members are also graduates of the Special Forces Freefall School and who conduct high-altitude, high-opening parachute operations, allowing offset parachute insertions.

d. Scout swimmer teams whose members are specially trained by the Landing Force Training Command (United States Marine Corps [USMC]) to conduct operations along coastlines, waterways, and inland waters. These teams conduct special insertion, reconnaissance, and security tasks often in support of amphibious operations.

e. Sniper teams, whose members are authorized by the table of organization and equipment (TOE), are used for both conventional and special operations. Along with the two-man sniper teams in the rifle company headquarters, each rifle platoon has two trained snipers.

f Demolition teams whose members are trained in advanced demolition and incendiary techniques and the use of special explosives,

g. Air defense teams trained in the use of man-portable air defense system (MANPADS) weapons. Such teams give the ranger battalion limited self-defense against air attack.

h. Liaison teams that are used by the ranger regimental commander to augment the existing command and control links from the employing unit headquarters to the ranger regiment. These teams are placed where they are most needed and may assist the higher headquarters staff in planning and target selection, as well as performing other liaison duties. See Appendix B for a detailed listing of the personnel and equipment included in the ranger liaison teams.

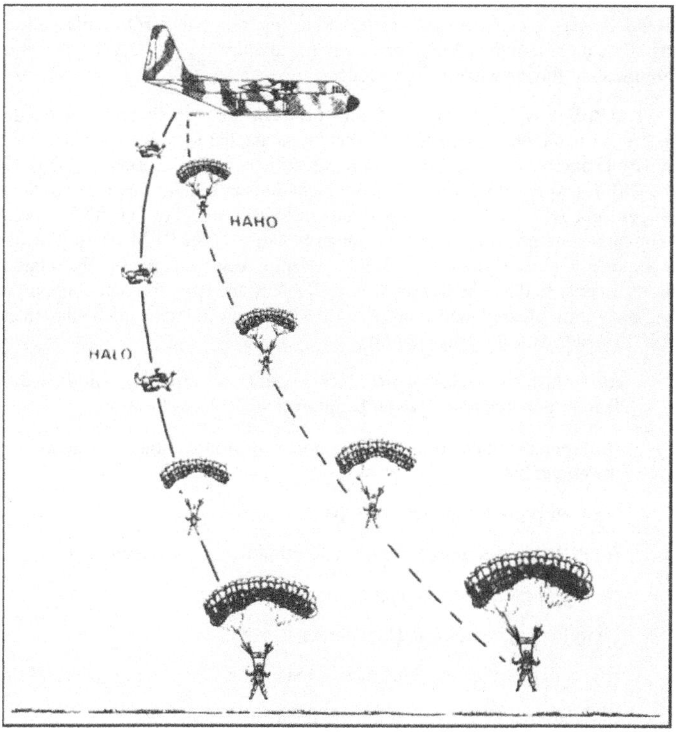

Figure 2-2. HALO and HALO parachute operations.

i. The ranger support element (RSE).

(1) The RSE is a special support element provided by host installation commanders at CONUS posts. It consists of elements and teams from either TOE units stationed at the installation or from the table of distribution and allowances (TDA) organizations under the command of the installation commander.

(2) The RSE provides the necessary support to ensure a timely and efficient deployment of the ranger battalion. During CONUS deployments, the RSE is responsible for both the departure and the reception support provided the battalion. During OCONUS deployments, the RSE is responsible for initial deployment and continuing support until the deploying ranger force is OPCON to the supported unified command or JTF. The responsible theater commander assumes support of the ranger battalion at the REMAB or the ISB if outside continental United States (OCONUS). On deployments that require the ranger battalion to move directly to the objective area from CONUS, the RSE continues support of the ranger battalion until released, and is prepared to accept considerable augmentation assets. The RSE provides the following support:

(a) Furnishing meals that are either served by a supporting unit in an established facility or taken to the ranger battalion at an isolated location.

(b) Issuing supplies from existing stocks, or from contingency items stored for use by the ranger battalion.

(c) Transporting supplies, food, personnel, and equipment.

(d) Palletizing equipment and ammunition for air movement and airdrop.

(e) Rigging vehicles and equipment for airdrop.

(f) Refueling vehicles and equipment.

(g) Issuing barrier and construction material for building of rehearsal sites.

(h) Helping in the construction of rehearsal sites and targets.

(i) Securing, storing, moving, and issuing ammunition and other Class V items.

(j) Maintaining ranger weapons and equipment, to include communications equipment. This is normally done by maintenance contact teams.

(k) Setting up and operating the RSE emergency operations center.

(l) Securing the REMAB if it is in CONUS.

CHAPTER 3

Deployment

Ranger units can quickly deploy by any means of transportation to any location in the world.

3-1. Readiness.

One ranger battalion is always in an advanced readiness condition as the Ranger Ready Force (RRF) available for immediate worldwide deployment. A second battalion is prepared to deploy later and a third battalion to follow. The regimental headquarters maintains command and control, liaison, communications, and reconnaissance elements immediately available for deployment. Higher status of readiness in response to specific world situations can be achieved.

3-2. Ranger ready force.

In order to maintain advanced readiness conditions, ranger battalions are rotated as the RRF.

a. Before taking over the duties of the RRF, battalions must do the following:

(1) Complete preparation for oversea movement (POM) qualification, to include an individual records check and required inoculations.

(2) Test fire all weapons, confirm battlesight settings, and repair or replace faulty weapons.

(3) Assemble and load on pallets the unit basic load (Classes I and V) and other supplies and equipment.

(4) Check unit-derived packing list components for accountability and serviceability.

(5) Prepare special operations troop listings.

b. During an RRF period, ranger units:

(1) Have a positive recall system for all personnel.

(2) Store enough air items for a battalion-size parachute assault.

(3) Conduct no off-post exercises, unless approved by the regimental commander.

c. Each unit keeps an alert roster and a telephone notification system for recalling personnel. The rosters ensure that the duty officer and the unit charge of quarters make a minimum number of calls. Each commander briefs personnel assigned to his unit on the recall system and tests it often.

d. Ranger units can maintain an increased readiness condition for long periods, but they should not be kept in that condition longer than needed. The effort to maintain this level of readiness will degrade the unit's training and effectiveness. If the needs of the operations dictate long periods at the highest readiness condition, provisions must be made for sustainment training. Also, added support will be required from the RSE.

e. The ranger regiment maintains a strong family support system when the unit is training or deployed on combat operations. Each member of the chain of command is responsible for family support.

3-3. N-hour sequence.

The N-hour sequence is a departure time sequence designed to help the ranger force complete all required deployment actions within a certain time. The length of the N-hour sequence depends on the readiness condition of the battalion before deployment. The N-hour sequence begins when the battalion is alerted and ends when the first elements depart. The N-hour sequence does not include travel time to the objective area.

3-4. Intelligence and security.

a. **Essential elements of friendly information.** Items of sensitive information and subject areas, if not secured, can be collectively analyzed by a hostile force. This analysis can reveal the intentions of a unit or compromise a mission. These items of information are designated as essential elements of friendly information (EEFI). They are never transmitted without encryption. The EEFI associated with a ranger unit alert normally are the mission, the use of staging bases, nature of the target, special terrain or weather conditions, and time limits. The ranger force commander must ensure that EEFI are not discussed outside of secure areas or made available to personnel without a security clearance and a need to know.

b. **Security plans.** Security plans for billets, marshalling areas, and out-loading areas must remain flexible to adapt to any situation. Certain items are always addressed in security plans. These are access control and physical security.

(1) **Access control.** Ranger units control access to their respective areas by allowing only authorized personnel to enter or leave. All entrances are secured and an armed guard with the access roster is posted. Access to telephones is restricted.

(2) **Physical security.**

(a) Equipment is open to theft and sabotage during the early phases of deployment. Units must provide guards to secure all unit and personal equipment.

(b) Weapons, ammunition, and sensitive items must be stored in a restricted area and guarded.

(c) Classified documents are open to compromise during deployment. Steps must be set up for maintenance, transportation, turn-in, and destruction of classified material. The most secure facility available must be used.

c. **Rear detachment security.** Security after deployment (maintained by the rear detachment) is as important as security before deployment. No information should be given to the news media from other than official sources. The rear detachment may continue to play a role in the unit cover and deception plan.

d. **Tactical operations center security.** Only personnel listed on TOC access rosters are allowed free entry into the TOC area. Personnel wanting to enter the TOC, who are not listed on the access rosters, are detained until their identity and purpose are proven. Such personnel are escorted at all times while within the TOC area.

e. **Marshalling and out-loading area security.** The ranger unit normally marshalls its personnel and equipment in the unit area, providing its own security force. The USAF security police provide this support at the departure airfield. On Army airfields, Army military police provide security. En route security is provided by the RSE.

f. **Personnel and unit equipment sanitization.**

(1) Before deployment, personnel are relieved of non-mission-essential material having intelligence value. This reduces the intelligence available to hostile forces if they capture a ranger force member.

(2) The ranger force commander sets the degree of sanitization IAW mission requirements.

(3) If strict sanitization is imposed, guidelines are issued as follows:

(a) The only personal items retained are identification card and tags.

(b) Unit insignia and markings are removed from uniforms and equipment.

(c) Personal items, such as wallets and photographs, are secured by the rear detachment commander.

(d) Personnel are briefed on how to compose and retain combat diaries. No personal diaries are allowed.

(4) The commander decides how much US currency is to be taken with the unit. '

g. **Area sanitization.** To avoid the compromise of classified or sensitive mission-related material, searches are conducted before and after leaving the host installation, REMAB, or ISB. These searches are held to locate and remove any material, indicators, litter, or other items that could compromise the ranger unit's mission. Unit commanders ensure that a thorough sanitization of their areas is conducted, to include dismantling rehearsal areas, destroying terrain models, erasing chalkboards, and emptying all classified and nonclassified waste containers. During their searches, commanders must pay close attention to emergency operations centers, tactical operations centers, and marshalling and loading areas. The ranger unit's intelligence officer coordinates support from local counterintelligence elements to search the area again after the ranger unit leaves.

3-5. Operations.

a. **Actions upon receipt of the OPORD.**

(1) Task-organize and equip units for special mission(s) IAW the warning order.

(2) Allocate/reallocate mission-essential equipment.

(3) Distribute augmentee personnel.

(4) Redistribute key personnel,

(5) Prepare a deployment time sequence.

(6) Prepare a movement plan.

(7) Prepare and maintain a current operation estimate.

b. **Advance elements.** A planning element is normally deployed to the REMAB/ISB or JTF headquarters in advance of the main body if the ranger force does not deploy directly to the objective area.

c. **Emergency operations center.** An EOC is setup by the ranger battalion headquarters. The EOC is responsible for relaying all reports to higher headquarters. It remains operational until ordered to cease.

d. **Air movement.** The ranger force always deploys by USAF aircraft. This is normally an administrative air movement to a REMAB or ISB. It may be followed by tactical air deployment from there. For some missions, tactical deployment begins at the unit's home station. Even if the movement is administrative, key personnel are cross-loaded. This is done under the direction of the operations section air officer based on aircraft allocations and load plans.

3-6. Logistics.

The following describes logistics readiness needs and steps for deployment.

a. **Supply.**

(1) **Class I.**

(a) Combat rations are normally palletized when the unit is RRF. These rations are to be used during the tactical operation. If the mission calls for deployment directly into the objective area, the palletized stock of rations is off-loaded in the unit area and issued as needed.

(b) Each ranger deploys with emergency combat rations.

(2) **Classes II and IV.**

(a) **Individual clothing and equipment.**

- Individual A bags are packed IAW a predetermined packing list, tagged, and marked. They are palletized, moved to the departure airfield by the RSE, and loaded IAW the air movement plan.
- Individual B bags are packed, assembled, and forwarded on order by the rear detachment.

- Individual environmental equipment is issued. Equipment shortages are replenished out of stocks maintained by the RSE.

(b) **Air items.**

- While the unit is on RRF, a contingency load of main and reserve parachutes is maintained by the RSE.
- Individual and unit air items are palletized by the RSE and shipped in bulk to the REMAB or ISB.
- For direct-deployment missions, air items and packing material are taken to the unit by the RSE.

(c) **General supplies.** The RSE maintains pallets of common contingency items and special equipment. Pallets are picked up by the out-loading support unit and moved to the unit area for issue or moved with the other pallets to the departure airfield.

(d) **Bulk supplies.** Barrier material is identified, relocated, palletized, and out-loaded IAW the air movement plan.

(3) **Class V.**

(a) The unit basic load of ammunition is secured, palletized, and loaded by the RSE IAW the air movement plan.

(b) For direct deployment missions, selected pallets are taken to the unit area and ammunition is issued. The convoy route is coordinated through the post EOC, and secured by the RSE.

(c) Designated items are rigged by the RSE for airdrop, if enough time exists before movement to the REMAB or ISB begins.

(4) **Class VI.** Personnel deploy with comfort items as needed.

(5) **Class VII.**

(a) All weapons and sensitive items not deployed are assembled and secured by the rear detachment personnel.

(b) Unit vehicles are prepared for airland or airdrop.

(c) Inoperable mission-essential equipment is repaired or replaced.

(d) Special equipment requests are forwarded to the logistics section as they are identified.

(6) **Class VIII.**

(a) Individuals deploy with one set of individual Class VIII items.

(b) Unit medics deploy with modified aidbags.

(c) Controlled medical items (less drugs) are palletized and deployed with section supplies. Controlled drugs are secured and issued to unit aidmen on order.

(d) Medical supply chests are maintained in the unit aid station.

(7) **Class IX.**

(a) Units deploy with initial prescribed load list (PLL) items.

(b) Critical PLL shortages are identified to the RSE for follow-on issue.

b. **Maintenance.**

(1) Upon alert, priority maintenance or replacement of key items of equipment is coordinated by the logistics officer. Only minimum mission-essential equipment is replaced.

(2) Items to be returned from supporting maintenance activities are identified, and pickup is coordinated by the property book officer.

c. **Out-loading coordination and support.**

(1) Upon alert, the RSE provides a liaison officer to the ranger logistics section.

(2) The support platoon leader (SPI,) attends the air movement conference and coordinates the following:

(a) Movement times (for both troops and equipment) to the departure airfield.

(b) Steps for weighing and joint inspection of the pallets and vehicles upon arrival at the departure airfield control group (DACG).

(c) Limitations on the pallet height and weight due to aircraft type, allowable cargo load (ACL), and availability.

(d) Changes in the planned number of pallets due to the mission requirements or airframe availability.

(e) Rigging and loading of the heavy drop and supply bundles.

(3) Empty pallets are placed in position for loading by the RSE in coordination with the SPL.

(a) Companies and staff sections prepare pallets.

(b) For airdrop operations, vehicles are prerigged at the unit location. On order, vehicles are driven to the heavy drop rigging site. The SPL and the officer in charge of the rigger detachment coordinate for a rigging site and the delivery of the required air items to that location. Rigging is done by the RSE, and supervised by a member of the ranger unit. Loading aboard the aircraft is IAW the air movement plan and supervised by RSE and USAF aerial port personnel.

(4) The DACG is provided by the host installation. Upon arrival at the DACG area, unit personnel and equipment are under the control of the DACG. As the deployment begins, a

member of the ranger unit is present at the DACG to ensure an orderly loading of personnel and equipment.

(5) The DACG plans for troop-holding areas in the event of delayed or aborted aircraft.

(6) Unit movement data are maintained by the air officer in the operations section.

3-7. Communications-electronics.

a. **Alert notification communications.** Alert notification of the ranger battalions by the regiment is done by secure communications means. Units in local training areas are informed by radio or through their assigned range field telephone. Units operating out of normal FM radio range during RRF periods set up a high frequency net, or other communications, to their higher headquarters.

b. **Telephone service.** Strict control is maintained over key telephones being used during unit alert. Care is taken to ensure that COMSEC and OPSEC violations do not occur. All telephones are assumed to be monitored; classified, sensitive, or official information is not discussed.

c. **Radio communications.**

(1) **Voice communications.**

(a) A secure radio net is set up early in the deployment. The ranger unit EOC, the installation EOC, and the RSE EOC operate in this net. A liaison officer from the ranger unit is placed at the installation EOC.

(b) The ranger support element liaison officer, attached to the logistics section, operates a secure voice station on a separate logistics net to the RSE EOC and the RSE commander.

(2) **Cryptographic material.** During RRF, COMSEC codes and enabling equipment are kept available to set up the secure voice net.

3-8. Remote marshalling base and intermediate staging base.

a. Ranger units can deploy from their CONUS base directly to the objective area. A more common method would be for the ranger unit to deploy first to either a REMAB or an ISB before insertion into the area of operations. (See Figure 3-1.)

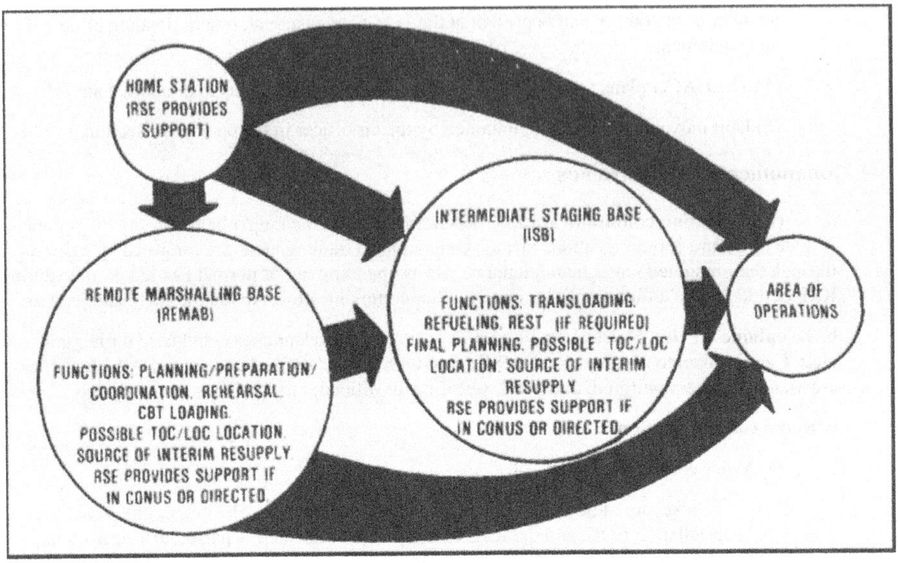

Figure 3-1. Deploying ranger units.

b. The REMAB is a secure base to which the entire ranger unit, including organic and attached support elements, deploys and where mission planning continues. The REMAB is normally within the geographical area encompassed by the command authority of the theater or JTF commander. This ensures that the CSS elements providing support to the ranger unit are operating within their normal area. It prevents or lessens out-of-sector support requirements for CSS elements.

c. The REMAB also provides a secure location for the ranger unit to conduct detailed planning and coordination with the controlling headquarters staff. In the REMAB, the ranger force commander conducts rehearsals, refines and modifies plans, determines PIR, and coordinates with the proper intelligence source to receive that information. It is in the REMAB that specialist augmentees to the ranger force are integrated into the unit, if they have not already joined. Specially trained supporting units, such as aviation and communications elements, also join the ranger force at the REMAB. The REMAB must provide--

(1) Access to the controlling headquarters staff.

(2) Physical security of billeting, planning, maintenance, and communications areas.

(3) Mess, billeting, latrine, and shower facilities for the ranger force and its supporting elements.

(4) Access to a C-141 or C-130 capable airfield, possibly with all-weather operations.

(5) Access to secure communications and processed mission intelligence.

(6) Access to rehearsal areas where sites can be built and live-fire rehearsals can be conducted.

(7) Access to the unit locations of major supporting elements, such as naval landing craft or Army aviation units.

(8) An external security force and an active counterintelligence agency.

(9) Vehicular transport for troop lift, equipment transfer, and administrative use.

(10) Access to maintenance support facilities.

(11) Medical support facilities to augment the ranger force medical personnel.

(12) Covered area to allow for packing of parachutes and rigging of airdrop loads.

d. The ISB is a more austere location. Elements of the ranger force deploy to an ISB to perform final planning, coordination, and task organization. The unit's organization and composition are finalized for movement to the objective area. The ISB is normally not occupied for long periods. Ranger units may deploy to the ISB from the REMAB or CONUS. Deployment to the ISB is more common when terrain or distance precludes insertion to the objective area directly from the REMAB or CONUS. Quick-reaction missions often involve the use of an ISB. Facilities needed at the ISB are limited to the following:

(1) Austere messing arrangements for the combat elements of the ranger force and its attachments.

(2) Austere billets or rest areas.

(3) Water point.

(4) Fuel for aircraft and vehicles.

(5) Areas for test firing of weapons.

(6) Covered and concealed areas for assembly of the ranger force and rigging of parachutes and door bundles.

(7) Austere airfield support facilities, possibly capable of all-weather operations.

(8) Remote location far away from civilians or traffic routes.

(9) Security and counterintelligence elements.

(10) Secure communications.

e. The REMAB and ISB should be located in an area similar in terrain and climate to the objective area. Time spent at the REMAB or ISB lets the ranger force begin acclimatization. Ranger units train yearly in all climates and emphasize physical conditioning. This allows acclimatization quicker than units not as well trained or conditioned. However, ranger units still need an acclimatization period if the terrain and climate of the objective area are different from that of their CONUS base. Not enough acclimatization means that the ranger force could be less efficient, especially when the objective area is hot, dry, or at a great altitude.

CHAPTER 4

Insertion, Extraction, Escape, and Evasion

Insertion is the entrance of a ranger force into a hostile area to execute a particular mission. It is the first critical phase of a ranger operation. These operations often involve deep penetrations of enemy territory by air, water, or land. This chapter emphasizes fundamentals and techniques employed by ranger units when their mission requires insertion and extraction. Extraction is conducted quickly after the mission to avoid casualties.

4-1. Intelligence.

a. Insertion plans are based on timely and accurate intelligence. The headquarters directing the operation must provide current intelligence information to the ranger force. The ranger regiment normally takes action to solicit required information from the controlling headquarters and from other agencies, if security permits. The type of intelligence needed varies depending on the specific mission, but normally includes the following:

(1) Current photographs from aerial or overhead reconnaissance.

(2) Detailed maps.

(3) Data on the population, terrain, and weather from area studies.

(4) Current enemy order of battle, enemy troop locations, and enemy capabilities.

(5) Details on the target area and infiltration routes.

b. To meet its need for current intelligence, the ranger force also relies on aerial reconnaissance and surveillance of the terrain over which it will move. Emphasis is on getting information on the enemy's ability to detect and engage the inserting forces. The location and abilities of air defense radar and weapons systems are critical. Terrain analysis is detailed and focuses on areas suitable for the insertion. Weather information must be detailed and current.

c. In getting information and intelligence data, OPSEC measures are stringent and must be enforced. Coordination with other agencies is kept within security constraints approved by the directing headquarters.

4-2. Deception.

There are always plans to deny the enemy knowledge of the ranger unit's insertion, or to deceive him as to the location or intent of the operation. Feints, false insertions, and other deception operations all add to ranger deception plans. Selection of unexpected means, time, place, and routes of insertion, as well as use of speed and mobility, give less reaction time to the enemy. Planning may include the use of diversionary fires to direct the enemy's attention away from the inserting unit. Deception techniques that may be used include--

a. Multiple airdrops, water landings, or both. This precludes the entire force from becoming engaged at the same time and place, if detected.

b. Dispersion of insertion craft (air or water) both in time and location.

c, Landing a force in an area closer to other potential targets than to the actual targets.

d. Leaks of false information and false messages.

e. False insertions or radio traffic to disguise actual insertions.

f. Diversionary actions such as airstrikes in other than the target areas.

g. Increased reconnaissance flights, either by manned aircraft or remotely piloted vehicles, over false targets.

4-3. Speed and Mobility.

a. Speed of action must be emphasized when the enemy can react quickly to the ranger force presence or when there are limited means of withdrawal. The operation must be conducted swiftly, allowing the ranger force to withdraw before the enemy can react.

b. Major problems facing ranger units are the need for speed and mobility, and the heavy loads carried by soldiers for sustained operations. Ranger units are required to be flexible and to be able to react to any problem within the objective area. However, if equipment is carried to cover every encounter, the load will become too heavy. This will reduce the speed and mobility of the ranger unit. The ranger force commander must always have a balance between the equipment carried and the mobility of his unit. He does this by making three different types of loads: fighting, mission, and existence. He decides what items will make up these loads, where they are carried, and what will be done with them upon contact with the enemy.

(1) The fighting load consists of those items of equipment, weapons, ammunition, food, and water that are common to all soldiers and are necessary for immediate use in combat. These items are normally carried and are retained at all times while in contact with the enemy.

(2) The mission load consists of those items of equipment or ammunition that are required by the mission, but are not normally carried by all rangers in a unit. The load may include spare batteries, demolitions, radios, antitank weapons, or mines. This load is normally spread out among members of the unit and is rotated often. It may be dropped upon enemy contact, but is normally carried to the objective and either used or emplaced there.

(3) The existence load consists of any items designed to provide protection from the elements: sleeping gear, changes of clothes, spare rations, or tentage. This load may or may not be brought into the objective area, depending on the tactical situation. If it is brought, it is normally dropped upon enemy contact or left in an assembly area and retrieved later.

4-4. Stealth.

Methods and steps chosen for insertion must emphasize stealth and try to avoid detection by the enemy. Multiple routes or methods of insertion may be used to preserve the ranger force if some soldiers or units are seen.

4-5. Suppression.

Effort must be made to suppress enemy detection devices, weapons systems, and command and control facilities. This may include electronic countermeasures or direct attack. This helps prevent the enemy from finding and engaging the ranger force en route to the objective area. Deception tactics add to suppression activities. If the ranger force is inserting by infiltration, suppression must not disclose friendly activities.

4-6. Security.

Emphasis on OPSEC prevents compromise of an impending operation. The operation must not be disclosed by overt rehearsals or training activities, the open use or procurement of special items of equipment, or the location of the marshalling area. Several measures can help in maintaining security:

a. Guarded and secure facilities for the headquarters and staff during planning.

b. Issue of operation orders only after units are moved to secure marshalling areas.

c. Alerts, rehearsals, and training to get the local people used to conditions that will exist during actual marshalling.

d. A secure facility set up to isolate units selected for an impending operation.

4-7. Night vision and electronic devices.

a. Rangers use night vision devices to detect and avoid the enemy's forces and his detection devices. These devices also assist in controlling and speeding movement and allow units to traverse almost impassable terrain.

b. The use of electronic homing devices permits precise navigation during reduced visibility. These devices may also be used with visual marking systems.

c. Passive night vision devices are used to assemble rapidly and to reorganize.

d. The use of active night vision devices in assembling, and the degree to which drop zones or assembly areas are "sterilized," depends on the following:

(1) Mission success may depend on not being seen during insertion. When a clandestine insertion is made, the time needed to get to the objective will normally be greater, the use of passive devices will prevail, and the sterilization of the area will be vital. Each ranger must be briefed on sterilization plans and techniques for erasing signs of the insertion.

(2) When detection is likely and the mission depends on getting to the objective before the

enemy can react, speed becomes crucial to the success of the insertion. Insertion may be near or on the objective. Rangers must get in and out quickly. They do minimum sterilization of the area and can use active RSTA devices.

4-8. Rehearsals.

Rehearsals should be close to the actual conditions of insertions and extractions. Security of rehearsals and rehearsal sites is important to the success of the operation. Rehearsals are held on terrain and structures like those in the objective area. If such structures are not available, they are built by the ranger force and the RSE with its supporting engineers.

4-9. Sand tables and terrain models.

Sand tables are useful to acquaint personnel with insertion sites and surrounding terrain. The use of sand tables and terrain models during the issuance of prejump orders and briefings enhances an orderly and speedy tactical assembly.

4-10. Communications.

Reporting during insertion and extraction missions is normally by exception, using an operations schedule.

4-11. Planning insertions.

a. **Centralized planning.** Planning for insertion of ranger units is centralized at the lowest level responsible for the overall operation. This is normally at the joint task force level, with detailed coordination required among all supporting forces. These forces may include other Army elements, USAF, United States Navy (USN), United States Marine Corps (USMC), other US agencies, and allies. Preparation must include detailed contingency planning for use of firepower, shock effect, speed to break contact (if detected), control measures, procedures, and planning. When planning takes a long time, intelligence must be continually updated through all available sources. This includes overhead photography and ground sources to ensure that the initial planning data are still current. Planning stresses the use of deception. Escape and evasion plans, external communication nets, and the location of caches must be known to leaders at all levels.

b. **Reverse planning sequence.** Insertion planning is done in the reverse planning sequence with the ground tactical plan being prepared first. Actions at the objective serve as the basis for the landing and assault plan, the plan for insertion and movement, the loading plan, and the plan for marshalling. All plans are interrelated. Some may be developed at the same time. Consideration of the factors of METT-T guides all planning. Special considerations that apply to these operations are:

(1) **Mission.** The nature of actions at the objective and requirements for special items of equipment and nonorganic personnel impacts on how the unit inserts. When the mission requires rapid deployment, the fastest method of insertion may be needed. In other operations, where success of the mission depends on maintaining secrecy, speed may be less important.

(2) **Enemy situation.** Enemy disposition, strength, and security measures along the route(s) near the objective area affect the method of insertion to be used. Planners must take into consideration whether the enemy:

(a) Uses RSTA detection devices or not.

(b) Has a strong air defense system or possesses air superiority.

(c) Has gaps between his defensive positions.

(3) **Terrain and weather.**

(a) **Topography.** Terrain such as mountains, swamps, or heavy forests favors land infiltration. Deserts, arctic regions, and open prairies favor air or water insertion. Areas not inhabited or with a friendly but scattered population favor land infiltration. A heavily populated area, or one which contains large enemy forces, favors air or water insertion. Hydrography--tides, currents, surf, reefs, and sandbars--must be considered when planning water insertion.

(b) **Weather.** Bad weather with reduced visibility and high winds favors land infiltration, and adversely affects air or water insertion. Limited visibility favors land infiltration; good visibility does not. Adverse weather aerial delivery systems (AWADS) lessens the impact of reduced visibility as a limiting factor on air infiltration. High winds may preclude airborne insertion.

(4) **Troops.** The number of men to be inserted and their level of training may be limiting factors. A need for special skills may call for use of nonranger unit personnel whose physical stamina and ability to perform many infiltration techniques may not equal that of the rangers. Whether supporting forces and special items of equipment are available must also be considered.

(5) **Time.** The main concern is whether the ranger operation is quick-response or deliberate-response. During a quick response mission, planning, preparation, and rehearsal time is short. The timing of the insertion is crucial with respect to weather, the enemy, and if an insertion craft is available.

4-12. Air insertion.

The most rapid form of insertion is by air. Rangers and equipment can be delivered by parachute (static-line or freefall technique), fixed-wing aircraft (airlanded), or helicopter (airlanded, parachute, or rappeling).

a. Suppression of enemy air defense along the insertion corridor may be necessary. The USAF will normally assume the major responsibility for suppressing enemy air defense artillery (ADA) capabilities that may interfere with the inserting forces. This is done by a variety of sophisticated electronic countermeasures applied against enemy equipment and by strikes against known enemy positions. Army fire support elements, special electronic mission aircraft (SEMA), and ground jammers may participate, as well as Navy aircraft, jammers, and gunfire.

(1) A main danger area is the forward edge of the battle area (FEBA) or frontier areas where

the enemy employs most of his sophisticated weapons systems and air defenses. Other dangerous areas are population or troop concentration areas behind enemy lines protected by point air defense systems.

(2) Since the majority of the enemy's detection devices and air defense weapons may be located near the point of entry, fire support, smoke screens (even at night), and suppressive measures may be critical. The use of special items of equipment may be needed to counter the enemy's RSTA effort.

(3) Long-range fires (artillery and naval gunfire) should be planned on known and suspected enemy antiaircraft locations and on prominent terrain features along the route(,s). Once beyond the FEBA, inserting units may be beyond the range of conventional indirect fire. More reliance will then be placed on air assets for fire support and electronic countermeasures.

b. There must be close coordination between the ranger unit and the transporting unit. What the transporting unit does during the insertion has a great impact on the success of the mission. To decrease the chance of detection, maximum use is made of periods of reduced visibility and tactical cover and deception. Drop zones and landing zones are chosen behind tree lines, in small clearings in forests, or on other types of terrain not easily seen.

c. All flight routes over enemy territory should be planned in order to avoid occupied areas and enemy air defenses. Routes should complement the cover and deception plan. They should avoid giving away final ranger objectives.

d. In-flight emergencies, mainly in deep-pedetration flights, should be considered, The ranger force receives a preflight briefing on the route to be flown and is informed of checkpoints along the route. Simple ground assembly plans for contingencies are set up before enplaning. In an emergency, the leader of the unit involved decides whether to continue the mission or abort. This is based on METT-T factors, contingency plans, and distance to the objective as compared to the distance to friendly territory. Contingency provisions for air and water rescue are made.

e. Deciding the DZ location is a joint effort. After receiving input from the air and ground commanders, the joint task force commander makes the final decision. During air assault operations, the ground commander is responsible for the overall process. After reviewing MEYF-T factors, it may be determined that surprise is more important than speed, or vice versa. This may determine the number, size, and locations of DZs and LZs.

f. Planning and preparation continue during the marshalling phase of any operation using air delivery for insertion. A review must be made of items of equipment that are crucial to the ranger force reaching its target area, accomplishing the mission, and returning safely. The need for special supplies and equipment, their transport, the distance to be traversed, the delivery means, and external support are all factors that enhance or reduce the chance for success. Once the means of insertion has been decided, logistics planning continues with emphasis on requesting and coordinating external support. The following must be considered:

(1) Detailed planning for mission logistical needs.

(2) Extra sets of key items of equipment to ensure redundancy and self-sufficiency of

inserting units.

(3) Plans for accompanying, planned, and emergency resupply. The ranger force should prepackage loads to meet specific contingencies. It should consider the use of supplies and equipment already in the operational area.

(4) In quick-response operations, only manportable items of equipment needed to accomplish the mission are carried. Plans and signals must be established for resupply of key items.

(5) Specific plans to evacuate casualties.

(6) All attached personnel and augmentation equipment must be properly prepared for movement.

(7) Provisions must be made for evacuating captured enemy equipment and personnel.

(8) All inserting units should be cross-loaded. Items to be considered in a cross-loading plan are the ground tactical plan, the assembly plan, the configuration of the DZ, the light route and formation of the aircraft, and the number of aircraft available. Cross-loading ensures that key leaders and equipment--for example, door bundles--are evenly distributed throughout the assaulting force. If one or more aircraft abort or are shot down, some key leaders and equipment will still arrive at the DZ. This permits mission accomplishment. Cross-loading aids rapid assembly on the DZ in support of the ground tactical plan. If a ranger is to assemble near the leading edge of a DZ, he is loaded on the aircraft to exit near the front of his stick.

(9) A bump plan is used if an aircraft should abort while still on the ground and there are no spare aircraft. The bump plan ensures that key leaders and equipment are not left behind when the main body takes off. If an aircraft aborts on the ground, key personnel move to another aircraft and replace nonessential personnel. This concept promotes success on the objective.

4-13. Airborne insertion.

a. During the planning phase of an airborne operation, there is no room for error in judgment. It can be planned and conducted well only if there is a constant exchange of information. All personnel involved must be informed of all changes in plans, times, and locations.

b. The reverse planning sequence is the key to the success of an airborne insertion. Most of the planning is based on the ground tactical plan, followed by the air movement plan to support it.

c. The USAF is responsible for flight planning, initial point (IP) selection, and crew procedures throughout the flight. The senior ranger in the aircraft must remain oriented and keep abreast of any last-minute changes. He must coordinate with the aircraft commander.

d. Most airborne insertions are made at either very high altitudes (HALO/HAHO) or very low. For combat operations, drop altitudes may be less than 500 feet above ground level (AGL), depending on the type of parachute used.

e. In airborne insertions, emphasis is on the use of special delivery or navigational techniques since normal insertion is during periods of limited visibility.

(1) **Adverse weather aerial delivery system.**

(a) A self-contained navigational system in some USAF C-130 aircraft is AWADS. The system allows the aircraft crew to fly a specific route to a DZ during darkness or bad weather. The system works with a Doppler radar that determines the ground speed of the aircraft, and an AWADS computer programmed with special flight route information (start point, route, checkpoints, wind speed and direction, and air speed). These two instruments show where the aircraft is in relation to the programmed checkpoints. This eliminates the need for airdrops under visual weather conditions.

(b) Only certain aircraft in formation are equipped with the actual AWADS device. Other aircraft are equipped with station-keeping equipment (SKE). This SKE consists of a computer that tells the aircrew where their aircraft is in relation to other aircraft in the formation.

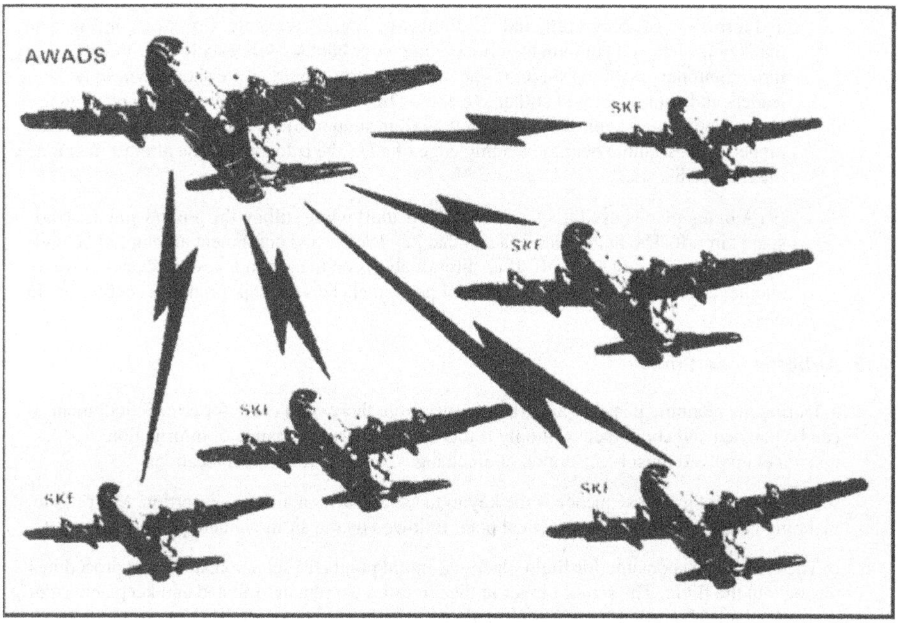

Figure 4-1. AWADS abd SKE aircraft formation.

(c) Airborne insertions using AWADS give the ranger commander the ability to airdrop personnel and equipment during low or zero visibility. This increases the safety of USAF aircraft. It protects them from visually tracked enemy antiaircraft guns and non-all-weather fighters. It also uses adverse weather as a combat force multiplier. Using AWADS, personnel and equipment may be inserted day or night without a pre-positioned USAF combat control team (CCT) or an Army assault team. Time length of the air train is shortened. The air corridor has to be cleared only once. This enhances the element of surprise. The CCTs are brought in with the assault elements. They are able to direct other low altitude parachute extraction systems (LAPES), container delivery systems (CDS), and air-land or heavy drop missions. Use of AWADS enables the ranger commander to conduct rapid, vertical reinforcement of units threatened by an enemy attack during bad weather.

(d) The use of AWADS requires detailed planning, rehearsal, and close coordination among the forces involved. Intense training of both USAF and Army personnel is required.

(e) Personnel must be briefed on terrain through use of photographs, maps, and terrain models. Limited visibility relies mainly on compasses and audible and visual assembly aids.

(f) During reduced visibility, essential heavy-drop serials may precede personnel aircraft. This reduces the time lag between personnel and heavy-drop chalks due to the need to move friendly troops off the DZ. Remaining heavy-drop serials or LAPES can be scheduled following P-hour.

(2) **Reconnaissance teams.**

(a) The operations of the main ranger force may be enhanced by use of reconnaissance teams. These teams are inserted ahead of the main body to provide early warning or to conduct reconnaissance of key objectives. If reconnaissance teams are required at a point along the flight path or near the main DZ, they may be cross-loaded on board any aircraft within the formation. They exit at planned times so as to land as close to their objective as possible. Should such teams be required at a point perpendicular to the axis of the main formation, one of the following is used:

- Another flank aircraft delivers teams to designated points before or with arrival of the main body.
- Selected aircraft deliver the main body, turn, and then deliver the reconnaissance teams.
- The teams are cross-loaded as complete elements on aircraft delivering other units whose flight path crosses the desired locations.

(b) Missions assigned to reconnaissance teams include:

- Providing local security for a USAF CCT, in which case, by joint doctrine, they form the Army assault team (AAT).
- Establishing surveillance or blocking positions on likely avenues of approach into the objective area or DZs.
- Locating and marking targets or assembly areas for the main assault force.

(c) Reconnaissance teams may be inserted by static line or HALO/HAHO operations.

(d) If the reconnaissance element is to locate a specific target or objective, it may be inserted up to 24 hours early. The longer the reconnaissance team remains near the objective area, the greater the danger of it being seen.

(e) In order for the information to be used by the ranger force commander, secure communications and a linkup plan must be prepared and rehearsed.

(3) **Assembly.**

(a) The ranger airborne force can conduct a quick assembly of combat power (weapons, personnel, and command and control elements) on the DZ. Quick assembly and reorganization is crucial because the unit is open to attack. Assembly must be fast and precise. Assembly areas and plans are set forth after considering all the factors of METT-T, mainly the location of the enemy, visibility, terrain, drop information, dispersion pattern, and cross-loading.

(b) The number of assembly areas depends on the location of the DZ, size of the assembly areas, and enemy's detection ability. Unit members move directly to the assembly areas without breaching security. Dissemination of information to all jumpers is the key to a successful assembly plan.

(c) For battalion-size or larger operations, general assembly areas are prescribed in the ranger force OPORD. Company commanders select and designate the exact location of their assembly areas. Although assembly area locations are terrain-dependent, they are generally on the DZ if the drop is at night and off the DZ if the drop is during daylight.

(d) Personnel chosen to carry assembly aids are cross-loaded in the center of their unit sticks. Strong, aggressive leaders should jump first and last in each stick. Upon landing, the leaders roll up their sticks toward the center, locate and move to the proper assembly aid. They may also carry or wear assembly aids. Individual rangers then follow these key personnel. During daylight operations, radiotelephone operators (RATELOS) can display color-coded pennants on their radio antennas to facilitate linkup with leaders. Key personnel may carry and jump with their own radios, and also display color-coded pennants on the radio antennas. Key personnel and RATELOs put their radios into operation before moving to assembly areas. For drops occurring near dawn or dusk, both daylight and night assembly plans are required.

(e) When tactical vehicles or door bundles are dropped, they are normally marked with panels, chemical lights, or strobe lights. The delivery parachutes are often color-coded.

(f) A ranger force should be able to assemble on an unfamiliar DZ, at night, within 45 minutes of the final aircraft pass. Because most ranger operations are time-sensitive, the unit normally moves from the assembly area to the objective when enough of the force has assembled to accomplish the mission. A straggler control point remains at the assembly area under the control of a designated leader. When everyone assembles, or at a specific time, the leader moves to link up with the unit. If available, a radio is left with the straggler control point.

(g) Assembly area security plans are prepared and disseminated. These plans must address hasty defense, sectors of responsibility, observation posts, fire support, and chain of command.

(4) **Contact on the DZ.** During the conduct of parachute insertion, rangers must be ready for enemy engagement at all times. Rangers are open to attack on the DZ before assembling. Just as contingency plans and chance immediate-action drills are developed, briefed, and rehearsed for other chance enemy contacts, rangers must be prepared to execute immediate-action drills for enemy contact on the DZ. The following actions are taken when there is enemy contact on the DZ:

(a) Rangers immediately assemble in fire team-size groups and fight off the DZ to designated assembly areas. Elements attempt to break contact and assemble.

(b) Assembly areas designated by commanders in the mission OPORD serve as primary rally points.

(c) Alternate assembly areas (normally 2 to 5 kilometers from the DZ) are designated in the mission operation order (OPORD). They are used by elements that cannot link up or assemble in the primary assembly area. These elements continue with the mission once assembled.

(d) During limited visibility, personnel should move directly and quickly off the DZ. They must avoid mistaking other friendly elements for enemy, which could cause firefights among friendly forces.

(5) **Reorganization.**

(a) The unit reorganizes according to plans, using designated assembly areas, aids, and identification markings for personnel and equipment. Replacements are named for key leaders or personnel who are lost or injured.

(b) Security teams assemble and move to their positions. Remaining units move quickly to their assembly areas, carrying supplies and equipment.

(c) Subordinate leaders keep the commander informed during assembly and reorganization.

(d) Runners may be used to report assembly status. This maintains radio silence. Designated personnel quickly recover supplies and equipment and move to the assembly areas.

(e) Reorganization is complete when all units have assembled or have been accounted for, and control has been established.

(6) **Pop-up technique.** This involves close terrain flying until the DZ is reached, then the aircraft quickly climbs to jump altitude. Special navigational equipment and aircrew training

are needed but may not always be available.

(7) **HALO/HAHO.** This involves high-altitude, low-opening and high-altitude, high-opening jumps with high performance parachutes designed to maneuver the jumpers to a specific point on the ground. During these operations, midair assembly procedures may be used.

(8) **Arctic.** Airborne insertion in extremely cold conditions is difficult. Times allocated for all parachute and equipment rigging must be increased. Rigging of individual equipment must be IAW FM 57-220. Ahkios can be rigged, packed inside a double A22 container, and dropped in platoon bundles. Because of the increased amount of individual equipment needed, allowable cargo loads for USAF aircraft are reduced.

(9) **Water.** Because of a lack of suitable land DZs, or to avoid detection, rangers may conduct airborne operations onto water DZs close to land. Procedures and techniques are similar to land DZ operations.

(10) **In-flight rigging.** On long-range operations involving extended flight time to the objective area, personnel may rig in flight. In-flight rigging requires detailed planning, coordination, of all jumpers. A rehearsal of these procedures should be conducted before deployment. Aircraft loads must be reduced to carry out in-flight rigging procedures. (See FM 57-220.)

4-14. Air-land insertion (fixed-wing).

a. **Planning factors.** Most of the planning factors that apply to airborne insertion operations also apply to air-land insertion operations. Most USAF aircraft cannot conduct air-land operations in total instrument weather conditions. The AWADS equipment is accurate enough for airdrop navigation, not air-land. Air-land operations are conducted, however, under conditions of extremely reduced visibility. Specially trained USAF crews, using sophisticated navigational techniques, can locate and land on small airfields in bad weather. As in airborne operations, ranger reconnaissance teams, as well as specially trained CCT elements, can jump before the air-land aircraft arrive. They emplace navigational and location markers to assist the pilots. Rangers may rely solely on air-land operations for insertion. This method does not provide for delivery of the force in the shortest time possible. Air-land insertions are normally in concert with an insertion of airborne elements.

b. **Time.** Since turnaround time is crucial, loads in aircraft should be unchained as the aircraft taxies. This also applies to personnel and seat belts. Aircraft should stop as close as possible to the assembly areas where personnel and equipment are off-loaded.

c. **Method.** Air-land insertions normally begin after a ranger Army assault team (AAT) and a USAF CCT have parachuted into the initial objective area and conducted reconnaissance, clearing, marking, and security operations. The ranger ground force commander is the arrival airfield commander. He has control over all arriving, taxiing, and departing aircraft (fixed-wing or rotary) through the CCT, which locates near him. The ranger force command and control element is usually on the first aircraft in and the last aircraft out.

d. **Security.** Air-land operations concentrate large numbers of men and equipment, which presents an open target to the enemy. The ranger ground force commander must ensure rapid off-loading, assembly, dispersion, and movement toward the objective.

e. **Communications.** The ranger ground force commander must have secure radio communications with the ground force, CCT, fire support elements, and higher headquarters. Wire should be laid to connect ground command and control locations.

f. **Operation duration.** Depending on the factors of METT-T, the aircraft may remain on the ground for extraction while the mission is being accomplished or take off at once. They may then orbit, pending recall, or return at a prearranged time.

g. **Extraction.** Extraction of the air-landed ranger force is normally planned and accomplished like an insertion, only in reverse order. A simple, yet effective, way must be found to see that no ranger personnel are left in the objective area.

4-15. Air assault insertion.

Ranger air assault (air-land, rappel, or combination) insertions normally take place during limited visibility. They quickly place the ranger force on, or close to, the objective. The ranger force (depending on the mission) normally plans on fighting when it hits the ground, completing the mission, and then being extracted. Air assault insertions need detailed planning, surprise, flexibility, speed, shock effect, concentration of combat power, and precise timing.

a. **Ground tactical plan.** The reverse planning process is important. The ground tactical plan, developed from the mission assessment, is the first planning area. All other planning begins from this point. The ranger commander and his staff decide stated and implied tasks, and request the proper helicopters for the mission.

b. **Fire support plan.** The fire support available to the ranger commander may be artillery, naval gunfire, mortars, attack helicopters, and USAF or USN tactical aircraft. Because of the timing of ranger air assault operations, preparatory fires are often not used. If they are used, enemy ADA sites normally have priority.

c. **Landing plan.** This plan puts the ranger force into the objective area at the time, place, and proper sequence to support the ground tactical plan. The ranger commander selects primary and alternate LZs and the priority for landing. The exact location of an LZ is determined after a detailed analysis of METT-T. Rehearsals are important. The ranger force must quickly assemble, reorganize, and accomplish the mission. The senior ranger on each aircraft ensures that the personnel on his aircraft offload at the correct LZ.

d. **Air movement plan.** The air movement plan provides for the coordinated movement of the ranger force into the zone of action in support of the 'landing plan. Flight routes, air movement tables, flight formation, altitude, and air speed should be considered. The ranger commander must have secure in-flight communications with his subordinate commanders.

e. **Loading plan.** This plan supports the air movement plan and provides for the pickup zone (PZ) setup and control. It sets priorities for the commander for loading personnel, equipment, supplies, and cross-loading. Air-loading table, aircraft loading formations, bump plan, and rehearsals should be considered.

f. **Staging plan.**

(1) This plan includes PZ selection and operation. The selection of PZs and LZs needs planning and coordination for best use of air assault assets. Site selection in or near the objective must be coordinated in person between the supported ranger commander and the aviation commander. The tactical situation is the key factor in PZ and LZ selection. Other factors are size of landing zones, surface conditions, ground slopes, approach and departure directions, prevailing winds, obstacles, communications, cross-loading, aircraft command and control, PZ and LZ markings, and rehearsals.

(2) The PZs are set up and secured about 15 minutes before touchdown of first aircraft. The fire support plan must support the other plans. All supporting fires must be coordinated with the air mission commander (AMC). Because of the depth of most objective locations, USAF tactical aircraft will be the main fire support method. Orbit locations for fire support aircraft must be planned.

g. **Other planning considerations.** These include escape and evasion, actions at the last LZ, assembly plan, downed aircraft procedures, control measures, forward area rearm and refuel point (FARP), weather delays, and deception plans.

4-16. Amphibious insertion.

Insertion by water may be by SCUBA, surface swimming, small boats, submarines, surface craft, helocasting, or a combination of these methods.

a. **Planning.** Planning for waterborne operations must be thorough. The plan is decided upon between the ranger commander, the transporting unit, and the joint headquarters directing the operation. Plans must be made for all types of enemy action and weather while on board the transporting craft. The transporting unit is given information on a need-to-know basis. Information that could endanger the operation may be withheld until the mission is under way. Initial planning includes:

(1) Time schedule of all events, from the beginning until the end of the joint operation, to be used as a planning guide. Timing for each event must be exact; the success of the operation depends on it. (See Figure 4-2)

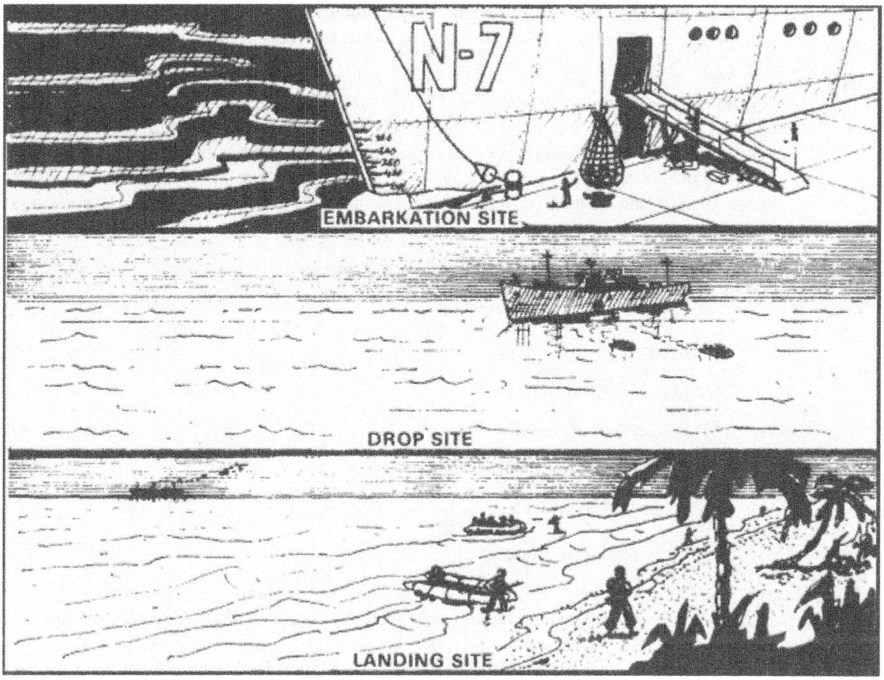

Figure 4-2. Waterborne insertion.

(2) Embarkation site, where troops and equipment are loaded on the transporting craft.

(3) Drop site, where the ranger force will depart the primary craft and load into smaller boats.

(4) Handing site, where the unit will beach its boats, or land directly from amphibious craft.

(5) Loading, when loads are lashed and waterproofed IAW unit standing operating procedures (SOPs). Inspections are important. A bump plan for key personnel and equipment must be developed.

(6) Cross-loading, which will help preserve the ranger force. location of leaders, crew-served weapons, and needed equipment are stressed.

b. **Landing site.**

(1) **Beaches.** Beaches are checked to find one that allows an undetected approach. Landing sites that cannot be approached from different directions are avoided. Stretches of defended or patrolled coastline should be avoided. If sand beaches are used, tracks and other signs of

operation that may compromise the mission should be erased. Rural, isolated areas are best. The area behind the landing site should provide an assembly area and concealed routes for exit.

(2) **Barriers.** Reconnaissance may find that good landing sites are not suitable because of barriers between the site and the objective. Such barriers may be patrolled roads, large areas giving little or no concealment, enemy obstacles or defensive positions, areas populated with enemy sympathizers, or an area used for enemy troop maneuvers and training. Unless a means of avoiding such barriers is devised, landing sites should be limited to those without barriers.

(3) **Other factors.** Beach selection should also consider:

(a) The scheme of maneuver.

(b) Surprise.

(c) Enemy dispositions.

(d) Distance to the target.

(e) Landing and exit sites.

(f) Trafficability of the beach.

(g) Cover and concealment.

c. **Routes.** The route the transporting craft takes to the drop site should be planned to deceive the enemy. A major change in the route immediately after the unit's debarkation could compromise the mission. The route should be the same as a route that would be used in some other naval operation--for example, minelaying or sweeping, or a naval patrol. An alternate route must be planned.

d. **Tactical deception.** There should be other plans to deny the enemy knowledge of the insertion besides the water approach route plan. Plans may include electronic countermeasures (ECM), diversions, feints, ruses, or demonstrations.

e. **Actions at the drop site.** Some operations may permit landing directly from the transporting craft. If not, primary and alternate drop sites must be set. The drop site should be at least 1,500 meters offshore to cover any noise in the loading and launching. If the enemy has a surface radar capability, a drop site should be chosen to avoid detection. The drop site may have to be several miles offshore or may need the use of ECM. The manner in which men, small boats, and equipment are loaded must promote fast action at the drop site.

f. **Navigation.**

(1) Ship-to-shore navigation (to the landing site) may be done by dead reckoning. The course may also be maintained by compass navigation, reference to a shoreline silhouette, radar, or prearranged signals from the shore.

(2) Boat handlers, navigators, and other experts are relied on in various phases of the

transport operation.

(3) Lead reconnaissance teams or scout swimmer teams may be used to plant radar, radio, or infrared buoys, or distinct, shielded lights. Boats may be equipped with radar reflectors or beacons to permit accurate tracking.

g. **Actions on the beach.**

(1) Boats hold offshore while ranger scout swimmers move to a covered and concealed position. After reaching the beach, they make a short listening halt. They check the beach and the area immediately beyond the beach for enemy activity. Swimmers signal the force "all clear," mark the landing site, and provide local security.

(2) When the boats land, chosen personnel move directly into covered and concealed security positions. They prepare to defend the landing site.

(3) If the ranger unit is not carrying equipment, boats may be moved off the beach and buried or camouflaged. If heavy equipment makes it hard to move the boats, chosen personnel start unloading at once. The remainder of the force secures the landing site. Each boatload moves into position under the control of a designated leader. Teams are positioned with the best cover, concealment, and fields of fire available.

(4) The ranger force leader sets up a command post from which he can control the operation. He may establish secure communication with his headquarters.

(5) Boats may be deflated and buried or camouflaged near the landing site or away from it. If the boats are to be hidden near the landing site, a team may be chosen to dig holes or cut brush for camouflage along with the unloading of the boats. A small team may be chosen to sweep the beach to erase tracks and drag marks.

(6) The ranger force should be assembled and prepared to move within minutes after the last boat lands. Every ranger must be briefed on the unit's plan should enemy contact be made upon landing.

h. **Insertion by air from ship.**

(1) Helicopters from a ship can launch the ranger force from over the horizon and move at a low level to the LZ.

(2) Helicopters may be vectored from ships to a landing zone. The aircraft may return to the ship while the ranger force conducts the mission, they may orbit offshore, or they may stay at the LZ. Other aspects of landing and assembling are the same as discussed for helicopter insertions.

i. **Helocasting.** This form of insertion combines helicopters and small boats in the same operation. It is planned and conducted the same as air assault operations, except that the LZ is in the water. While helicopters are moving at low levels and low speeds, rangers drop small boats and jump into the water. They then assemble, climb into the boats, and move to shore. Ranger elements are usually assembled and loaded into the small boats within 20 minutes of the last ranger entering the water. This type of operation is normally conducted during limited visibility.

j. **Fire support.** Fire support may be ship-to-air, ship-to-shore, air-to-ground, or ground-to-ground. Specific missions include:

(1) Suppressing enemy air defense.

(2) Neutralizing enemy field artillery within range of boat or helicopter routes.

(3) Employing on-call fires against shoreline targets.

(4) Destroying enemy aircraft.

k. **Contingency planning.** The following must be planned for:

(1) Actions on enemy contact.

(a) En route.

(b) Helocast site.

(c) Flares.

(d) Aerial attack.

(e) Small arms fire.

(f) Indirect fire.

(2) Bump plan with priority to key leaders, crew-served weapons, and radios.

(3) Downed craft procedures.

(4) Escape and evasion.

(5) High surf.

(6) Adverse weather.

(7) Separation.

l. **Rehearsals.** The ranger force must rehearse all aspects of the amphibious insertion. This includes boat launching, paddling, beaching and assembly, boat commands, and capsize drills.

4-17. Land infiltration.

Land infiltration to the objective may be the best way to complete a mission. This is normal when the enemy has total air control or effective air defenses in the operational area to be penetrated, or when the objective is close to friendly territory. Rangers can infiltrate over any type of terrain, in any climate. However, thick forests, swamps, and broken or steep terrain offer the best chance for success.

a. **Planning.** There are many methods of infiltrating by land, and each has its good and bad points. Plans are designed to get the infiltrating force to the objective area without being seen. Plans should include:

(1) Selecting concealed routes for surprise and protection. Primary and alternate routes are based on detailed map and aerial photo study, ground reconnaissance, and data from agent reports.

(2) Avoiding obstacles, populated areas, silhouetting (when forced to cross obstacles or ridge lines), enemy positions, main avenues of approach, and movement along often used routes and trails.

(3) Selecting the time of infiltration to make use of reduced visibility, such as darkness, rain, snow, and fog, and periods of reduced enemy alertness, such as early morning, holidays, and times following protracted movement or combat. The time of the infiltration is important during critical phases (crossing borders or passing through enemy troop concentrations or populated areas).

(4) Using speed and stealth. All rangers must know routes, linkup points, alternate linkup points, time schedules, danger areas, and actions on enemy contact and at linkup points.

(5) Centralizing planning and decentralizing execution. Centralized coordination must ensure that units are acting IAW cover and deception plans. Control should be given to the leaders of the infiltrating groups. Within their designated zones, these leaders select their own routes and formations. Leaders control the rate of advance. When the enemy has electronic detection devices, radio communications must be kept to a minimum. Key leaders normally go with the group that is using the fastest lane or the lane least likely to be detected.

(6) Seeding infiltration routes to the assembly area with unattended ground sensors. These sensors can help the infiltrating force by reducing the chance of surprise contact during the infiltration, and by helping in controlling the forward progress of the units.

(7) Planning for fire support, which must include coverage of routes, danger areas, assembly areas, objectives, and PZs.

(8) Rehearsing actions at the objective, actions on enemy contact, actions at danger areas, and movement techniques.

b. **Movement.** Land infiltration is conducted using one of three techniques:

(1) **Movement in small groups along one axis.** All members of the force use the best route. Small groups are harder to detect, easier to control, and do not compromise the total force if detected. This technique may take too long and it needs an assembly area or linkup point. If the lead group is detected, other groups may be ambushed.

(2) **Movement in one group.** This technique has no reassembly problem. The same route is used by all, making navigation easier. A large force can fight out of a dangerous situation easier than a small one. It has no coordination problems between other infiltrating units. This form of movement is easier for the enemy to detect. If the force is detected, the mission may be endangered.

(3) **Movement in small groups along several routes at the same time.** This technique avoids putting the total force in danger, is less likely to be seen, forces the enemy to react in

many locations, and makes it harder for the enemy to determine the size of the force or its mission. Some groups travel over poorer routes and have problems of assembly, control, and support; the capture of personnel may endanger the mission. Regardless of which technique is used, ranger units normally move to reduce the chance of enemy contact.

c. **Assembly area.** Following movement into the general area of operations, infiltrating units converge on the assembly area or linkup point. Assembly may begin at once after the passage of lines, or it may take place at any time before reaching the objective. The infiltrating force must remain undetected before assembly. If an infiltrating unit is seen at this time, the enemy may learn the location of the assembly area, and the objective. It may take action to destroy the infiltrating force.

d. **Linkup point(s).**

(1) Selection characteristics include terrain that--

(a) Is free of enemy troops.

(b) Offers cover and concealment from enemy observation and patrols.

(c) Does not restrict movement of friendly units.

(d) Accommodates the unit.

(e) Is easily identified.

(f) Offers escape routes to alternate linkup points.

(g) Offers concealment from air attack.

(2) Action sequence for occupying the linkup point.

(a) Halt short of the point.

(b) Establish local security.

(c) Conduct a reconnaissance of the linkup point.

(d) Occupy the linkup point.

(e) Emplace observation posts and listening posts.

(f) Link up with other groups.

(g) Conduct final preparations.

(h) Continue the mission.

(3) Security may be enhanced with the aid of devices such as night vision goggles and night sights. Unattended ground sensors may be placed on likely avenues of approach into the linkup point. Normally, only passive observation devices are used to assist in avoiding detection.

(4) Plans must set a time schedule for actions at the linkup point. All unit members must move in quickly, consistent with security, so that the unit may proceed to the objective rally point (ORP) and on to the objective.

4-18. Stay-behind technique.

A ranger force can be positioned behind enemy lines by employing the stay-behind technique. In this method, a ranger force allows itself to be bypassed as the enemy advances. This is done to allow the ranger force to perform a specific mission. Stay-behind operations may require the concealment, or cache, of supplies before the enemy bypass. Rangers may need to use captured enemy supplies. The location of the assembly area and the plan for extraction are important parts of stay-behind plans.

4-19. Extraction.

The ranger force is normally extracted after the mission is completed. The enemy may also force the unit to abort its mission and to call for extraction. Contingency plans must be made to extract a force under enemy pressure, as well as one that is not in contact. Primary and alternate extraction sites are planned and coordinated with all ,porting forces. Ranger operations are normally conducted against deep targets. Extraction may begin with land movement but normally it includes the use of boats or aircraft. Plans may include primary I alternate means of extraction.

a. **Terrain.**

(1) The terrain is important in choosing the point of extraction - for example, whether or not the extraction site offers good cover from enemy direct-fire weapons and positions from which the extracting force can suppress the enemy. Tidal conditions and PZ size must be considered.

(2) Unlikely locations such as swamps, jungles, or mountain areas are often used for extraction. Specially trained air and naval crews, using the latest extraction devices, make such terrain useable.

b. **Enemy.**

(1) Extraction is best conducted without enemy pressure. However, such pressure could develop during the operation. Detailed plans are made for emergency or contingency extractions forced by the enemy as well as clandestine extractions not under enemy pressure. If under pressure, units will have to use planned fire support to suppress enemy weapons. At the beginning of the extraction with security posted to provide early warning, the unit may start to thin out the same as with a night withdrawal. Ground security may be relied on until the last unit withdraws under cover of suppressive fires.

(2) If enemy contact is made during an extraction, the commander must decide whether to reinsert units or break contact and move to another extraction point. Fire support and suppression are important and include suppressing enemy air defense by deception, ruses, or decoys.

(3) The ranger unit may have special weapons systems or equipment to aid in the extraction. Shoulder-fired antiaircraft missiles may be used to defend against enemy aircraft threatening

the extraction.

c. **Extraction by air or water.** Extraction by air or water is preferred when the resources are available and their use will not endanger the mission. Other factors that favor their use are when--

(1) Long distances must be covered.

(2) Time of return is essential.

(3) The enemy does not have air and naval superiority.

(4) Heavily populated hostile areas block land movement.

(5) Ranger forces are burdened with prisoners of war,casualties, or critical items of equipment.

d. **Extraction by land exfiltration.** Stealth and evasiveness are key elements of exfiltration. This method of extraction is favored when--

(1) The enemy has air superiority.

(2) The enemy is able to prohibit air or water extraction.

(3) The distance to friendly lines is short.

(4) The terrain provides cover and concealment (for movement on foot) and limits enemy mobile units.

(a) Multiple routes of exfiltration may be used to prevent destruction of the force if soldiers or units are detected. Captured enemy vehicles and equipment may be used to assist in the exfiltration.

(b) The ranger force may exfiltrate in one body or in small groups. Exfiltrating in small groups avoids the time delay of assembly.

(5) The exfiltrating force is lightly equipped and not burdened with captured personnel or material.

(6) The exfiltration route passes through an area occupied by friendly civilians or guerrilla forces who can assist the movement.

(7) Areas along exfiltration routes are not inhabited.

(8) The enemy force is dispersed or is not able to concentrate against the exfiltrating force.

4-20. Escape and evasion.

a. A vital part of all premission planning is the development of en-route and post mission E&E plans. Such plans enhance survival of the ranger force, and the transporting air or boat crews. The ranger force commander is responsible for--

(1) Helping in the development of an E&E plan in coordination with all supporting agencies.

(2) Ensuring that all members of the ranger force and the supporting boat or air crew are briefed on the E&E plan.

b. Each E&E plan is unique. Each situation has unique problems associated with E&E. The E&E plan devised by the ranger force commander must address these problems, while profiting from the individual abilities and training of the rangers and their supporting air and boat crews. The following considerations apply to E&E plans devised during ranger operations:

(1) The plans enhance survival of personnel who can no longer accomplish their assigned mission. The senior ranger in a group must decide if there are any remaining missions that the group can accomplish. If not, then he must try to evade and escape enemy capture if unable to link up and be extracted with the rest of the ranger force.

(2) Because of the depth of penetration behind enemy lines, most successful E&E plans may involve either air or water movement away from enemy-held territory. The air and boat crews supporting ranger operations are highly skilled and dedicated. If there is any chance of a successful rescue, these forces continue to try to retrieve rangers escaping from an enemy-held area.

(3) Rangers possess stamina and endurance. They can move a great distance (especially at night) and over rugged terrain to reach an area where they can rendezvous with search-and-rescue (SAR) aircraft or boats. Rangers are also resourceful and highly trained in communication, stealth, and camouflage. This, coupled with their self-discipline, means that successful E&E can be conducted over a great distance, through heavy concentrations of enemy, and over a long period.

(4) Escape and evasion plans for ranger elements should include avoiding contact with the local civilians. However, the aid of friendly insurgent forces may be enlisted. The ranger E&E plan may also include the use of other special operations forces E&E networks that are in place behind enemy lines. These nets must not be compromised by the volume of evaders.

c. Escape and evasion plans should be based on three phases:

PHASE ONE: That portion of the insertion following entry into the hostile zone where personnel survival and recovery is the only thing considered. Should an aircraft or boat carrying rangers be disabled during phase one, the ranking survivor assumes command. He must consider E&E planning and the experience and skills of other survivors. He should be concerned with the security of the mission, caring for wounded, and moving away from enemy activity while contacting SAR elements for recovery.

PHASE TWO: That portion of the insertion that is near the objective and could permit the ranger element to successfully pursue its mission. Although part of the E&E plan, actions by the ranger force during this phase become a forced land infiltration rather than an E&E. The senior surviving ranger must decide if enough personnel and equipment can be moved to the objective area in time to accomplish the mission. If so, he must proceed, taking along any surviving air or boat crew members. If not, he may proceed directly to the extraction point and wait, or he may contact SAR elements for recovery.

PHASE THREE: That portion of the operation after mission accomplishment. It is a hard time to escape and evade since the enemy is alert and reacting to the ranger forces. The confusion caused by the violence of the attack, coupled with the rangers' ability to move quickly cross-country away from the objective area, offers the best chance of a successful escape and evasion. The surviving ranger elements that cannot be extracted can do one of the following:

(1) Move over land at once to a PZ or beach for extraction by SAR elements.

(2) Move to an area where reentry into friendly lines can be made.

(3) Move to a neutral or nonaligned country where either internment or return to friendly control can be effected.

d. Any plan for E&E should provide for the issue to each ranger of an area map, a covert means of recognition, and a small compass. An emergency ration may also be included and any items needed for the climate in that objective area. The ranger force commander must balance the need for other survival equipment against the added weight and bulk involved.

CHAPTER 5

Strike Operations

The ranger regiment conducts strike operations in support of the Air Land Battle. Ranger strike operations are integrated into an overall plan designed to destroy, delay, and disorganize the enemy. They may also cause him to divert a major part of his combat forces and power to rear area security. Strike operations also help create an environment in which other special operations forces can operate. Strike operations can complement air power in disrupting enemy forces in depth. They allow the theater commander to regain the initiative and lead to decisive action.

5-1. Application.

a. The ranger regiment can conduct strike operations either as deliberate-response or quick-reaction missions. It can be inserted into the area of operations by land, sea, or air.

b. Strike operations can be conducted throughout the depth of the battlefield in support of conventional or unconventional warfare. Multiple strike operations can have a synergistic effect and can create situations that help friendly forces seize and maintain the initiative.

c. Strike operations include raids, personnel and equipment recovery, and interdiction of lines of communication. They are best accomplished by ranger units who use conventional techniques and special skills. Strike operations vary in size. When well augmented and supported, the ranger regiment can conduct strike operations deep in enemy territory using up to three battalions. Strike operations normally are limited in time and may be conducted unilaterally or in conjunction with other special operations forces or allied forces. They are normally ended by the extraction or exfiltration of the ranger force.

5-2. Raid operations.

a. A raid is a strike operation conducted behind enemy lines against strategic objectives, targets of high tactical value, time-sensitive targets, or key personnel and facilities in enemy rear areas. It normally involves a swift penetration of hostile territory to confuse the enemy or destroy his installations. Ranger units can conduct deep penetration raids when given the appropriate assets for insertion.

b. A strategic raid by the ranger regiment would be directed by the NCA and conducted under the OPCOM of a unified or JTF commander. An operational raid would be directed by the unified or JTF commander and conducted by that command directly or under the OPCOM of a corps commander. Typical targets would include:

(1) Command, control, communications, and intelligence centers at front level or combined arms army level.

(2) Nuclear and chemical weapons storage sites and delivery means.

(3) Key logistical centers, warehouses, ammunition complexes, or fuel pumping centers supplying logistical support to a specific front.

(4) Air defense and air traffic control integrating centers and air defense weapons sites.

(5) Key power generating and transmitting facilities, hydroelectric dams, and irrigation systems.

(6) Key ports or rail complexes.

(7) Key installations or facilities, such as airstrips, buildings, bridges, dams, or tunnels.

c. The size of the ranger raid force is determined by a detailed analysis of METT-T. The unit committed to the ground phase of the raid mission is kept as small as possible. It may be as small as a ranger company or it could include all the battalions of the ranger regiment. The force must be tailored to complete the mission quickly, violently, and with few casualties. The size of the force may also be affected by the augmentation of specially trained personnel needed for that mission.

d. Raids are normally conducted in five phases. (See Figure 5-1).

Figure 5-1. Five phases of a raid.

PHASE 1. The ranger force inserts or infiltrates into the objective area.

PHASE 2. The objective area is then sealed off from outside support or reinforcement, to include the enemy air threat.

PHASE 3. Any enemy force at or near the objective is overcome by surprise and violent attack, using all available firepower for shock effect.

PHASE 4. The mission is accomplished quickly before any surviving enemy can recover or be reinforced.

PHASE 5. The ranger force quickly withdraws from the objective area and is extracted.

e. This paragraph discusses the raid after insertion/infiltration. All considerations discussed in Chapter 4 must be applied in order for the unit to get deep behind enemy lines. Upon receipt of a raid mission, the following basics are emphasized:

(1) **Maximum use of intelligence information.** The gathering and dissemination of information must be continuous and provided to the raid force even while en route to the target area. To ensure mission accomplishment, the ranger unit must be kept informed of the latest enemy developments in the objective area to prevent being surprised. Only then can the full combat power of the rangers be concentrated at the decisive time and place.

(2) **Plan development.** The reverse planning sequence and the planning steps set forth in FM 101-5 are followed in the plan development.

(3) **Coordination.** Coordination is through the operational headquarters with friendly units (such as higher and supporting Army or joint headquarters, intelligence agencies) for fire, RSTA, special equipment, and personnel and logistical support.

(4) **Rehearsals.** Rehearsals help validate all aspects of planning for the raid and ensure precision in execution-They allow changes to be made in the plan before it is carried out.

f. A successful raid is ensured by--

(1) Launching the raid at an unexpected time or place by taking advantage of darkness and other periods of limited visibility, and moving over terrain that the enemy may consider impassable.

(2) Avoiding detection through proper movement techniques and skillful camouflage and concealment to include taking advantage of natural cover of the terrain, and using sophisticated equipment to detect and avoid enemy forces.

(3) Timing the operation as close as possible.

(4) Using all available support, both organic and nonorganic, to include use of special weapons such as Air Force "smart bombs" and artillery cannon-launched guided projectiles, with the ranger unit using laser target designators.

(5) Performing quick, violent, precise, and audacious actions that focus full combat power at the decisive time and place.

(6) Disengaging quickly upon mission completion.

(7) Withdrawing swiftly using planned routes and including a deception plan.

g. Four functions are normally performed by ranger tactical formations when conducting a raid. Each subelement is organized and equipped to do a specific part of the overall mission. Depending upon the specific mission, nature of the target, enemy situation, and terrain, these four functions are as follows:

(1) The command group controls movement to and actions at the objective. This unit normally consists of the ground commander, other subordinate leaders in the raid organization, and communications to support these leaders.

(2) The security element, whose organization is determined by the mission of the raid force, size and type of enemy force and its mobility and state of alert, terrain and avenues of approach into the area, and the time needed to seal off the objective area. The security element may perform the following:

(a) Secure the ORP.

(b) Give early warning of enemy approach.

(c) Block avenues of approach into the objective areas.

(d) Prevent enemy escape from the objective.

(e) Provide overmatch for the units at the objective and suppressive fires for their withdrawal.

(f) Provide short-range air defense.

(3) The support element provides the heavy volume of fire needed to neutralize the objective. Because fires from this unit are violent and devastating, they must be closely controlled to ensure the precision needed. On order or as planned, fires are lifted and shifted to cover the maneuver of the assault element by suppressing enemy fire from the objective or aerial fires. The support element may also be given specific locations to cover by fire in support of the security element if an enemy quick-reaction force moves toward the objective area. These may include routes into and out of the objective site, key terrain features, or installations adjacent to the main objective. Once the assault has been completed, or on order from the raid force commander, the support element displaces to the next planned position. Organization of the support element is determined by the following:

(a) Size of the objective, the geography of surrounding area, and the enemy threat (to include air) in the area. This element should be able to neutralize the objective (when supported by air or naval gunfire) and to lift or shift fires either when the assault is launched or when ordered by the raid force commander.

(b) Mission of the assault unit.

(e) Suitable firing positions.

(d) Size and nature of the enemy force in the objective area and those enemy forces capable of reinforcement at the objective.

(e) Fire support from other units (air strikes, naval gunfire, surface-to-surface missiles, and artillery fire).

(4) The assault element seizes and secures the objective and protects demolition teams, search teams, prisoner-of-way teams, or other teams.

(a) The organization of the assault element is always tailored to the mission. Each target area must be examined carefully. To assault, seize, and destroy an installation, position, or equipment, the assault element could be organized into one or more assault teams. The element's mission is to overcome resistance and secure the objective and to destroy the installation or equipment. Other specialized teams may also be needed. For example, sniper teams could be needed to remove key sentries. To capture prisoners, liberate personnel, and seize or destroy equipment, the assault element could be organized into assault teams, prisoner teams, search teams, medical teams, demolition teams, or breach teams.

(b) To destroy a point target or installation in a heavily defended area where the USAF cannot get close en to be effective, the assault element might be organized with one small team equipped with laser target designators. From covered and concealed positions, members of the team could then provide guidance for USAF delivery of laser-guided munitions from a safe distance.

h. The site chosen for the raid LZ or DZ must support the planned actions at the objective. There are two options when choosing sites:

(1) The ranger force can land on or near the objective and seize it before the enemy can react. (See Figure 5-2.) This avoids forced marches over land carrying heavy combat loads. If there is no suitable landing area near the objective, or the enemy has a strong reaction force nearby, this option is not favored.

Figure 5-2. Airborne landing on the objective.

(2) The ranger force can land unseen far from the objective. It then assembles, reorganizes, and moves into an ORP near the objective. (See Figure 5-3.) The objective is seized after security and support elements are in place. This option makes coordinated action easier by setting up control of small units before engaging the enemy.

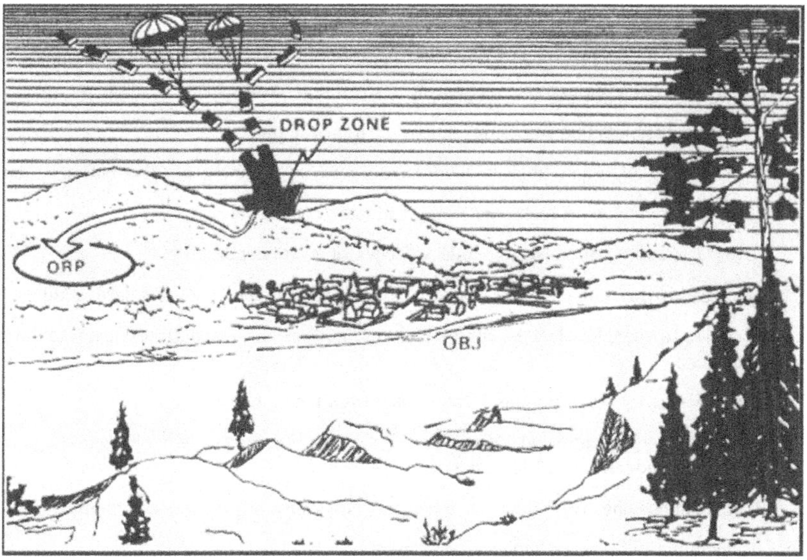

Figure 5-3. Airborne landing distant from the objective.

5-3. Interdiction operations.

a. AirLand Battle doctrine set forth in FM 100-5 is a shift in emphasis toward keeping the initiative in battle. The deep battle seeks out enemy elements separated from the main battle area and attacks them before they can be deployed. Interdiction of selected enemy forces by attacking them or blocking chokepoints hinders, delays, or interrupts the enemy's use of his lines of communication; it denies him the use of certain areas; it destroys industrial facilities, military installations, and equipment; and it stops or slows his advance along a particular route of approach. The ranger regiment's ability to conduct interdiction is vital to execution of the AirLand Battle doctrine.

b. Interdiction by ranger forces includes blocking or channeling approaches, holding key terrain for a limited time. and destroying facilities.

c. Ranger forces conduct interdiction missions that supplement those conducted by long-range Army and USAF weapon systems. Large-size ranger forces are normally used only when other systems fail or are not available, and the target is of such priority that it must be attacked. Small ranger laser target designator teams may work in concert with USAF or USN strike aircraft to direct terminally guided weapons systems. Rangers should be directed against crucial interdiction targets that disrupt, confuse, or delay the enemy and allow friendly forces to gain the initiative at the forward line of own troops (FLOT).

d. Typical interdiction operations are--

(1) Destroying rail lines, switches, rail yards, and rolling stock.

(2) Destroying bridges, overpasses, tunnels, and truck parks.

(3) Closing mountain passes or routes in restricted terrain.

(4) Cutting oil or natural gas pipelines and destroying pumping and distribution systems.

(5) Destroying dams, locks, spillways, and flood control gates.

(6) Destroying rail transload points, switches, and yard engines.

(7) Destroying shipyards, dry docks, wharfs, lighters, cranes, and stevedore facilities.

(8) Destroying electrical generation facilities, power lines, transformers, and grid monitoring centers.

(9) Destroying water pumping stations and filtration systems.

(10) Destroying radio and TV stations, phone lines, microwave terminals, satellite receiving stations, and telephone exchanges.

(11) Ambushing enemy forces and personnel moving along lines of communication.

e. Considerations when using rangers for interdiction are:

(1) Insertion and extraction.

(2) Support resources available.

(3) Ranger force vulnerability.

(4) Target identification.

(5) Threat environment.

(6) Other or supporting means of interdiction.

(7) Command and control.

(8) Time and magnitude of the effect on the enemy forces at the forward edge of the battle area.

(9) Collateral effects of target destruction.

f. Interdiction targets are not identified and attacked at random. They are part of an overall deep attack plan to destroy or cripple an entire system. Interdiction is based on the assigned deep campaign plan that directs which targets are to be attacked, by whom, the results desired, and the priorities of attack for specific systems. Based on his plan, the ranger force commander selects those ranger elements best suited to conduct the attack. Target selection, which requires detailed intelligence and planning, is based on:

(1) **Target criticality.** A target is critical when its destruction or damage would have a major influence on the enemy's ability to conduct or support operations. Each target is considered in relation to other elements of the particular target system designated for interdiction. The criticality of a target changes with the situation. For example, when the enemy has few locomotives, a railroad bridge is less critical.

(2) **Target accessibility.** The ranger force must be able to insert into the target area. Terrain, security controls, location, and means of insertion are considered when studying a target for accessibility. systems. Based on his plan, the ranger force commander selects those ranger elements best suited to

(3) **Target restorability.** The enemy's ability to restore a damaged facility to normal operating capacity affects target selection. The enemy's repair and maintenance ability and his reserve stocks must be considered.

(4) **Target vulnerability.** A target is vulnerable when it is open to attack by means available to ranger forces. Vulnerability is affected by the nature of the target such as type, size, disposition, and security. It is also affected by the means available to attack it such as explosives, incendiaries, and special devices.

g. Key to the interdiction process is the ability to locate and evaluate target systems for attack and the timing of the attack. A target system is a series of related functions that serve a common purpose.

(1) A target is one element, installation, or activity identified for attack such as a locomotive, train, bridge, or prison.

(2) A target complex is several targets in the same general area, such as a railway marshalling yard, an airfield, or port and dock facilities.

(3) A target system consists of an industrial system and its sources of raw materials; the rail, highway, waterway, or airway systems over which these materials are transported; the source of power and method of transmission; the factory complex; and the means by which the finished product is transported to the user. The targeting process is usually accomplished in the special operations force cell at EAC and the battle coordination element (BCE) at corps level.

h. Interdiction techniques include ambush, emplacement of conventional demolitions, target designation for laser-guided munitions, and emplacement of beacons or sensors.

NOTE: The enemy surface-to-air threat normally precludes the use of an airborne platform for direct observation and designation. This may be accomplished by the on-ground ranger force.

i. Through interdiction, ranger units are an ideal force to contribute to the Army's AirLand Battle operational concept. This includes taking the initiative, moving quickly to strike deep in the rear of the enemy, and defeating the enemy force beyond the FEBA. Planners and commanders must ensure that the ranger units are used on missions that are synchronized to contribute to the overall deep battle campaign. Strikes against critical targets must combine with efforts of other assets to take the initiative from the enemy, limiting his flexibility, initiative, and momentum.

5-4. Personnel and equipment recovery operations.

Recovery operations are specialized raids to liberate imprisoned or detained personnel or to return certain items of equipment to friendly control. These operations include recovering and extracting downed or hijacked aircrews and political or military leaders. Rangers use surprise and combat power to overwhelm resistance before harm can come to the detainees or prisoners.

a. **Planning.** The planning of personnel and equipment recovery operations is the same as for the raid. The main difference is that the ranger force commander must plan for the extraction of the recovered personnel and the loading and extraction of sensitive equipment. Personnel and equipment recovery raids are often executed as quick-response missions. They normally require the ranger force commander to plan and execute the recovery quickly.

b. **Augmentations.** This type of strike operation often requires that the ranger force be augmented with nonranger personnel with special skills or abilities. Examples are medical personnel, technical experts, mechanics or crew members trained in repair and retrieval operations, linguists or translators, and other special operations force elements.

c. **Organization.** The organization of the recovery force is the same as for a raid. Sometimes special teams are needed to perform certain missions involved in the recovery. This mission is conducted aggressively, usually ending in close combat during darkness or other limited visibility conditions. Reconnaissance teams may be inserted ahead of the ranger force to reconnoiter the objective and to locate the detained personnel.

(1) All planning and execution takes place as described for the raid. The emphasis must be on detailed, timely intelligence.

(2) Medical teams must be available to care for the detainees.

(3) The ranger personnel must be adept at searching and clearing of buildings.

(4) The ranger force must--

(a) Locate the specific building or area where the friendly detainees are being held.

(b) Gain entry into buildings, use stun grenades, and avoid obstacles and booby traps.

(c) Engage all enemy personnel using selective-kill techniques.

(d) Locate, identify, and secure detained personnel.

(e) Move personnel to a secured extraction site.

(f) Leave no detainees or rangers behind in the objective area.

d. **Execution.** The key to success is surprise and speed. The enemy must be assaulted when least expected. The ranger force must ensure that friendly detainees are not harmed during the assault - either by rangers or enemy personnel.

e. **Extraction.** The ranger force commander should plan for the extraction point to be as close as possible to the point of recovery. This prevents the detainees from having to move cross-country or for the ranger force to transport bulky equipment long distances. For small groups of detainees or small items of equipment, the extraction point can be further away. Recovery operations may use any method or combination of methods of insertion and extraction (see Chapter 4). Close planning and coordination is normally required with US Army, USAF, or USN aviation for both delivery and evacuation of the target area.

CHAPTER 6

Special Light Infantry Operations

Ranger units are normally employed on strike operations. However, broad US strategic needs demand a wide range of other military abilities. The ranger regiment may conduct special light infantry operations when larger conventional forces could not be used. The regiment conducts these operations using light infantry doctrine, methods, and techniques.

These operations include, but are not limited to, movement on foot, military operations on urbanized terrain (MOUT), passage of lines, ambush, and reconnaissance. Ranger units do not normally conduct planned operations against enemy armored forces. Battlefield surprises or METT-T considerations may cause ranger units to engage armored forces for short periods. The ranger regiment can conduct a conventional light infantry defense for short periods. A defense is not normally conducted adjacent to other friendly forces. It is normally a perimeter defense from which the unit then breaks out from encirclement and links up with friendly forces.

6-1. Movement.

a. Most missions involve movement on foot in tactical operations. Ranger units infiltrating behind enemy lines seek to avoid chance enemy contact. They choose the movement that allows them to retain security and control. To avoid loss of surprise and initiative, casualties, and mission failure, ranger units normally--

(1) Avoid chance enemy contact, if possible.

(2) Move on covered and concealed routes.

(3) Avoid likely ambush sites and other danger areas.

(4) Practice camouflage, noise, and light discipline.

(5) Maintain all-round security.

(6) Make unavoidable enemy contact with the smallest element possible.

(7) Retain the initiative to attack at the time and place of the ranger unit's choice.

b. In planning the movement of his unit, the ranger commander must consider all the elements of METT-T and determine how they will affect the movement. He must determine the route, formations, security measures, methods of command and control, location of key leaders and weapons, and action to be taken on chance enemy contact.

(1) **Routes.** The ranger force commander conducts a careful route analysis, using the factors of observation, concealment, obstacles, key terrain, and avenues of approach (OCOKA). He chooses a route that maintains the security of the force while ensuring surprise. He tries to choose a route that will avoid enemy contact and speed movement. The commander makes a detailed route reconnaissance using maps, aerial photographs, and reports from persons familiar with the area. If possible, he reconnoiters the route himself or uses a reconnaissance team. He may have that team act as guides or mark the route using covert marking devices.

(2) **Formations.** Ranger units use the movement techniques found in current light infantry doctrinal publications. The traveling, traveling overmatch, or bounding overmatch technique is chosen based on the likelihood of enemy contact. The ranger force normally tries to avoid detection and contact with the enemy while moving. The commander chooses a movement technique that ensures that if unexpected contact were made, it would be with the smallest element possible and not the entire formation. The distance between moving ranger elements depends on the terrain, visibility, and enemy situation. March interval is normally great enough to allow each succeeding element either to deploy or change direction if enemy contact is made. Distances are close enough that each element could quickly assist the element to its front.

(3) **Security.** Security is to the front, rear, and flanks when enemy contact is likely. A front security element is placed well forward with adequate communications with the main body. The security element is used to detect the enemy and warn the main body. The main body then either changes direction or hides until the enemy is past. If the enemy does detect the main body, the security element may be used to attack the enemy from the flanks or rear. Flank security elements are placed next to the route of march. They move forward either by alternate or successive bounds. An alternative is to have the security element moving adjacent to the column along routes paralleling the direction of march. Rear security is handled the same as front security. Plans are made for the rear guard to support the main body during chance enemy contact. This is done either by maneuver or by furnishing supporting fire.

(4) **Command, control, and communications.** All available communication is used, consistent with OPSEC and movement security, to assist in maintaining control during movement. March objectives, checkpoints, and phase lines may be used to aid in control. The number of reports is reduced, normally only exception reports are needed. The ranger force commander should be well forward in the formation but may move throughout as the situation demands. Communications with security elements are mandatory. Operations security often precludes the use of radios. Therefore, connecting files, runners, and visual signals are normally used. Detailed planning, briefing, rehearsals, and control are valuable if there is enemy contact. Alternate plans are made to cover all possible situations.

(5) **Weapons.** Rangers are placed within the formation to allow a mix of their individual weapons to cover the front, flanks, and rear of the formation. Rifles, grenade launchers, and light automatic weapons cover all sectors. Crew-served weapons such as machine guns, antitank weapons, and mortars should be given designated locations within the formation. This allows the element leader to employ them quickly. They should be placed near the lead element, but not so close as to be engaged along with it. Crew-served weapons are used to

provide overmatch, especially when employing the bounding overmatch technique of movement.

(6) **Rehearsals.** The ranger force, expecting movement as part of an operation, conducts the following combat drill rehearsals:

(a) Conduct movement.

(b) React to enemy contact.

(c) Maneuver under fire (fire and movement).

(d) Cross danger areas.

(e) Break contact.

(f) Conduct a hasty ambush.

(g) Make a temporary halt.

(h) Establish a perimeter.

(i) Establish a rally point.

(j) Take actions upon consolidation.

(k) Use arm-and-hand signals.

(l) React to flares and indirect fire.

c. After a ranger force is inside enemy territory, it must be alert to avoid detection before reaching the objective.

(1) Should the ranger force security element become aware of enemy presence, it alerts the main force. The main force tries to move away without alerting the enemy and continues the mission.

(2) If the enemy does make contact, the security element then engages the enemy, trying to deceive him as to the size of the ranger force and its direction of movement. After breaking contact, the security force divides into small groups and moves to a rally point to rejoin the main force.

(3) The ranger force does not fight en route to the objective unless there is no alternative, then it quickly breaks contact and moves away.

(4) Ranger units may need to set up a temporary defensive position for resupply, evacuation of wounded, planning, or extraction. One technique that can be used is the temporary perimeter defense.

(5) If the moving ranger force comes under indirect fire, it moves quickly out of the impact area and continues the mission.

(6) The most effective means of combating an ambush is through the immediate return of a

high volume of fire by all weapons. This requires alertness, discipline, dynamic leadership, and rehearsed plans.

(a) Regardless of the method of movement, all weapons should be positioned for immediate use. A well-executed ambush is hard to counteract. However, the effectiveness of an ambush can be reduced by the immediate return of fire. Enemy weapons may be silenced, thus creating weak points in the ambush position. Smoke and high-explosive (HE) grenades are effective in executing counterambush actions. They create confusion, provide a screen for movement, and disrupt the ambush plan.

(b) When all elements of a ranger force are not trapped within the ambush, the elements that are free to maneuver should initiate an immediate flank or rear assault against the ambush force. A flank assault permits better coordination with personnel in the ambush. Supporting fires are provided by those weapons that cannot be hand carried and fired in the assault.

(c) Specific actions against a near and far ambush are as follows:

- **Near ambush.** Elements caught in the kill zone return fire at once and move out of the kill zone. Elements not in the kill zone help by maneuvering and conducting a flanking attack. The unit assaults the ambush position.
- **Far ambush.** Those ranger elements not in the kill zone set up an overmatch position and lay down a base of fire for the elements in contact. The elements in contact move out of the kill zone and prepare to assault the enemy, using individual movement techniques and fire and movement.

(7) The reorganization after enemy contact involves the use of rally points, plans for local security, reorganization of the force, evacuation of dead and wounded, and movement based on the unit mission.

6-2. Ambush operations.

a. **Definition.** An ambush is a surprise attack, from a concealed position. It is used on a moving or temporarily halted enemy to destroy or capture him and his equipment. An ambush is a useful tactic because--

(1) Small, well-trained, disciplined forces, with limited weapons and equipment, can destroy much larger enemy forces.

(2) It reduces the enemy's overall combat effectiveness by destruction and harassment of his forces.

(3) Enemy morale and effectiveness suffer heavily at little cost to the force executing the ambush.

b. **Execution.** A successful ambush must be executed with precision, violence, speed, and audacity of execution. For success, ambush operations must emphasize the following:

(1) **Surprise.** Surprise, more than any other single aspect, enhances the value of an ambush. Surprise increases the potential for inflicting damage on the enemy with less danger to the

ranger force.

(2) **Coordinated firepower and shock effect.**

(a) Coordinated firepower is used for maximum shock effect. Massive volumes of accurate fire, explosives, and mines, coupled with an aggressive attack, breaks the enemy's spirit to fight back. Surprise increases shock effect and the chances for success. Shock effect can cover unexpected defects in an ambush--for example, ambushing a much larger force than expected.

(b) All weapons must be sited with interlocking fires in the kill zone and along likely avenues of entrance or exit. Mortars should be used if the terrain permits. Tripods and traversing and elevating mechanisms are normally used with machine guns to lock in fires. All riflemen use firing stakes to mark left and right limits, and elevation stakes. There is a tendency to shoot high in an ambush--especially at night. The M203 grenade launchers are sited to cover the dead space and routes of escape. When the rules of engagement permit, plastic bags containing CS crystals may be taped to the front side of Claymore mines.

(3) **Control.** Control is essential.

(a) Leaders must have contact with all members of their unit to alert them to the oncoming enemy. Leaders should not move around the ambush site during this crucial period. A method used to alert members can be to tie strings or vines to rangers' legs or arms. By a series of light tugs, all members of the ambush can be alerted to enemy presence.

(b) The leader must initiate the ambush with a casualty-producing device. A bank of Claymore mines on a double-ring main is an excellent device to spring an ambush. Other good techniques are to use a 90-mm recoilless rifle firing antipersonnel (APERS) rounds or a machine gun. All must be under the direct control of the ranger leader. Whistles or pyrotechnics must not be used. They will allow the enemy time to react.

- As soon as the enemy is hit, he reacts. The ambush force has only a few seconds to destroy the enemy in the kill zone. He will quickly recover from the initial shock and leave the kill zone--either with a direct counterattack or withdrawal, Subsequent fires and other banks of Claymore mines must be planned.
- The leader always executes the ambush, except when a member of the ambush knows he has been discovered. He then has the authority to execute--with killing fire, not by yelling.
- The cease-fire must be controlled by the ranger leader. A whistle or other device may be used to get attention, and then cease-fire is signaled.

(4) **Security.** The flanks and rear of an ambush site are open to counterattack. Flank and rear security may be enhanced by -

(a) Echeloning in depth.

(b) Designating sectors of observation.

(c) Positioning of RSTA devices.

(d) Enforcing noise and light discipline.

(e) Having a good withdrawal plan.

(f) Securing routes of withdrawal.

(g) Executing with speed and violence.

(h) Positioning security forces to sea] off the ambush area.

(i) Having good camouflage.

(5) **Simplicity.** A simple, direct plan improves the chance of success. The ambush plan must be clear yet concise to offer the greatest likelihood of success. For example:

(a) Mission statements for security, support, and assault elements must be clear, concise, and direct.

(b) Tasks to be performed by the ambush elements should be easy to understand.

(c) Contingency plans should be simple.

(d) Routes into positions and withdrawal routes should not cross. They should be the shortest, most secure routes.

(6) **Training and self-discipline.** Precise and violent execution is how ranger units win against enemy forces. All advantages must be exploited. Discipline must be strict. There must be no sleeping, talking, eating, or smoking in the ambush site. If an ambush is to be set up for long periods, then the elements of the ambush must be pulled back at set times to the ORP for rest. Extended ambushes of 24, 36, or 48 hours require six- or eight-hour shifts. It may take a company to man an extended platoon ambush position. Tired troops cannot man an ambush well. Other vigorous operations cannot be performed all day and have an alert ambush all night.

(7) **Maximum use of specialized equipment.** This includes the most sophisticated weapons, communications equipment, and RSTA devices. The effectiveness of any mission is enhanced by the use of this equipment.

(8) **METT-T.** Every ambush mission must be evaluated by the ranger commander IAW METT-T to design an ambush that works.

(a) **Mission.** Clearly state the mission--for example, stop enemy resupply, ambush enemy armor columns, or capture personnel or equipment.

(b) **Enemy.** State the type of enemy units--for example, guerrillas, main force, logistics carrying parties, or armor units. What is the size of the force, its habits when attacked, and type of support? What is the enemy night vision ability and expected time of movement?

(c) **Terrain.** Is it wooded, mountain, jungle, or urban? Some good ambushes can be set in a city or suburb. Are the routes available to the enemy suitable to his activity?

(d) **Troops.** Squads and platoons are the basic ranger ambush forces. An ambushing force should be able to engage a force two to three times its size--but evaluate enemy and terrain, A platoon ambushing an enemy tank company in daylight in open terrain is risky. But engaging that same column at night or in a forest may give the ranger platoon an advantage.

(e) **Time.** How long will it take to emplace the ambush? A complex, lethal ambush with plenty of "dirty tricks" takes hours to set up; a hasty ambush on a trail takes 15 to 30 minutes. How long is the ambush to be in place? What hours of the day, or at nighttime only? How soon must it be set up? Do not set up an ambush in darkness--even if it is to be executed in darkness. Try to allow every ranger to '6see" and "lay" in his weapon in daylight. Time is important.

c. **Organization of the ambush force.** A ranger unit conducting an ambush must be task-organized to perform the following functions: assault, security, and support. The ambush forces should be task-organized according to the TOE--by platoons, squads, and fire teams. The TOE should not be changed to create smaller elements for an ambush. The TOE formations may be reinforced with machine gun, recoilless rifle teams, or a 60-mm mortar squad.

(1) **Assault.** The elements assigned the assault mission either move directly into their positions or through a release point. The mission may include any combination of the following:

(a) Conduct the main assault.

(b) Halt an enemy's motorized column or any moving target,

(e) Kill or capture personnel.

(d) Recover supplies and equipment.

(e) Destroy vehicles and supplies.

Search teams are not always used. The ranger leader must decide how and when to use search teams. When rangers leave the security of their well-chosen, concealed ambush position, they are subject to the fires of the enemy who may also be hidden and ready. If it is at night, do not throw tripflares or shoot illumination to light the search area as this will illuminate and expose rangers to the enemy. Always assume there is hidden enemy--the ambush will not kill them all. Night observation devices (NODs) or a red filtered flashlight taped to M16s should be used to quickly search in ranger buddy teams. If the return fire from the enemy is great or if the ambush missed the main body, then the leader may choose to break contact and leave without searching the kill zone.

(2) **Security.** The elements assigned the mission of security may move to their positions

directly or by way of a release point. Their missions may include any or all of the following:

(a) Secure flanks, rear, or ORP.

(b) Provide early warning.

(c) Seal off the kill zone to prevent the enemy from escaping or reinforcing.

(d) Assist in executing the ambush.

(e) Cover withdrawal of main ambush force.

(3) **Support.** The units assigned a support mission provide fires that may include employment of--

- Heavy automatic weapon fires.
- Antitank fires.
- Mortar fires.
- Mines.
- Flame munitions.

d. **Ambush site.**

(1) **Choose an ambush site.** When choosing an ambush site, all sources of information must be used to enhance surprise, exploit the enemy's weak points, and take advantage of the terrain. An analysis of the terrain is also based on the other factors of METT-T to include maps, aerial photos, and, if possible, personal reconnaissance. Emphasis is on--

(a) Natural cover and concealment for the ambush force.

(b) Routes of entry and withdrawal (at least two) that should be direct and easy to reach.

(c) Good observation and fields of fire.

(d) Harmless-looking terrain.

(e) Few enemy escape routes.

(f) Limited enemy reinforcement ability.

(g) Nearby assembly or rendezvous area.

(h) Terrain that will canalize enemy into kill zones, and natural obstacles to keep him there.

NOTE: Try to select a site covered by friendly supporting indirect fires.

(2) **Take advantage of the terrain.** Emphasis must be on exploiting all natural cover and concealment afforded by the terrain. Site the ambush and individual positions based on the terrain rather than trying to adapt the terrain to a fixed geometric design.

(3) **Restrict enemy movement.** Restricting enemy movement by natural or man-made

obstacles should also be planned.

e. **Types of ambushes.**

(1) Ambushes have two categories for ranger purposes--area ambush and point ambush.

(a) Area ambushes may be set up by ranger platoons, companies, or battalions. They are used to interdict enemy movement in a given area or inflict casualties on his forces. Area ambushes consist of a series of point ambushes. The size and location of the ambushes are dictated by the METT-T analysis.

(b) Point ambushes are set at the most ideal location to inflict damage on the enemy. Such ambushes must be able to accept the enemy force from more than one direction. The ranger force levels may enable it to execute an ambush in two or three main directions. The other directions must be covered by security that gives early warning of enemy attack.

(2) The basic ambush is linear (Figure 6-1). An ambush is set on an expected avenue of approach. This ambush can accept contact from three basic directions--left, right, and front. The rear is secured by a security team.

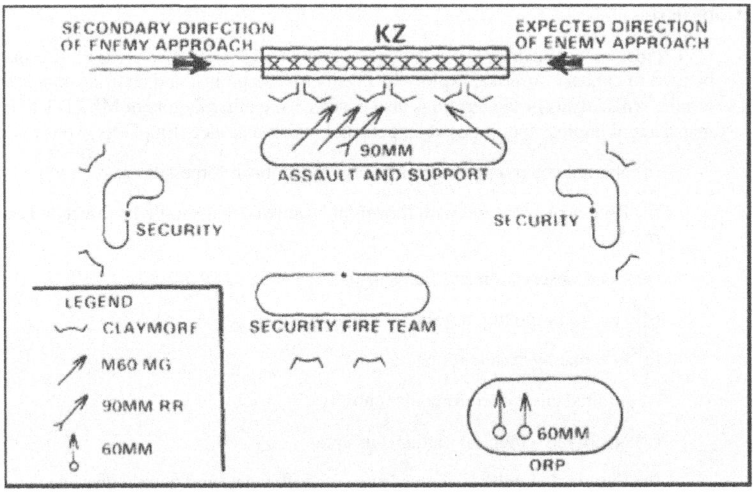

Figure 6-1. The linear ambush.

(3) A variation of the linear ambush is the L-shaped ambush (Figure 6-2). The L-shaped ambush is formed with the base (bottom) of the L perpendicular to the expected enemy direction of advance. This is a good ambush for a road, jungle trail, or an area where the enemy is canalized and his approach route is known. The L-shaped ambush can handle an enemy coming from the expected direction (toward the base of the L) from the front (stem

of the L). It is less effective against an enemy formation that comes from the opposite direction. The ambush must be executed when the enemy main body has cleared the base of the L. The flank security must protect the rear of the base of the L. In the L-shaped ambush, weapons must be sited to avoid direct or ricochet fire into friendly forces. The security elements must be considered as fire fans are being planned.

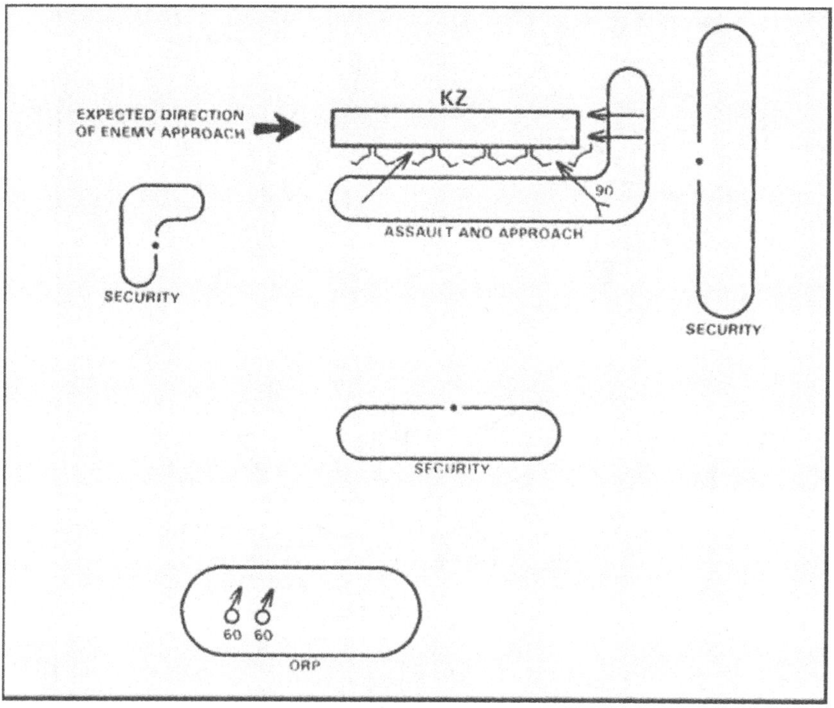

Figure 6-2. The L-shaped ambush.

(4) Other ambushes (the Z, the X, the V, and so on) are all variations of the linear and L formations (Figure 6-3). The more advanced ambush formations are difficult. Units must be expert in the basics before they are ready to employ such techniques.

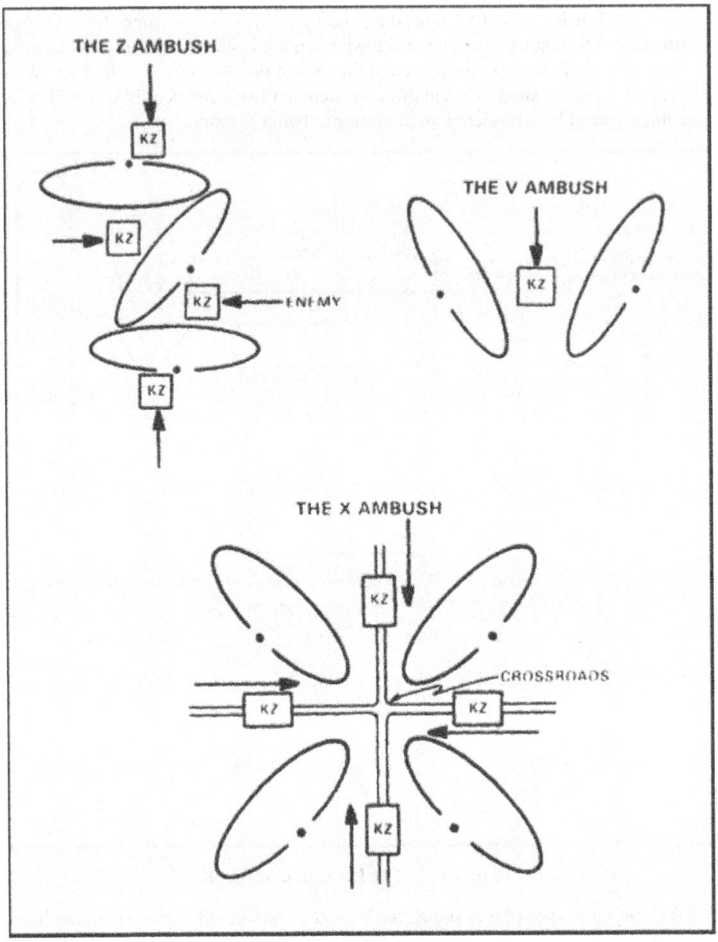

Figure 6-3. Other ambush variations.

(5) Another type of point ambush is the mechanical ambush (Figure 6-4). It consists of Claymore mines set in series with a double-ring main. It is command detonated or detonated by a triggering device activated by the enemy. Mechanical ambushes are normally manned. Rangers prepare to engage the enemy with direct fire if the mechanical ambush does not detonate or if it is wholly or partially ineffective. Mechanical ambushes are an effective way to interdict a large area using a small ranger force. If the mechanical ambush is effective and

rangers do not reveal their presence, the enemy is confused. This has a devastating effect on his morale and effectiveness.

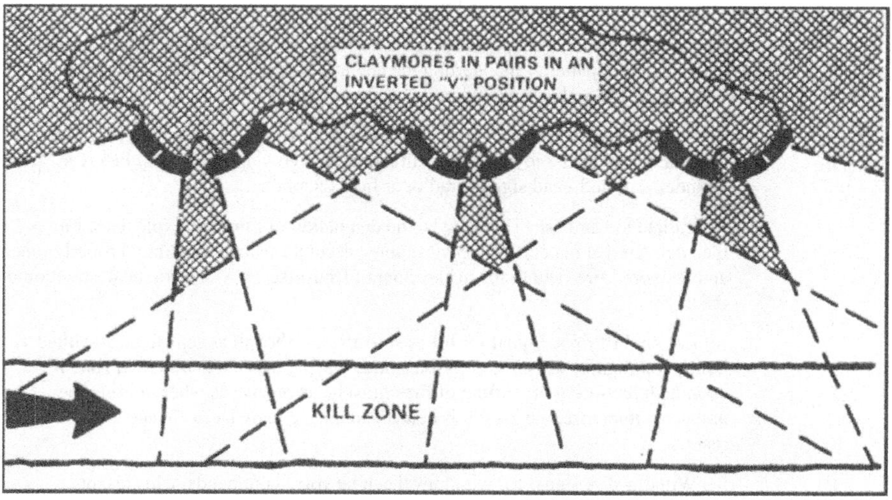

Figure 6-4. The mechanical ambush.

f. **Execution of the ambush.**

(1) **Occupation.** Stealth and security are important factors.

(a) Position security teams and early warning detection devices first.

(b) Use the rest route to main ambush position consistent with security.

(c) Quickly occupy the ambush position and set up communications and signaling devices.

(d) Position key weapons (automatic and antiarmor).

(e) Rig Claymore mines, tripflares, and booby traps.

(f) Ensure that all weapons are correctly positioned. Assign sectors of fire to provide mutual support and cover dead space.

(2) **Camouflage.** Each man must be hidden from the enemy, During preparation for the mission, each man camouflages himself and his equipment, and secures his equipment to prevent noise. At the ambush site, positions are prepared with minimum change in the natural appearance of the site. All debris resulting from preparation of positions is concealed to prevent evidence of occupation.

(3) **Movement.** Movement is kept to a minimum. The number of men moving at a time is closely controlled. Every man is as quiet as possible, especially at night. Light discipline is rigidly enforced at night. Smoking is forbidden.

(4) **Signals.** Audible and visual signals, such as whistles or pyrotechnics, must be changed often to avoid setting patterns and alerting the enemy. Three or four simple signals are needed to execute the ambush. Signals are used to--

(a) Provide early warning of an enemy approach. A signal by the security force to alert the patrol leader to the correct direction of enemy approach may be given. This includes arm-and-band signals, radio, or field telephone.

(b) Initiate the ambush. This may be the detonation of mines or explosives, Fire is then delivered at once in the heaviest, most accurate volume possible. Properly timed and delivered fires add to the achievement of surprise, as well as to the destruction of the target.

(c) Lift or shift fires, Signal for lift or shift fires, if the kill zone is to be assaulted. Voice commands, whistles, or pyrotechnics may be used. When the kill zone is assaulted, the lifting or shifting of fires must be as precise as when starting the ambush. Otherwise, the assault is delayed and the enemy has a chance to recover and react.

(d) Withdraw. A signal for withdrawal can be voice commands, whistles, or pyrotechnics.

(5) **Objective rally point.** The ORP is located far enough from the ambush site so that it will not be overrun if the enemy manages to attack the ambush unit(s). Existence loads are normally left in the ORP. Routes of withdrawal to the ORP are scouted (when possible by each man), Withdrawal routes should provide cover and concealment for the unit and hinder enemy pursuit. Withdrawal routes are a main consideration in the selection of the ambush site. They may be the key to survival after execution of the ambush. On signal, the ambush force quickly (but quietly) withdraws to the ORP. If pursued, the withdrawal may be by bounds, with mines or hasty ambushes used to delay pursuing forces.

(6) **Ambush variety.** If one method of ambush is used predominantly, the enemy will develop an effective defense against it and will be affected less by the shock of the ambush since he knows what to expect. More than one method of ambush should be used. No single method will fit all combinations of terrain, equipment, weather, and enemy capabilities. Variety should also be used when signaling - for example, use multiple signals, both audible and visual. Use weapons fire, mines, and RSTA when possible and vary signals to avoid compromise.

(7) **Swift action.** Speed in the execution of the ambush and the withdrawal should prevent enemy reaction forces from engaging the ambush force. Speed is often a shield against casualties and failure. When there is contact with reaction forces, speed may enhance quick disengagement. Ranger units must not only perform better than the enemy but also move faster than the enemy can react.

g. **Successful ambush.** Emphasize the following:

(1) Intelligence to ensure the enemy is ambushed at a time and place when he least expects or is least prepared to fight. This is preferred during reduced visibility to achieve surprise.

(2) Detailed planning and thorough training and rehearsing of all elements in all phases of the ambush. This ensures maximum shock effect through swift, precise execution.

(3) All available RSTA devices to permit daytime effectiveness at night when moving, shooting, or detecting enemy movement. To avoid detection, active RSTA devices should not be used until after the ambush has been triggered.

(4) All available firepower with emphasis on antiarmor, area and automatic weapons, and precision-guided munitions.

(5) Speed to achieve surprise and enhance security of the force.

(6) Cover, concealment, and overall protection afforded by the terrain when moving or when occupying ambush positions.

6-3. Antiarmor operations.

a. Antiarmor operations are not primary missions for ranger forces. They lack antiarmor firepower and have limited antiarmor ability. Ranger units can, however, perform limited antiarmor missions on close terrain by interdiction or ambush. The antiarmor weapons they do have are man-portable, but the ammunition is too heavy to carry many rounds cross-country. A ranger battalion has ten Dragon missile launchers and nine 90-mm recoilless rifles. It is also equipped with light antitank weapons (LAWs) and laser target designators (LTDs) that can guide Copperhead rounds.

b. Every ranger must be skilled with the limited antiarmor systems available. Proper selection, fortification, and camouflage of antiarmor ambush sites, route selections, withdrawal procedures, and initiative are important. Armor kill zones (AKZs) are normally set up after infiltrating the ranger force behind enemy lines under limited visibility.

c. A ranger antiarmor force is organized into four elements: command and control, support, security, and armor-killer teams (Dragons, LAWs, 90-mm recoilless rifles). These elements function the same as a raid force. The exception is that the assault force is the armor-killer team with the mission of destroying enemy armor, and it will probably not enter the AKZ. After the ambush occurs, the ranger force withdraws to its ORP. Indirect fires (if available) are then brought to bear on the AKZ.

d. Ranger units may conduct antiarmor ambushes and interdiction operation by use of laser-guided Copperhead rounds or air-delivered precision-guided munitions. Air-delivered family of scatterable mines (FASCAM) can be used to provide the ranger force security against attack by armored forces.

6-4. Military operations on urbanized terrain.

a. The ranger regiment normally tries to avoid the manpower-intensive, high-casualty battles associated with MOUT. However, due to the nature of strike operations and the types of targets selected, the regiment may conduct MOUT for limited periods. As in all ranger operations, the ranger force commander tries to surprise the enemy, attack him from unexpected directions, and destroy him with a violent, precise attack. The maze of cellars, rooms, sewer systems, rooftops, and alleys common to urban areas presents many opportunities for the ranger force.

b. Offensive operations by a ranger force exploit the centralized planning and decentralized execution aspects of ranger unit training. Individual action and initiative are always channeled to accomplish the commander's intent. The ranger unit's offensive tactics are based on quick action and boldness. Attacks by the ranger force are short and sharp. Execution is marked by swiftness, daring, timing, and surprise. The enemy's sleeping habits, eating habits, and relief times are watched. launched almost always at night or under smoke screens, attacks are often made without mortars or artillery, relying on surprise for success. Snipers and sharpshooters are used to shift the balance of battle.

c. Deliberate defensive operations are not good ranger unit missions. However, if a ranger unit must defend for a short period, the close terrain of urban areas allows it to function well against a stronger force. The buildings in a city break up advancing enemy formations and force them to move along streets. The ranger force executes an active defense anchored on strongpoints, such as a building or a group of buildings. It launches surprise counterattacks against the enemy's flank and rear. The aim of ranger defense is to give the enemy no respite and to make him feel threatened by a sudden, deadly attack. Ranger units should be withdrawn from defensive positions in urbanized areas as soon as possible. Prolonged MOUT and the associated casualties quickly degrade a ranger unit's combat effectiveness.

6-5. Perimeter defense.

The Perimeter defense may be conducted to control key terrain, gain time or economize forces.

a. **Planning.**

(1) **Warning order.** The order must be issued soon after mission receipt. It should contain the mission, time schedule (including rehearsal and inspection times, mission time, and place for OPORD), uniform and equipment, and special instructions.

(2) **Operation order.** The leader normally gives the OPORD on the ground to be defended. Locations and sectors of fire for squads and crew-served weapons should be designated. If time permits, leaders walk the terrain with subordinates as the OPORD is issued. The OPORD covers security, communications, signals, target reference points (TRPs), armor kill zones, and priority of engagement rules for antiarmor weapons. The OPORD also states the priority of work.

(a) Set up security and communications.

(b) Position crew-served weapons and personnel.

(c) Clear fields of fire.

(d) Prepare fighting positions.

b. **Organizing.**

(1) Squads are positioned with overlapping and interlocking sectors of fire and observation.

(2) The M60 machine guns are positioned either with an assigned final protective line (FPL) of fire or with a primary direction of fire (PDF). The PI)Fs are employed on covered avenues of approach such as gullies. The M60 FPLs should interlock across the unit front.

(3) The 90-mm recoilless rifles and Dragons are positioned with frontal cover to engage attacking armored vehicles from the flank.

(4) Riflemen are positioned to support and protect crew-served weapons. Riflemen are assigned FPLs that interlock at a point just beyond hand-grenade range.

(5) Dead space is covered with Claymore mines, antipersonnel mines, and M203 fires.

(6) Antitank mines, positioned to provide close-in protection from armored vehicles, are covered by LAW fires.

(7) Leaders plan indirect fire for long-range suppression of armored vehicles and short-range protection from dismounted assault. Mortar or artillery final protective fires (FPF) are placed on the most dangerous dismounted avenue of approach.

(8) Range cards are prepared for all crew-served weapons. Squad and platoon leaders prepare sector sketches and fire plans.

(9) The ranger force camouflages everything from ground and aerial observation.

(10) Leaders designate alternate and supplementary positions.

c. **Executing.** Leaders control the distribution and concentration of fires in their sectors. They constantly consider the overlap of fires with units to their left and right.

(1) Squads and crew-served weapons engage targets in sectors assigned by platoon leaders.

(2) Each weapon is used in its best role:

(a) Rifle, against enemy infantry within 300 meters.

(b) Machine gun, against enemy infantry within 1,000 meters.

(c) Grenade launcher, against infantry within 350 meters and armored vehicles within 200 meters.

(d) Light antitank weapon, against armored vehicles within 200 meters.

(e) Recoilless rifle, against armored vehicles within 400 meters.

(f) Dragon, against armored vehicles within 1,000 meters.

(3) Most dangerous targets are engaged first.

(4) Squads shift fires as needed or as directed by platoon leaders.

(5) The platoon masses fire on selected targets while maintaining coverage of the entire platoon sector.

(6) Platoon leaders control fires.

(7) After repelling an enemy assault, the platoon consolidates and reorganizes the defense.

6-6. Breakout from encirclement.

A breakout from encirclement is conducted when units, operating behind enemy lines, find themselves cut off from friendly forces and surrounded by superior forces. Given ranger unit missions, the chance of operating as an isolated force behind enemy lines is great. The breakout is characterized by determining enemy weak points, deception, massing of combat power, and a direct attack for a violent and timely breakout.

a. **Characteristics and fundamentals.** The breakout from encirclement involves the following characteristics and fundamentals:

(1) Set up a hasty defense. Adjust the perimeter as needed and make contact between subunits. Assign each a specific sector.

(2) Plan fire support. Plan FPFs on likely dismounted avenues of approach. Mass antiarmor weapons on the most dangerous mounted avenue of approach.

(3) Conduct reconnaissance. Identify enemy weak points. Locate enemy armored forces and the routes they must use. Locate enemy automatic weapons.

b. **Planning.**

(1) Organize all support personnel into maneuver elements--for example, company headquarters, mortars, and antitank. Organize the ranger force into a rupture force, reserve, support force, and a rear guard.

(2) Issue a fragmentary order. It should contain current local enemy situation and cover all pertinent areas of the five-paragraph field order.

(3) Select rally points.

c. **Execution.**

(1) Reduce the defense perimeter. This is done slowly until the unit is massed for the breakout attempt.

(2) Create a diversion. This is done either by fire from the rear guard force or by a limited attack in a direction other than that of the main breakout attack.

(3) Break through the defense with the rupture force. Hold the shoulders of the breakthrough.

(4) Move the reserve through to exploit the breakthrough.

(5) Move the support troops through with casualties. Have them collect casualties en route.

(6) Have the rear guard follow the support troops. It may have to fight a delaying action and use fire and movement to break contact with the enemy.

(7) Consolidate and reorganize following the breakout.

(8) Have the rupture force break contact with the enemy once the rear guard joins.

6-7. Linkup operations.

During operations behind enemy lines, ranger units may be required to link up with other friendly ground forces. Linkup may occur in airborne, air assault, amphibious, or shore-to-shore operations. It may occur during the breakout of an encircled force or during the convergence of separate forces. It may involve conventional or unconventional US forces, allied forces, or friendly guerrillas. Rangers may be required to link up with other rangers, partisans, or friendly ground forces that are conducting offensive operations.

a. **Planning.** Ranger commanders plan detailed linkup operations to avoid friendly forces firing on each other. Coordination and planning of communications, recognition signals, fires, command relationships and responsibilities, and control measures are essential. Coordination should be done before the start of the operation.

(1) Site. The linkup site should be easy to recognize, have cover and concealment, be located away from prominent terrain the enemy might use, be defensible, and provide multiple access and escape routes. An alternate site should be planned in case of compromise.

(2) **Communications and recognition signals.** Frequencies, call signs, codes, visual signals, and alternates for each of these should be planned before departure of friendly lines. Radios may be used to report the location of each unit to the site, and occupation and securing of the site. A system of mutual recognition must be devised to keep friendly patrols from firing on each other. This may include visual signals such as arm bands, panels, colored lights, or RSTA devices.

(3) **Fires.** The headquarters directing the linkup sets up fire control measures and priority of fires. The most commonly used control measure is the restrictive fire line (RFL), beyond which one force may not fire without coordination with another force. The RFL may be adjusted as two units draw closer together. Such an adjustment needs close coordination and should be planned before the operation. Successive phase lines between the two units can serve as on-order RFLS, provided that neither unit comes too close to the RFL before it is shifted.

(4) **Command relationships and responsibilities.** These are set up by the headquarters directing the linkup. Ideally, liaison personnel are exchanged before the operation. The key to success is liaison and guides.

b. **Execution.** Each member of the ranger force must understand the plan. Rehearsals are necessary in order to avoid confusion during the linkup. One uninformed ranger can start firing and cause enough confusion that friendly forces could be identified as enemy. Actions must be quick since units are open to enemy attack while they try to link up.

(1) **Moving units.** A linkup between moving units is hard. As the units move closer to each other, the chance of them engaging one another increases. Therefore, the linkup units must adjust their movements to each other and continuously coordinate on a designated secure radio net. If possible, one or both units should come to a short halt before linkup.

(2) **Stationary units.** Moving units must know the positions of the stationary units. They must keep stationary units advised of locations. Stationary units guide the moving units to the linkup point by radio. The stationary units must be ready to accept and position the moving units smoothly and quickly.

6-8. Passage of lines.

A passage of lines is an operation in which one unit moves through another unit that is stationary and disposed in a tactical formation on a FEBA. Ranger forces may conduct a passage of lines to get behind the enemy. Movement in forward unit areas must be controlled, coordinated, and kept to a minimum. This avoids conflict with friendly troops or the activation of RSTA devices. Ranger units treat the positions of forward units as danger areas. They must be assumed to be under enemy surveillance in all weather or visibility. The ranger force is open to attack during the passage. Personnel and units may be more concentrated than is desired. Fires of the stationary unit may be masked for a limited time. The passing ranger unit may not be in a good position to react to enemy contact. Detailed reconnaissance and coordination are crucial to ensure that the passage is conducted quickly and smoothly.

a. **Principles.**

(1) **Coordination.** The ranger force commander, or his representative, must coordinate departure and reentry of friendly front lines (FFL). This is done either directly with the units through which the rangers pass or through designated staff agencies. Items to be coordinated include fire support, passage lanes, control measures, guides, communications, intelligence, CSS, and actions on enemy contact.

(2) **Reconnaissance.** The ranger force commander should make a ground reconnaissance of the area through which the unit will pass. He should pick the initial rally point (IRP) at this time. The reconnaissance should include observing the area just forward of the friendly unit for routes, danger areas, and obstacles. During reconnaissance or coordination, rangers should wear the same uniform as the troops in the forward unit. This will avoid attracting the enemy's attention.

(3) **Security.** Maintain security to avoid contact with the enemy while departing or reentering. It is important that the ranger force make no enemy contact at this time. Control is hard if a fire fight starts just forward of an FFL.

b. **Departure of friendly front lines.**

(1) Designate an IRP.

(2) Maintain security. Use an appropriate movement formation when departing the FFL.

(3) Do not move within the friendly unit's area without a guide.

(4) Make final coordination with the friendly unit commander.

(5) Count the ranger force members as they depart.

(6) Make a security and listening halt.

c. **Departure of friendly units.**

(1) The ranger force arrives at the forward unit and is met by a guide. The guide leads the rangers to an assembly area chosen by the ranger commander during initial coordination with the forward unit commander.

(2) No one should move, either singly or as a unit, anywhere in the FFL without a guide.

(3) The ranger commander makes final coordination with the forward unit commander. He determines if any changes have taken place since initial coordination and learns of any recent enemy activity.

(4) Before leaving to coordinate, the ranger commander gives instructions (called a contingency plan) for what should be done while he is gone. These instructions state where he is going, who is going with him, how long he will be gone, what to do if he does not return, and actions to be taken if there is enemy contact.

(5) On the commander's return from final coordination, he may issue a fragmentary order (FRAGO) to cover any changes.

(6) Three common threats and techniques for countering them during departure from the FFL are:

(a) **Ambush and chance contact.**

- The ranger force commander must avoid enemy contact while departing.
- The ranger force goes to the friendly side of the departure point where a point team is dispatched to see that the area forward of the barriers is clear.
- The point element should check the area out to the first covered position or an area large enough to allow the unit room to maneuver if it is hit while departing.
- The distance varies with the size of the ranger force and the terrain.
- The point team notifies the ranger commander when the area is clear. The rest of the unit is brought forward.

(b) **Indirect fire.** If enemy indirect fire is falling, the ranger commander should not halt his unit after final coordination at the forward command post. Instead, he should move it quickly through the FFL. The point element should be far enough ahead of the main body to provide security from ambush or chance contact.

(c) **RSTA.** If the enemy has radar, sensors, or night vision devices, they may be offset by the following countermeasures:

- Use a well-hidden departure point, such as a reverse slope or a dense woods.

- Infiltrate the unit through the departure area and have the rangers rendezvous in a defilade.
- Pass through during rain, fog, snow, or darkness to help conceal the passage.
- Employ electronic countermeasures.

(7) Have a security and listening halt after the ranger force has moved out of sight and sound of the FFL. This is a short halt to see if enemy is in the area, and to accustom the unit to the sights and sounds of the battlefield. Halt in a position that covers the unit from chance friendly small-arms fire. If the security and listening halt must be in an exposed position, have each man lie flat on the ground.

(8) During movement, have frequent halts to see that the unit is not being followed and that no enemy is in the area.

d. **Reentry of friendly front lines.**

(1) Set up and occupy a reentry rally point (RRP).

(2) Maintain security at the RRP and at the reentry point.

(3) Use as few personnel as possible to locate the reentry point while the bulk of the unit remains in the RRP.

(4) Meet a guide at the reentry point. Normally, a forward-of FFL password is used since it may be overheard by enemy.

(5) Count the ranger force members to ensure no infiltrators follow the unit, especially during reduced visibility.

(6) Give the friendly unit commander a spot report, providing combat information of immediate tactical value.

e. **Reentry of friendly units.**

(1) The unit is moved into a rally point near the reentry point. This rally point should be on a prominent terrain feature where the ranger commander can pinpoint his location, with respect to the reentry point.

(2) The forward unit is alerted by radio that the ranger unit is ready to reenter. A code word is used for security and brevity. The code word must be acknowledged by the forward unit before the ranger force reconnoiters the reentry point. This shows that a guide has been sent to the reentry point and is waiting for the unit.

(3) If the ranger commander is certain of the reentry point location, he moves the entire unit directly to the reentry point. He has the point element coordinate the reentry.

(4) An effective method of locating the reentry point is through the use of a thermal imagery device. It detects the body beat of the reentry guide, even though he may be well camouflaged or his location masked by smoke. If ground surveillance radar is available, it can be used to vector the ranger unit to the reentry point.

(5) The reconnaissance party should never move parallel to friendly barriers or move around the wire. If the reentry point cannot be found by the initial reconnaissance or if the reconnaissance party finds only the barrier wire, the unit should notify higher headquarters and move to another rally point to wait until daylight. The unit should not stay in a rally point from which a radio transmission has been made.

(6) When the reentry point is found, the ranger commander can bring the unit forward. If the reentry point is easy to find, he can call by radio for the unit to come forward.

(7) The guide leads the unit through the barriers to the security position coordinated with the forward unit commander. The unit halts in the security position, remaining in movement formation. The ranger commander then gives a spot report to the forward unit commander. The ranger commander tells him anything of intelligence value or of immediate tactical use to the forward unit.

6-9. Reconnaissance platoon operations.

a. The reconnaissance platoon organic to the ranger regiment consists of a platoon leader, a senior reconnaissance sergeant, and three five-man reconnaissance teams that include a communications specialist in each team (see Appendix A). This platoon is responsible for short-duration reconnaissance of objectives for the ranger force commander. This reconnaissance is routinely in support of future strike or special light infantry operations. The teams can also be used by the ranger force commander to conduct deep target acquisition and designation, or to confirm or disprove information reported by other sources. Rarely will teams be used to attack enemy targets.

b. The reconnaissance teams of the ranger regiment do not replace (or perform the same reconnaissance tasks associated with) long-range reconnaissance patrol (LRRP) units or long-range surveillance units (LRSUs) that may be organic to the corps or division. These units are organized, trained, and equipped for long-term passive surveillance and reconnaissance in enemy territory. The ranger reconnaissance teams are trained and organized to support the immediate intelligence-gathering needs of the ranger force commander. They are not to act as a human intelligence (HUMINT) collection asset for other operational headquarters.

c. The reconnaissance platoon headquarters trains, equips, and controls the employment of the three reconnaissance teams. It operates the communications stations needed to receive reports from deployed teams. In the absence of the S2, it debriefs team members upon extraction or exfiltration. It can provide one or more teams to support each ranger battalion on independent operations. The reconnaissance teams can--

(1) Infiltrate the objective area by parachute (HALO, HAHO, or static lines), helicopter, fixed-wing aircraft, SCUBA, small boat, foot, or other means.

(2) Remain undetected in the vicinity of the objective area up to five days.

(3) Perform reconnaissance operations employing a full range of night observation devices, infrared detection devices, unattended sensors, and photographic equipment.

(4) Perform demolition target analysis.

(5) Operate small watercraft and inflatable boats.

(6) Emplace unattended ground sensors, omni-directional navigational beacons, hand-emplaced expendable jammers, and electronic target designation devices.

(7) Collect combat information to satisfy priority information requirements and mission-essential elements of information. Teams report that information by use of long-range, secure, burst-transmission communications equipment.

(8) Perform DZ selection, marking and reception duties.

(9) Report objective area weather conditions.

(10) Perform highly selective, limited attacks or ambushes when so tasked.

(11) Link up with the main body of the ranger force in the objective area, or escape and evade the enemy in order to return to friendly lines.

(12) Act as part of the AAT during airborne operations.

d. There are many limits of the reconnaissance teams.

(1) Mobility is normally restricted to foot movement in the area of operations.

(2) Use of radio and other active electronic or optical devices makes the teams open to enemy detection.

(3) All supplies and equipment needed by the team must be carried on the initial insertion, since resupply increases the probability of detection by the enemy.

(4) Conduct of insertion and extraction operations deep in enemy-held territory is difficult due to the need for secrecy, timeliness, security, and accuracy of location.

(5) Medical treatment of team casualties is limited to individual first aid. Casualty evacuation is difficult unless linkup with the main ranger force has been achieved.

e. The training of reconnaissance team members parallels that given to all members of ranger units with emphasis on infiltration, communications, and reconnaissance skills. Extensive training is also given in survival, evasion, resistance, and escape (SERE) and advanced first aid. Training emphasis is placed on actions at the reconnaissance site. The reconnaissance teams must conduct their mission and remain unseen. Every team member is trained to consider the need for information balanced by the likelihood of detection. Detection by the enemy compromises future operations.

6-10. Reconnaissance planning.

a. Detailed planning ensures that the reconnaissance mission is successful. It is also vital to the survival of the reconnaissance team. This planning is conducted by the ranger force planning group assisted by elements of the JTF staff. The supporting air or naval commander, navigator, and electronic warfare officer may aid the ranger force staff in the planning.

b. Upon receipt of the mission, the team selected begins an intense preparatory phase at the REMAB or the ISB. There the team receives its initial mission briefing. The items covered allow the team to complete its detailed planning. The reconnaissance platoon leader reviews the team plan. He ensures compatibility and integration into the ranger force commander's plan for employment of the main force. Staff specialists at ranger regiment and battalion--such as those concerned with intelligence, weather, or communications--provide help.

c. Planning for the use of reconnaissance teams is conducted within the framework of the ranger planning sequence for the entire mission. There are five planning considerations for the employment of reconnaissance teams:

(1) Operations security must be considered in all staff efforts to provide maximum protection for the reconnaissance teams. This includes intelligence, CE, logistics, administration, and maintenance. Operations security must be integrated throughout every ranger force operation. It consists of four main categories of security measures: signal security, physical security, information security, and deception. All are related and must be considered at the same time for each operation.

(2) Some missions may require rapid deployment to the operational area, thereby dictating the quickest method of insertion. However, others may depend on maintaining total secrecy, regardless of time.

(3) The transportation means selected for the delivery and recovery of the teams depends on the mission. This selection is also based on the abilities, limits, and availability of mission support platforms.

(4) The distance to and from the objective area must be considered, as well as the distance from the insertion point to the actual target. Reconnaissance teams can move great distances over rugged terrain. However, this movement does require time and limits the team's ability to carry bulky or sensitive equipment.

(5) Consideration must also be given to intelligence information available concerning the objective area which includes, but is not limited to--

(a) **Enemy situation.** The enemy threat--his abilities, disposition, security measures, detection, and air defense systems - affects the selection of the means for delivery or recovery.

(b) **Terrain.** Land formations must be considered in selecting the method of infiltration and exfiltration. Terrain affects the selection of altitudes, approach and exit routes, landing areas, drop zones, and beach landing sites. Air infiltration routes that provide terrain masking are desirable in static line parachute operations.

(c) **Weather.** Seasonal weather conditions affect team delivery and recovery. Factors to be considered are temperature, precipitation, visibility, clouds, and wind. If para-SCUBA techniques are used, high surface winds and their effect upon surf conditions or period of reduced visibility may preclude the use of parachutes, inflatable boats, or swimming as entry or recovery techniques. These same conditions favor land infiltration and exfiltration, The AWADS reduces the impact of visibility as a limiting factor for air infiltrations.

(d) **Astronomical conditions.** Periods of sunrise and sunset, moon phase, moonrise and moonset, and twilight must be considered.

d. A reconnaissance team mission must be specific and support the mission. Information concerning the target may not be gathered because the risk of detection by the enemy is so great. Detection may preclude the conduct of the planned operation. If more than one mission is assigned, priorities are set, All team missions must be coordinated to avoid other friendly forces operating near the target. Ranger forces conduct operations at such depth behind enemy lines that no other US or allied forces are normally present. There is a chance that a friendly indigenous guerrilla force may be operating near the objective, but ranger reconnaissance teams do not normally make contact with them.

e. The reconnaissance platoon leader joins with the intelligence and operations sections of the ranger force headquarters in the initial planning for reconnaissance team operations. The method of operation while moving, communication procedures, reporting, surveillance procedures, and actions on enemy contact should be included in the platoon's SOPS.

f. The ranger force intelligence officer normally prepares the reconnaissance plan IAW the guidance provided by the controlling headquarters. He is helped by the operations officer and the reconnaissance platoon leader. Selected team members and a member of the unit providing the insertion and extraction means are briefed on the mission early in the planning phase. They should participate in the detailed planning to follow. The essential details of a reconnaissance plan normally include--

(1) Areas to be observed and recommended areas or positions from which this can be done. These positions are determined in advance of the insertions of teams. They are based on a study of terrain, road and rail nets, enemy order of battle, delivery means, operations plans of the controlling headquarters, and the desires of the ranger force commander.

(2) Load plans and procedures. This includes both insertion and extraction. Alternate sites are chosen for possible use. Primary and alternate DZs are selected, if applicable.

(3) The flight plan. This includes insertion and extraction flight routes and alternate routes.

(4) The movement plan to and from a patrol base. This includes the times, routes,. and destinations when individuals or small groups are required to operate away from the patrol base.

(5) The fire support plan. This includes the location and times of planned air strikes in support of the main body, and all planned fires by the main assault force. The reconnaissance teams should know the frequencies and call signs for tactical air support.

(6) The timing and ground tactical plan for execution of the main force operation. This includes disposition of the main force, special mission teams from the ranger force, or any other special operations force elements.

(7) The communications plan. This includes frequencies, reporting schedule, emergency reporting procedures, and alternate communications plans. It also includes the ground-to-air

communications plan and the friendly recognition signal to be used during linkup with the main assault force or other friendly forces.

(8) Plans for use of technical specialists or special equipment.

(9) The plan for the treatment and evacuation of sick or wounded team members from the objective area.

(10) The E&E plan. This includes routes out of the objective area, passwords, and alternate extraction zones or methods far from the objective area.

(11) Mission-specific elements of information desired by the ranger force commander, and any special or on-order missions the team must conduct.

g. The team leader uses specific steps in planning, preparing, and executing missions. These procedures are comprehensive, yet adaptable to any situation. Three of the most important steps are:

(1) **Issuance of warning order.** This action gives the team maximum time to prepare for the mission. It should be issued as soon as the team leader has been notified of a mission and updated after he has made his tentative plan.

(2) **Issuance of OPORD.** The team leader conveys the information and instructions needed to accomplish the mission. He requires a briefback from each team member.

(3) **Rehearsals.**

h. The specific steps used when planning missions are as follows:

Step 1. Study the mission.

Step 2. Plan use of time.

Step 3. Issue initial warning order.

Step 4. Study terrain and situation.

Step 5. Make tentative plan.

Step 6. Organize.

Step 7. Update warning order.

Step 8. Coordinate (continuous throughout).

Step 9. Consult aerial photographs and map updates.

Step 10. Complete detailed plan. (To be combined with paragraphs 3, 4, and 5 of the operation order.)

Step 11. Issue operation order.

Step 12. Supervise (throughout), inspect, rehearse.

Step 13. Execute mission.

i. A workable E&E plan must be developed. Such a plan enhances the survival of the reconnaissance team members. It ensures that they are psychologically prepared to conduct the mission, by knowing there is a system and plan for returning them to friendly lines. The E&E plan must include a section describing the actions of the reconnaissance team members should their insertion craft be disabled en route to the objective area. It should also include what to do if the team is unable to link up with the main ranger force, or if the main ranger force has not been inserted into the objective area.

j. Rehearsals and briefbacks are best for finding flaws in procedures or planning. Thorough coordination of all procedures is essential. The exact type of aircraft or other means that will be used for insertion should be used for training. Rehearsals should be conducted under terrain, astronomical, hydrographical, and meteorological conditions close to those to be met on the operation. The team leader conducts a briefback with the platoon leader. This lets the platoon leader decide if the team is ready for the mission. It also allows him to make changes.

6-11. Execution of insertion.

a. Reconnaissance teams normally conduct blind insertions. Blind insertions are amphibious, heliborne, or parachute drops onto unmarked DZs without help from a DZ-marking party or a CCT. Blind insertions often depend on favorable light and weather conditions in the objective area. Favorable conditions for the reconnaissance team often means low-lying cloud cover, fog, and darkness. The teams are able to operate during reduced visibility by the extensive use of night observation devices. The team's extensive training and high level of land navigation skills allow it to quickly traverse rugged terrain and not be detected.

b. Team insertion into the objective area often needs a cover and deception plan. Team members can conduct HALO and HAHO operations. Teams can be inserted into areas not normally acceptable as a DZ for conventionally trained parachutists. By the use of HAHO techniques, reconnaissance teams can assemble in the air and glide to a landing at a location far from the drop point.

c. If reconnaissance teams need special or bulky equipment that cannot be carried on the individual team member, it can be airdropped, hidden near the DZ, and retrieved later. Such container drops are often used during low-level static line insertions. The high-altitude airdrop resupply system (HAARS) can be used to deliver such items at the same time as reconnaissance teams are inserted by HALO.

CHAPTER 7

Combat Support

Combat support is used to enhance the combat power of the maneuver companies. Combat support elements may be organic or nonorganic. They can conduct a variety of missions in providing indirect and direct fire support, security, intelligence, and maneuver and fire control.

7-1. Intelligence support.

a. The success or failure of a ranger mission often depends on the accuracy, detail, and timeliness of intelligence information. Both deliberate and quick-response missions need specific information that can be obtained only by a multidiscipline collection and analysis agency. An active interface with all levels of the intelligence system helps get the needed information quickly.

b. The theater or JTF commander is responsible for providing intelligence support. The 1st SOCOM also provides information and analyzed intelligence to the ranger force commander. The main source of processed intelligence is the Echelons Above Corps Intelligence Center (EACIC), similar to the division level military intelligence (MI) all-source production center (ASPC). The EACIC provides intelligence, security, and electronic warfare (EW) interfaces (see Figure 7-1). It is through the EACIC that ranger commanders can request support from national, joint, combined, or CONUS intelligence analysis systems.

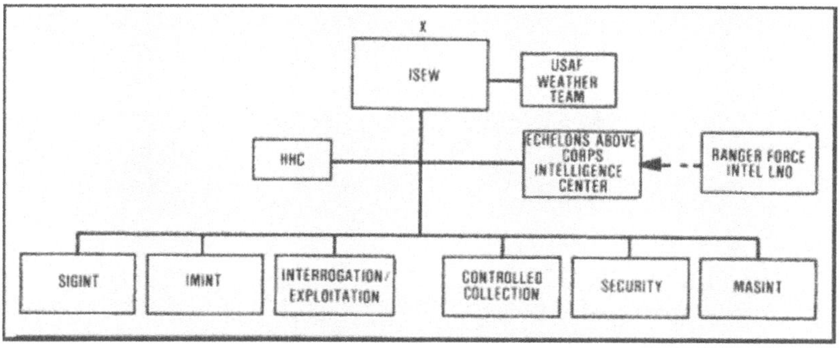

Figure 7-1. Intelligence at echelons above corps.

c. Intelligence requirements for ranger operations are unique, highly sensitive, and compartmented. The intelligence officer of the 1st SOCOM and the appropriate 1st SOCOM regional planning cell coordinate with theater intelligence agencies to provide needed information.

d. The ranger regimental staff and the ranger battalion staff have personnel in their intelligence sections to correlate and analyze information from many sources.

e. The reconnaissance platoon, organic to the ranger regiment, maintains surveillance of the objective area before the insertion of the main ranger force. It reports any last-minute developments that would affect the ground tactical plan. The platoon can also be tasked to determine any mission-specific items of essential information that the ranger force commander needs to complete his plan. This platoon is a highly specialized unit that is reserved for use by the ranger force commander. It is not a HUMINT collection agency for use by the theater or JTF commander. Reconnaissance teams normally link up with the ranger force in the objective area and are extracted with the main body. If this linkup cannot be made, they will exfiltrate enemy territory. They may use escape and evasion nets set up by other special operations forces or Department of Defense (DOD).

f. The ranger regiment normally places a liaison team at the headquarters of the command having OPCOM of a ranger force. This team provides command and control communications and liaison. It puts an intelligence liaison officer (LNO) at the EACIC or the appropriate ASPC (see Figure 7-2). This provides coordination to ensure that the intelligence needs of the ranger force are relayed to the intelligence processing center, and that the analysis is on the specific needs of the ranger force commander. This LNO coordinates with another located at the corps or EAC targeting center. Close coordination between the regiment, the intelligence analysis center, and the targeting center is essential to effective employment.

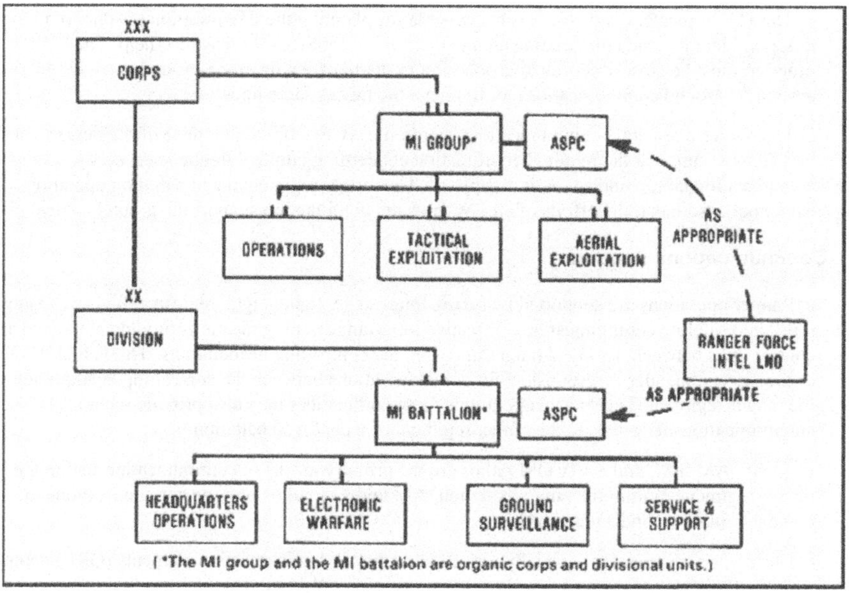

Figure 7-2. Intelligence source at corps and division.

7-2. Electronic warfare support.

a. Depending on the nature of the target and enemy abilities, a ranger mission may need support from EW units. This is mainly true during insertion. Electronic warfare operations disrupt or destroy the enemy's command and control of his forces and weapons systems, and retain friendly use of the electromagnetic spectrum. Electronic warfare is also used to support deception operations. They mislead the enemy by manipulation, distortion, or falsification of indicators, and persuade him to react in a manner prejudicial to his own interests.

b. Active jamming and chaff dispersal can prevent enemy early-warning radar from detecting the airborne force or determining its route. False transmissions can aid deception plans or feints used to help the ranger force's entry behind enemy lines. Selective jamming and imitative deception can disrupt the enemy command and control nets. This prevents the enemy from reacting to the presence of the ranger force in time to prevent mission accomplishment, or slows and interrupts his deployment of reaction forces.

c. Ranger elements, or other military forces, may conduct closely coordinated direct attacks against enemy EW sites during a ranger force's insertion or extraction. Not only does this reduce the enemy's ability to direct EW against the ranger force command, control, and communication system, but it misleads and confuses the enemy as to the true ranger objective.

d. The JTF or theater commander is responsible for planning the EW program to support a ranger mission. Planning and coordinating elements from the 1st SOCOM may also help. The EW program must be closely coordinated so as not to disrupt friendly air-to-ground or ground-to-ground communications, or to reveal the ranger force intentions.

e. The ranger regimental communications-electronics (CE) officer and those of the ranger battalions ensure that the required coordination concerning jammed frequencies, codes, authentication tables, and so on, is completed. The ranger regimental and battalion operations personnel ensure that the efforts of all EW units are combined to support the ground tactical plan.

7-3. Communications support.

a. Ranger operations are supported by secure long-range, lightweight, real-time, high-frequency (HF), and satellite communications. Effective long-range communications provide command and control links between deployed ranger units and the controlling headquarters. The JTF, ARFOR, or theater commander is responsible for communications between the controlling headquarters and the ranger regiment. The special operations force commander may also provide secure communications terminals to the ranger regiment or a deployed battalion.

b. Secure AM, FM, and SATCOM radios are the primary means of communication within the ranger regiment. Within the ranger battalion, AM and FM radios provide communications to company, platoon, and squad level.

c. During certain missions, specially trained and equipped quickreaction elements (QRE) deploy with the ranger force to provide secure communication links to the special operations force commander. These JCSE elements operate on either SATCOM or tactical satellite terminal (TACSAT) channels or through an airborne communications relay platform (see Figure 7-3). Depending on the mission, an airborne battlefield command and control center (ABCCC) aircraft or a joint airborne communications center/command post (JACC/CP) may be used. Their communications systems can operate at all levels of the national chain of command to permit a quick response to the tasking authority.

Figure 7-3. Airborne communications relay platform.

d. When only one ranger battalion is employed and the regimental headquarters is not the controlling headquarters, the ranger regiment provides a liaison cell to the controlling higher headquarters. This liaison cell includes a communications element from the regimental communications platoon. It can provide secure SATCOM, teletype, and facsimile transmission support to the ranger force in the objective area.

e. If two or more ranger battalions are employed, the ranger regimental headquarters deploys and acts as the command and control headquarters. The regimental communications platoon would then provide another communications link to the special operations force commander.

f. The ranger regimental CE officer ensures that the necessary communications links are set up and coordinated. The many communication means and channels available provide for effective control of a deployed ranger force. However, communication means must be closely coordinated at all levels of command to control the complex operations of a ranger mission. Planning and coordination with supporting aviation, transport, fire support, medical, and logistical elements before an operation are vital to efficient communications. The regimental CE officer must consider the communication systems linking the ranger force and other services. Air-to-ground and ship-to-shore communications are vital and must be set up early in ranger operations.

7-4. Fire support.

a. **Planning.**

(1) The ranger regiment has limited organic fire support assets. The six 60-mm mortars in each ranger battalion (two in each rifle company) provide the only organic indirect fire

available to the battalion commander. Each rifle company also has three 90-mm recoilless rifles and three Dragon missile systems that provide a small measure of organic direct fire support.

(2) The missions assigned to the ranger regiment and its subelements normally require fire support from organizations outside of the regiment. To plan, coordinate, and control these fires, the ranger regiment is organized with personnel dedicated to these tasks.

(3) The ranger regimental headquarters has a fire support element consisting of one fire support officer (FSO), two noncommissioned officers, and two fire support specialists. They prepare the fire support annex to the regiment's operation order. They ensure that all available fire support is planned to best support the regimental commander's ground tactical plan. They act as the primary coordinator for all external fire support, such as artillery, armed helicopter, close air support, naval gunfire, and short-range air defense. They also coordinate with each ranger battalion FSO to ensure unity of effort. The fire support element coordinates the efforts of air force, naval, and marine liaison officers when they are attached.

(4) The regimental commander may assign part of the fire support element to a committed ranger battalion as an attachment to help in the integration and planning of fires. The ranger battalions request through the fire support element that more or different fire support means be integrated into the ground tactical plan.

(5) The ranger battalion headquarters has a fire support team headquarters consisting of the battalion's FSO, two noncommissioned officers, and two fire support specialists. The FSO is the prime fire support coordinator for the ranger battalion commander. He maintains coordination with the regimental FSO and any attached fire support coordination elements, such as the USAF tactical air control party or naval shore fire control party. The FSO performs the following:

(a) Advises the ranger battalion commander on all fire support matters.

(b) Recommends allocation of fire support.

(c) Prepares fire support plans.

(d) Assigns target numbers.

(e) Processes target lists from the ranger company fire support teams to eliminate duplication.

(f) Monitors and functions as net control on the ranger battalion fire support coordination (FSCOORD) net.

(g) Reports changes in the status of fire support units to the ranger battalion commander, staff, and FISTS.

(6) When the ranger battalion is operating a single TOC, the FSO and his element are normally located there. When the ranger battalion is operating two TOCS, one TOC is normally manned by the FSO, the fire support sergeant, and a fire support specialist. The

senior fire support sergeant and the other fire support specialist would then monitor the FSCOORD net at the other TOC. The FSO ensures that any attached fire support elements also divide when needed to operate in a two-TOC configuration.

(7) The ranger battalion fire support team headquarters has three fire support teams assigned to it. These are normally allocated one to each of the ranger rifle companies. The company FSO and his team--

(a) Locate targets and request and adjust surface-to-surface fire support (mortar, field artillery, and naval gunfire).

(b) Plan fires to support the company ground tactical plan and prepare target lists.

(c) Coordinate fire support requests through the FSO for surface-to-surface and air-to-surface fires.

(d) Report battlefield information.

(e) Control air-to-surface fires of helicopters and prepare to assume control of fires from fixed-wing aircraft.

(f) Coordinate airspace use within the ranger rifle company operational area.

(g) Advise the ranger company commander on all fire support capabilities, limitations, and methods of employment.

(h) Inform all fire support units of target priority changes.

(i) Designate targets for laser-guided munitions.

(j) Assume operational control of, and administrative responsibility for, all augmentation fire support personnel (TACP, shore fire control party, and attack helicopter liaison teams).

(8) The company FSO is normally with the ranger rifle company commander. The FOs and their RATELOs are normally with the ranger rifle platoon leaders.

b. **Coordination.**

(1) Coordination of fires is a continuous process. It involves implementing the fire support plans of the ranger force commander and reacting to changing tactical situations. The success of a mission often hinges on how well fires are coordinated during the operation.

(2) The fire support planners within the many echelons of the ranger force become fire support coordinators during the actual execution of the operation. They use standard fire control coordination measures.

c. **Organic fire support.**

(1) **Indirect fires.**

(a) The 60-mortars constitute the only organic indirect fire support available within the ranger regiment. The mortar squads in each ranger rifle company normally operate as a section under the control of the mortar section sergeant. However, they can operate alone for short periods.

(b) The 60-mm mortar is used to destroy enemy equipment and light vehicles and to attack enemy personnel. Its main advantage is the ease with which it can be brought into operation against the enemy. This ease of employment allows the ranger force commander to concentrate combat power at the critical time and place. This helps to surprise and destroy the enemy.

(c) Mortar fires are employed by the ranger rifle company commander and coordinated by the company FSO. They are normally requested and adjusted by the FIST observers. The mortars are normally emplaced in a central position, and their fires are controlled by the fire direction center (FDC) using the indirect-fire method. The mortars may also use the direct-alignment or the direct-lay method.

(2) **Direct fires.**

(a) Each ranger battalion is authorized 90-mm recoilless rifles and Dragon missile launchers. These weapons constitute the heaviest organic direct-fire ability in the regiment. The antitank section in each ranger rifle company may operate as a complete section, but normally individual weapons are assigned throughout the company sector wherever their fires can best support the mission. These weapons can destroy most enemy armored vehicles and can penetrate the flank and rear armor of all enemy main battle tanks. The 90-mm recoilless rifle has a multiple projectile round that is effective against exposed enemy personnel. The HE round is highly effective against buildings and bunkers.

(b) The main disadvantage of direct-fire antiarmor weapons is their weight. These weapons, and their ammunition, are man-portable but place a strain on the unit when moving across rugged terrain.

(c) The ranger battalion is trained in the use of light antiarmor weapons, such as the US M72A2 LAW and the Warsaw Pact rocket-propelled grenade (RPG). The following provide protection for the ranger force against attack by enemy armored forces:

- The employment of light antiarmor weapons at all organizational levels.
- The placement of the 90-mm recoilless rifles or guided-missile systems on the most likely avenues of enemy-armor approach.
- An aggressive, offensively oriented employment.
- The ability to move into and across terrain impassable to armored vehicles.

d. **Field artillery fire support.**

(1) Many ranger unit operations are out of the range of supporting field artillery fires. Whenever such fires are available, they are planned for and integrated into the ranger force ground tactical plan.

(2) Coordination and exchange of call signs, frequencies, and target lists occur before insertion of the ranger force. Unless required by the tactical situation, ranger units do not normally receive augmentation forward observers from supporting field artillery units. The degree of mental and physical training necessary to execute certain ranger techniques requires more preparation than field artillery units are able to provide. The FIST from the ranger battalion requests and adjusts field artillery fires for the ranger force.

(3) Field artillery fires can be used to support the ranger force even if the objective area is out of range. Field artillery cannon fire and multiple rocket launch system (MRLS) fire can be used to suppress enemy air defenses. This helps the ranger force as it crosses the FEBA during airborne or air assault operations.

(4) Field artillery fires can be used to support the exflltration of ranger elements as they approach friendly lines. These fires can contribute to the deception plan and add combat power to feints used to support ranger operations.

(5) When in range of the objective area, field artillery units can be used to emplace field artillery delivered FASCAM to enhance the security of the ranger force.

(6) Copperhead rounds fired by 155-mm field artillery units can be terminally guided by the ranger FIST forward observers (FOs). They can attack hardened point targets or enemy armored vehicles by using a man-portable laser target designator (see Figure 7-4).

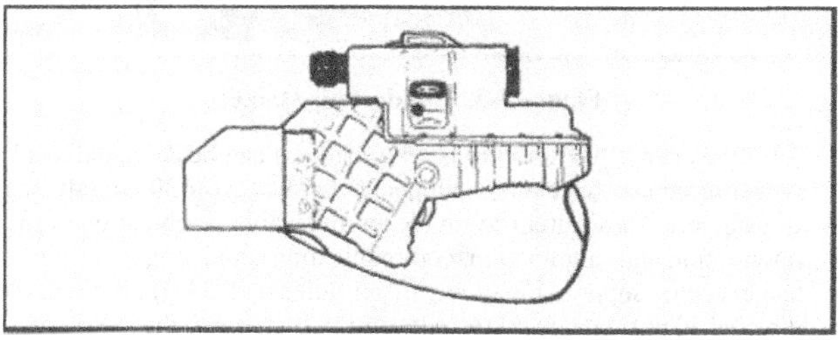

Figure 7-4. AN/PAQ-1 laser target designator.

e. **Aerial fires.**

(1) Aerial fire support is usually the prime means supporting the ranger force due to the distance behind enemy lines at which most ranger operations take place. Aerial fire support can be provided by either fixed-wing aircraft or helicopters.

(a) Fixed-wing aerial fire support may come from USAF, USN, or USMC units. The type of unit providing support, the aircraft, and the mix of ordnance carried all affect the fire support planning and coordination process. Some aircraft have a night and all-weather strike ability enabling them to support the ranger force during any level of visibility. Operations during poor weather that limits visibility to less than 3 nautical miles are still somewhat restricted. The ranger fire support coordinators must ensure that the correct aircraft are requested and employed effectively on the enemy. The

TACP directs and adjusts aerial fires in the objective area.

(b) The ranger force can use ground laser target designators to pinpoint targets for air strikes, as well as electronic navigation aids to permit nonvisual air strikes (beacon bombing). The ranger rifle company FIST or the TACP can control a laser-designated standoff air strike (see Figure 7-5).

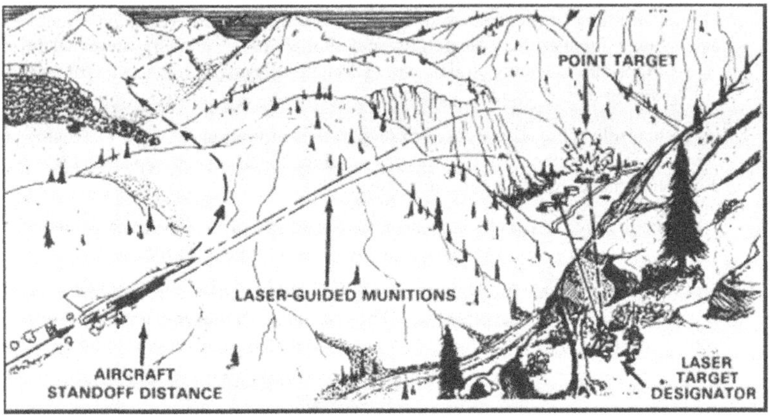

Figure 7-5. Standoff air strike.

(c) If the enemy ADA ability is not great or it can be degraded to a low level, the ranger force uses specially equipped and armed AC-130 aircraft (see Figure 7-6) for fire support. These aircraft provide an invaluable combination of firepower, night observation and illumination, communications, and long loiter time. A well-planned and executed suppression of enemy air defense (SEAD) program, coupled with ECM directed against enemy ADA units, normally permits the use of AC-130 aircraft.

Figure 7-6. AC-130 aircraft.

(2) The attack helicopter, armed with a mix of antitank guided missiles (ATGM), 2.75-inch rockets, 20-mm cannon, and 40mm grenade launchers, is accurate and responsive aerial fire support. The ranger force commander plans to use all sources of aerial fire to help accomplish his mission. However, the nature of ranger operations may preclude the extensive use of armed helicopters due to their limitations. Some attack helicopters are limited in range and lack an all-weather ability. They are restricted during operations at night by a lack of sophisticated night vision devices. They may lack the speed to go with the ranger force on some heliborne insertions. Attack helicopters may be used to escort and assist the ranger force as it crosses the FEBA. They may also be used to conduct feints and demonstrations to cover the insertion of the ranger force. As the MV-22 Osprey and the AH-64 Apache enter service, more use will be made of these aircraft to support ranger operations beyond the FERA, (See Figures 7-7 and 7-8).

Figure 7-7. MV-22 Osprey.

Figure 7-8. AH-64 Apache attack helicopter armed with Hellfire missiles.

(a) If attack helicopters are used to support a ranger operation, planned indirect fires are normally delivered along entry and exit corridors. Attack helicopters approach and depart the objective area using nap-of-the-earth flight profiles.

(b) Fires from armed helicopters are normally requested and controlled by the company FSO or one of his FOs, operating on a special ground-to-air net. The laser target designator may be used to precisely identify targets for the AH-64 Apache. Friendly unit locations may be marked by smoke, panels, lights, mirrors, or infrared sources.

f. **Naval gunfire.**

(1) During amphibious assaults or operations near a coastline, the ranger force may receive indirect fire support from naval gunfire. Normally, a ranger battalion would be supported by either a destroyer or a cruiser in a direct support role. However, the type and importance of the mission, the type of targets, the ships available, the hydrographic conditions, and the enemy capability determine how many and which type of ships are provided to support the ranger force.

(2) Naval gunfire is characterized by large volumes of highly destructive, flat-trajectory fire. Planned strikes in support of the ranger force may also include surface-to-surface missile fire. Some naval guns can fire a laser-guided projectile much like the Copperhead. This projectile, called a Deadeye round, is terminally guided the same as the Copperhead. Deadeyes can be guided either by an attached spotting party from the ANGLICO or by ranger LTD teams.

(3) When ranger units are being supported by naval gunfire, a shore fire control party is attached to the ranger force, This party normally consists of a liaison team and several spotter teams (forward observers). The liaison team is integrated into the operations of the fire support element (FSE) at the ranger battalion TOC. The spotter teams are attached to the ranger rifle companies. The shore fire control party LNO is the ship's representative to the ranger force commander, through the FSO. The spotter teams request and adjust fires from surface vessels. They can also request and control air strikes by carrier-based aircraft.

(4) The LNO and the spotter teams operate in the ground spot net, communicating with the ship by high frequency (HF) radio to request and adjust naval gunfire. The spotter teams communicate with the LNO using very high frequency (VHF) radios. The LNO also can communicate with aircraft using ultrahigh frequency (UHF) radios.

(5) Coordination and control measures that apply to naval gunfire are the same as for field artillery, with the addition of two terms, which are:

(a) Fire support area (FSA)--a sea area within which a ship may position or cruise while firing in support. It is labeled as FSA and numbered by a Roman numeral - for example, FSA VII.

(b) Fire support station (FSS)--a specified position at sea from which a ship must fire. This is very restrictive positioning guidance. It is labeled as FSS with a Roman numeral--for example, FSS VII.

g. **Air defense artillery.**

(1) Army air defense artillery units are rarely used to directly support the ranger force since

ranger operations normally take place deep behind enemy lines. However, it is possible that during the insertion or extraction phase of an operation, ADA units may support the ranger force as it crosses the FEBA.

(2) The ranger regiment's fire coordination element and the FSO in the ranger battalion are mainly concerned with adequate coordination of ADA fires. This prevents friendly units from being engaged. It allows the integration of ADA fires into the scheme of maneuver. The fire support planners ensure that enough coordination has been made so that ADA units are aware of the ranger force and know enough about its flight paths, routes, and altitudes to prevent engagement by friendly units. When conducting this coordination, care must be taken to pass only needed information to the ADA unit. To aid coordination, an ADA officer may be attached to the ranger force.

(3) The plan for a heliborne or airborne insertion and extraction of the ranger force must include the correct identification-friend-or-foe (IFF) codes, the flight profiles, the times of FEBA crossings, and the estimated number of aircraft making the crossing. There must also be a plan for crossing by straggler aircraft and battle-damaged aircraft that cannot reply to IFF interrogation. If possible, the routes selected should avoid concentrations of friendly ADA units.

(4) During an aerial crossing of the FEBA by a ranger force, the regimental headquarters may assign a liaison officer to the headquarters controlling the ADA fires in the crossing area. The LNO is then responsible for close and continual coordination. This ensures that the crossing is made without friendly ADA units engaging the ranger force while it is exiting or entering friendly lines.

(5) During ranger operations, the primary source of air defense fires is shoulder-fired antiaircraft missiles carried by specially trained individual rangers. These weapons provide the ranger force commander a limited defense against attack by enemy aircraft. The decision to take these weapons into the objective area depends on the factors of METT-T. Ranger operations should be completed before the enemy can react with ground forces. However, enemy aircraft may be able to react in time to threaten the ranger force. These lightweight missiles offer a highly effective, short-range counter to enemy air attack.

(6) The ranger force commander must define the weapons control status and rules of engagement for all short-range air defense (SHORAD) teams under his control. Priorities of target engagement must also be determined to prevent multiple engagements of a single enemy aircraft. The ranger battalion FSO and the attached ADA officer are responsible for the coordination of ADA fires within the objective area. Standard ADA weapons-control status and rules of engagement may be modified by the ranger force command.

7-5. Aviation support.

a. **Planning.**

(1) The ranger regiment may need extensive aviation support on some missions. The ranger battalion is organized and trained to move quickly to the objective area by air and to be resupplied by airdrop for short periods.

(2) Often the decision to execute a mission depends on the amount and type of available insertion and extraction aviation support. Normally, the theater commander or JTF commander allocates the type and amount of aviation assets to support a ranger operation. The theater or JTF commander's staff is responsible for the planning and coordination concerning aviation support. The ranger force commander must play an active part to ensure mission success.

(3) Due to the extensive training requirements and specialized operational techniques needed to support special operations, special air operations units in the Army and the USAF have been equipped and trained for the mission. These forces train with the ranger regiment and jointly develop tactics, techniques, and procedures of employment. These special units operate diverse aircraft that have unique capabilities. These help the ranger force commander to effectively employ his units. Command of these special units is maintained at a very high level in the chain of command. They are assigned to support a specific operation. Then they return to their parent unit's control.

b. **Coordination.** In addition to the staff of the theater commander or JTF commander, the coordination of aviation support occurs at all levels in the ranger force. These operations often are conducted in phases. A separate air mission commander is then designated. The AMC is responsible for the completion of each phase of the aviation mission, as well as the close and continuous coordination with the ranger force, commander. The ranger regimental air operations officer and the movements control officer are responsible for the planning, integrating, and controlling of aviation support.

c. **US Air Force.**

(1) The USAF provides special aviation support to the ranger regiment and other elements of the special operations force. The USAF has elements that are equipped with MC-130 Combat Talon aircraft (see Figure 7-9) and AC-130 Spectre gunships, as well as HH-53 Pave Low, UH-60, and UH-1 helicopters. These aircraft are specially equipped to conduct long-range infiltration, exfiltration, attack, and search-and rescue operations. The USAF also provides C-141B and C-130 aircraft for strategic and tactical airlift of the ranger force. There are specially trained USAF crews that can operate using AWADS and SKE. This gives them an excellent means of inserting the ranger force during reduced visibility.

Figure 7-9. MC-130 Combat Talon aircraft.

(2) The AFFOR commander provides the airlift control element (ALCE) support at the departure airfield. The ranger air operations officer ensures that coordination of aircraft load limits, air movement times, specific load plans, and manifests is conducted in an effective and timely manner. The air operations officer and the movements control officer ensure that a departure airfield control officer (DACO) from the deploying unit is designated to provide face-to-face coordination at the departure airfield.

(3) The USAF normally provides an airborne battle command and control craft. This is a specially equipped C-130-type aircraft that provides communications and data system down-links to ground stations.

(4) The terminal guidance needed for assault landings and air traffic control in the objective area is provided by a specially trained and equipped CCT. This element has habitual training and operations relationships with the ranger units. The USAF is also responsible for the specially trained and equipped weather teams that can be sent into the objective area with the CCT.

(5) The USAF Tactical Air Command (TAC) includes fighter, interceptor, ground attack, electronic countermeasure, early warning, reconnaissance, and airborne command and control squadrons, any of which may be used to support the ranger force. The Tactical Air Command provides the results of aerial reconnaissance to the AFFOR, who then processes it for use by the theater or JTF commander's staff. This reconnaissance may be by low-level photo, imaging infrared, side-looking airborne radar (SLAR), low-light-level television, or overhead photography.

(6) The Military Airlift Command (MAC) equips and trains special long-range SAR aircraft and crews. These aircraft, mainly helicopters, are equipped with sophisticated navigation, observation, and communications equipment, as well as extended-range fuel tanks and in-flight refueling capabilities (see Figure 7-10). The ranger regiment uses these aircraft for certain missions.

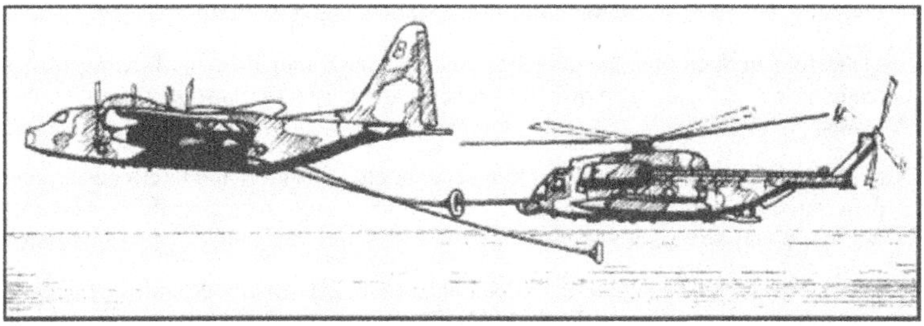

Figure 7-10. In-flight refueling.

d. **US Navy and US Marine Corps.** Both the USN and the USMC have units equipped and trained to support ground forces. The type of aviation support normally provided to the ranger regiment is close air support against targets in the objective area, or as suppression missions against enemy air defense installations. Long-range helicopters from the USMC may be used to support an amphibious operation or to insert special teams during ranger operations. The USN also has a search-and-rescue ability that may be used to support the ranger regiment.

e. **Special operations aviation units.**

(1) The US Army has specially equipped and trained SOA units tasked to support the ranger regiment, as well as other elements of the special operations force. These SOA units have a habitual training association with the ranger regiment and use dedicated aviation assets to -

(a) Insert, extract, and resupply the special operations force.

(b) Conduct armed escort, reconnaissance, surveillance, and electronic warfare in support of the special operations force missions.

(c) Provide airborne command, control, and communications for special operations force elements.

(d) Provide general support aviation during peacetime and contingency operations.

(2) Penetration of enemy airspace is normally done during limited visibility at very low altitudes. Penetrations are planned to maximize the use of existing terrain and exploit blind spots in enemy air defenses. On-board navigational aids are crucial to mission success. There normally is not any terminal guidance to the selected LZ or DZ. Deception operations of a tactical or operational nature help SOA units penetrate to the objective and reenter friendly lines. Advanced helicopters and short-takeoff-and-landing aircraft provide the help needed to do these missions.

(3) To survive enemy air defense systems and prevent mission compromise, SOA units avoid detection. Detected aircraft may reveal the location of supported special operations force ground forces or may be destroyed by enemy ADA fires. The SOA normally avoids detection by operating at night without illumination. This condition requires flight using sophisticated night vision devices. The SOA also operates in other conditions of reduced

visibility, such as clouds, fog, rain, snow, smoke, and dust storms.

(4) Before an operation, the ranger force commander may need to place reconnaissance teams near the objective. Army SOA assets can be used to transport ranger reconnaissance teams.

(5) Elements of SOA provide the ranger regiment with short and intermediate range penetration, resupply, and extraction capabilities. Long-range aircraft must be provided from USAF special operations assets.

(6) The ranger regimental or battalion commander is normally provided a liaison aircraft while the ranger force is in the REMAB. This ensures that he can travel quickly to the dispersed headquarters of the agencies providing support for the mission. This aircraft and crew are provided by the theater or JTF commander from EAC general support aviation assets. It may be a helicopter or fixed-wing aircraft, depending on the location of the REMAB to a suitable landing field.

(7) Army helicopters and fixed-wing aircraft may be used to provide SEMA aerial reconnaissance; radar, infrared and photographic imagery; and electronic intelligence support to the ranger force commander. This support is normally provided by the aerial exploitation battalion of the theater Army military intelligence group.

7-6. Engineer support.

a. The ranger regimental and battalion operations officers plan and request engineer support for ranger operations.

b. The most common engineer-type tasks needed during a ranger operation are the destruction of enemy facilities, equipment, and material by use of conventional explosives; and the interdiction of major lines of communication. To reduce the need for external support, the ranger regiment trains selected individuals in advanced demolition techniques. All members of the ranger regiment are well trained in the basic employment of conventional demolitions. Therefore, there is no need to attach combat engineers to conduct normal demolition missions. If specialized skill or knowledge is needed for a specific mission, qualified personnel can be attached.

c. During the occupation of the REMAB and the rehearsal for an operation, the ranger regiment needs construction engineer support to build simulated target areas. The simulated target areas should be camouflaged to prevent the enemy from recognizing the target. The construction engineers augment the ranger unit's self-help construction efforts. Limited-facilities construction and improvement may also be needed at the REMAB to provide security and isolation. The theater commander or JTF commander provides the construction engineer support.

d. Certain missions, such as those of interdiction, may need engineer special weapons teams. The teams are provided by the theater commander or JTF commander, or they may come from other special operations force assets. They are attached to the ranger force for a specific mission and released at the completion of the extraction and debriefing phase.

e. Some operations may need the ranger force to use captured enemy heavy construction equipment. Selected members of the ranger regiment are trained to operate such equipment.

However, construction engineer personnel may be attached to the ranger force for this mission.

CHAPTER 8

Combat Service Support

Combat service support consists of the logistical and administrative effort to maintain the battalion's ability to fight. The commander forecasts his needs for CSS. He employs CSS units to ensure accomplishment of the mission.

8-1. Ranger support element.

a. The ranger support element is a non-TOE organization formed from elements of CSS units collocated with CONUS-based ranger units augmented by selected personnel from the host installation. The mission of the RSE is to provide all the CSS needed to outload and deploy a ranger unit from CONUS. The personnel designated as part of the RSE have the same alert notification, reaction, and security requirements as members of the ranger unit they support. The RSE is normally built around a command and control group provided by a major CSS unit on the ranger unit's host installation. The remaining CSS elements and teams are then designated. An RSE consists of--

(1) An emergency operations center liaison group that provides timely coordination between the ranger unit and the functional elements of the RSE.

(2) A command and control group that coordinates and directs the actions of the functional elements.

(3) A supply section responsible for the timely issue of all Class II, III, IV, and VII items. This section maintains a stockage of contingency items marked for use by the resident ranger unit. It also issues items out of common stocks.

(4) A transportation section moves equipment and personnel from the unit area to the departure airfield and picks up equipment from local storage sites. It may also operate material-handling equipment and trailers to support loading of aircraft.

(5) An ammunition section responsible for the requisition, storage, issue, loading, and security of the ranger unit's basic load(s) of conventional munitions.

(6) A mess section provides all Class I support to the ranger unit. This includes operating a dining facility, providing food to remote sites, and drawing and issuing operational rations.

(7) A load team loads and unloads all equipment and supplies for the ranger unit. This team prepares pallets for loading, documents hazardous or dangerous cargo, and loads equipment pallets on USAF aircraft. It can also provide limited security for loads.

(8) A communications and electronics section provides secure communications among the RSE subelements, ranger unit, EOC, and DACG.

(9) An engineer platoon that is responsible for the construction of targets and rehearsal areas. This element of the RSE may or may not be activated, depending on the mission given the ranger unit. If the ranger unit deploys from CONUS to the objective area, this platoon has a key role to play in the premission preparation.

(10) A maintenance team whose primary mission is to ensure that the vehicles used by the RSE remain operational. It also provides predeployment maintenance and Class IX support for ranger unit vehicles and equipment.

(11) A medical team that provides unit-level medical care while the organic ranger medical section completes its outloading preparations. This team also provides any Class VIII supplies that the ranger unit may need.

(12) A security team responsible for security at the EOC, ranger unit location, and departure airfield. This team enforces stringent physical security precautions and maintains a constant counterintelligence (CI) surveillance of the outloading procedure. The security team provides armed guards for the protection of ammunition convoys, equipment movement, and vehicles.

(13) A rigger detachment that maintains, packs, rotates, and issues the personnel and equipment parachutes required by the ranger unit. It maintains enough stock of personnel parachutes to conduct a battalion-level parachute assault. The riggers lend technical skill to the rigging and aerial delivery of bulky, unusual, or delicate equipment.

b. The primary mission of the RSE is to provide the support necessary to quickly load and deploy the ranger units. The theater commander provides CSS to the ranger force once it has arrived at the OCONUS REMAB or ISB. The transition of support from the RSE to the CSS assets of the theater commander may be delayed for certain subelements in order to ensure a smooth deployment and enhance mission accomplishment. These subelements are released when the in-theater CSS assets can provide the support. Retained subelements of the RSE may be released from support requirements all at once, released in increments, directed to continue to provide support until mission completion, or augmented by in-theater CSS assets. The maintenance, medical, security, and rigger elements are those most likely to continue providing support OCONUS. This is likely if deployment is to an austere ISB pending immediate employment.

c. Once a ranger unit has been deployed OCONUS and has closed to a REMAB or ISB, the theater commander's staff coordinates CSS. If 1st SOCOM elements have deployed to the theater or operational area, the special operations support battalion (SOSB) of the 1st SOCOM coordinates CSS. Ranger logisticians at regiment and battalion level are mainly planners with limited logistics operations ability. Ranger logistical and personnel officers coordinate with CSS elements who provide such services.

d. The Theater Army Materiel Management Center (TAMMC) is the point of contact for information concerning in-theater logistical assets. In a developed theater of operations, a series of theater Army support groups will have been set up. One of these groups is assigned the mission of supporting the ranger force. The 1st SOCOM may also deploy certain assets available to provide CSS to the ranger force. The many CSS sources, the rapidly changing nature of SOF operations, and the unique CSS requirements of ranger units combine to make providing logistical and personnel support a challenge. The ranger unit's logistics officer must be included in the early phases of planning and deployment.

8-2. Logistics and personnel.

a. The echelons-above-corps CSS structure provides support to the ranger regiment as directed by the theater Army (TA) commander. In a low-intensity contingency involving a corps or less, there may be no echelons-above-corps logistics and personnel support elements. The corps support command (COSCOM) involved may be able to meet all needs from its own sources. However, in a combined operation in a high-intensity environment, the theater commander assembles a large CSS organization to provide support to the TA. This CSS base normally provides for all the ranger unit's requirements. During contingency operations in an area where there is no TA, the JTF commander provides the CSS elements. Until the buildup of these elements is complete, the ranger regiment must be prepared to operate from austere facilities with only the supplies it brings from CONUS.

b. The ranger regiment is normally assigned to the theater command for CSS. The TA commander, as directed by the theater commander, provides CSS to the ranger regiment and sets up support priorities. The EAC CSS organizations are under the command of the TA commander and consist of subordinate area and functional commands. The TA plans, coordinates, and executes its CSS functions through area-oriented support commands and specialized subordinate organizations. During the early stages of conflict, the TA may be organized as shown in Figure 8-1.

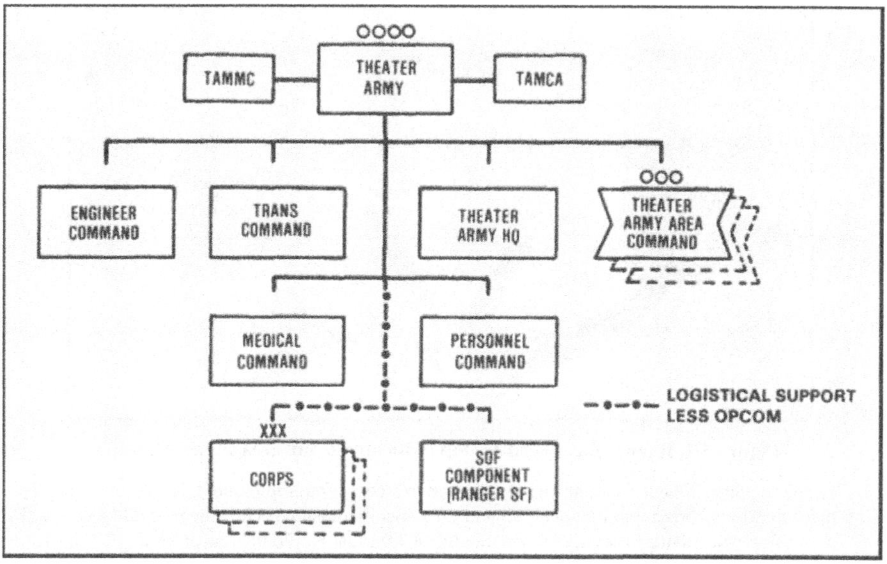

Figure 8-1. Ranger force relationship to theater Army.

c. The Theater Army Area Command (TAACOM) is the key CSS operator in the communications zone. It functions as a major subordinate command under the theater Army. The primary mission of the TAACOM is to provide direct support CSS to units such as the ranger regiment located in or passing through its assigned area. While located in an OCONUS REMAB, the ranger regiment receives most of its logistical support from the TAACOM.

d. The TAACOM area support group (ASG), which is subordinate to the TAACOM, is structured with operating units to provide direct support CSS to units located in or passing through its assigned geographical area of responsibility (see Figure 8-2). Area support groups are usually located along the lines of communication. The ASG is the prime source of CSS for the ranger regiment. It contains the units that actually provide services requested by the ranger unit logistics officer.

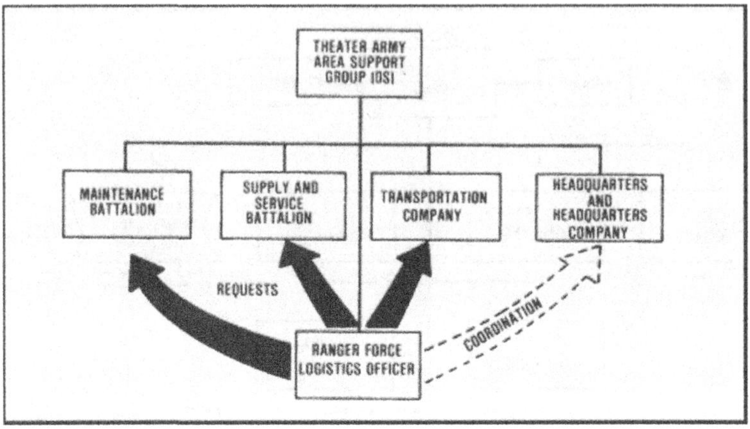

Figure 8-2. Ranger force relationship to theater Army area support group.

e. The ranger regiment is not normally assigned to a corps for employment. However, it may be placed under a corps' operational command for a specific mission. The ranger unit logistics and personnel officers must coordinate with the COSCOM that would then be providing CSS. (See Figure 8-3.) Corps support groups are major subordinate elements of the COSCOM. They provide command and control of COSCOM companies and battalions, which provide direct support (DS) and general support (GS) supply and field services to nondivisional units, such as the ranger battalion. (See Figure 8-4.)

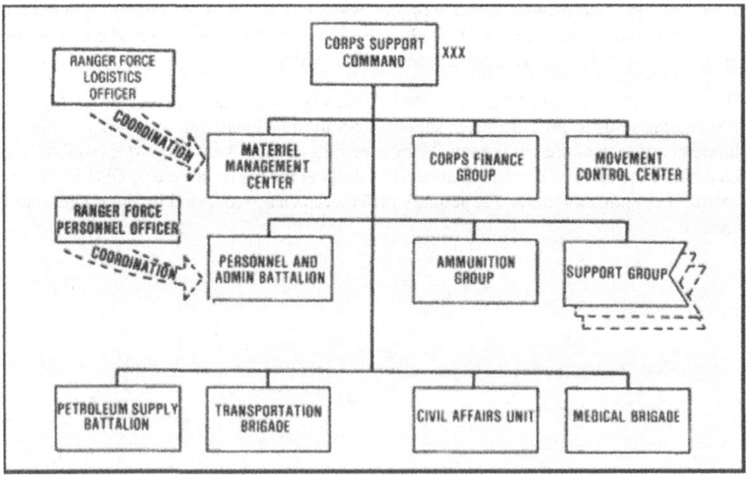

Figure 8-3. Ranger force relationship to corps support command.

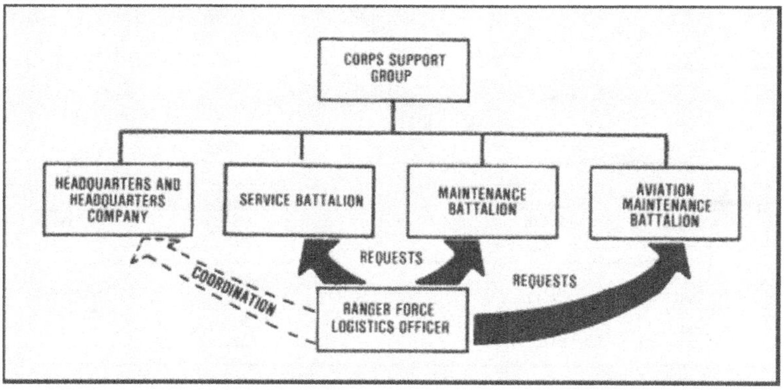

Figure 8-4. Ranger force relationship to corps support group.

f. Normally, the ranger regiment receives its logistical support from the corps support group responsible for the geographical area in which the REMAB or ISB is located. The corps support group provides all supplies except Classes V and VIII. The medical brigade provides Class VIII supply and medical maintenance support. The ammunition group provides Class V support. Financial and personnel services are provided by the corps finance group and the personnel and administration battalion, respectively.

g. Transportation needs are met by the transportation brigade. The ranger unit places ground transportation requests with either the Movement Control Center (MCC) or a designated movement control team. All airlift supply requirements are passed IAW FM 100-27. Standard and nonstandard supplies are normally requested from the supporting direct support unit (DSU). If authorized by the corps commander, the ranger regiment can place requirements for nonstandard items on the appropriate division within the corps MMC.

h. The COSCOM MMC (see Figure 8-5) performs supply and maintenance management for all classes of supply. The ranger unit logistics officer may coordinate with the MMC division chiefs or operations section.

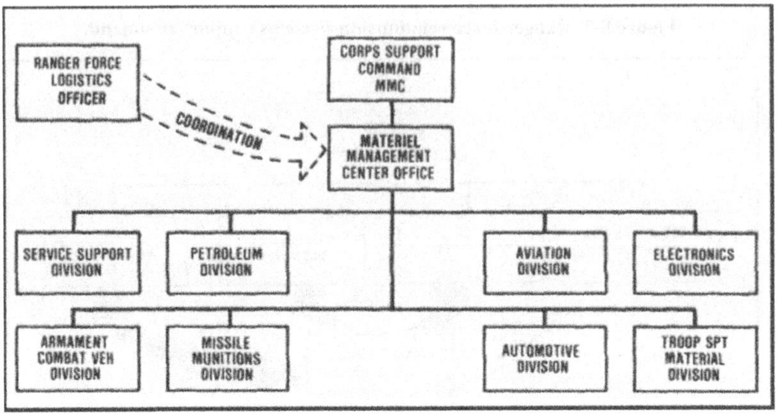

Figure 8-5. Ranger force relationship to COSCOM MMC.

i. Critical logistics operations, such as those supporting ranger and other SOF elements, need intensive management. They may be assigned to special staff sections that act as special operations force logistics managers. If designated, a logistics manager for special operations force elements would be a major point of contact for logistical support for the ranger regiment.

j. The ranger regiment may also receive logistical support from the SOSB. Normally, the SOSB provides administrative and logistical support to the Headquarters, Army Special Operations Forces (ARSOF). It complements the support provided by the Army component in the theater of operations by coordinating, augmenting, or assisting CSS forces. The SOSB provides unique quick-reaction support to the ARSOF or deploying SOF elements.

k. The ranger regiment does not make inordinate demands on the logistical system. However, a ranger unit may need certain nonstandard or low-density equipment. The concurrent need for nonstandard Class IX items requires management by logisticians to ensure support.

(1) Ranger battalion mess sections are responsible for Class I operations at the REMAB. The ranger regimental headquarters, which has no mess section, either collocates with one of the battalions or is supported by a mess facility operated by a nearby unit. It may share this facility on an area basis. The ranger unit personnel officer coordinates with the supporting DSU, which initiates the issue of Class I supplies.

(2) Class II, IV, and IX support to the ranger force is normally not of a great volume. It may consist of some nonstandard items. Depending on the need, the item, and the security, the SOSB may get some nonstandard items directly.

(3) Class III support needed by the ranger regiment is small and limited to standard items. The normal source of supply for Class III is the supply and service company providing area support.

(4) Class V support needed by the ranger regiment is small. It may include requests for nonstandard items of ammunition and explosives. The ranger logistics officer passes requests for standard items to the ordnance company (conventional ammunition) of the ammunition battalion assigned to the TAACOM ammunition group. Requests for nonstocked items are passed through the TAACOM Materiel Management Center or 1st SOCOM channels. The ranger unit must contact the ammunition company early to ensure a continuous supply of ammunition.

(5) Class VII supplies needed for the ranger regiment are often nonstandard or in excess of TOE. (See Appendix C for an equipment density listing.) All nonstandard or excess requests must be coordinated through the TAMMC or 1st SOCOM channels.

(6) Class VIII supplies for the ranger regiment are normally standard items supplied by the area support Class VIII point. The ranger regiment personnel officer sets up the Class VIII supply points of contact through the ranger battalion's medical officer.

(7) Aerial delivery equipment is provided by the airdrop equipment repair and supply company of the theater Army support group (TASG) (GS). There is normally one such company for each TAACOM. This unit supplies and repairs aerial delivery equipment. It does not pack and rig supplies for aerial delivery. The ranger regiment's requirements for aerial delivery supplies normally consist of standard items. Certain low-density items, such as HALO or HAHO parachute equipment, may also be provided through 1st SOCOM channels.

(8) The supporting motor transportation group of the transportation command is responsible for unit movement of the ranger regiment. The ranger regiment does not have any organic transportation assets. It depends on outside support whenever motor movement is needed.

8-3. Resupply.

a. Resupply operations for ranger units are normally covered during the planning and coordination phase of the mission. Ranger units try to complete their mission quickly and then be extracted or exfiltrate from the objective area. Ranger units carry into the objective area the equipment and munitions they need. They do not normally remain long enough to need extensive resupply. Certain operations may require the use of bulky or heavy items of equipment that cannot be carried by personnel. If the ranger force is airlanded on or near the objective, one or more of the aircraft may contain palletized supplies. These can be quickly off-loaded for future use.

b. If the ranger force is inserted into the objective area by parachute assault, aircraft can also drop resupply loads either just before or just after the personnel drop. These loads might include extra water, ammunition, demolitions, or barrier material. During peacetime, preloaded airdrop pallets are stored at CONUS depots for such contingency use. These mixed loads can be quickly moved to a departure airfield and inserted along with, or following, the ranger force. By dropping a resupply of water, food, equipment, and munitions with the ranger force insertion, the heavy-drop aircraft do not have to conduct a separate penetration of the enemy air defense zone. The ECM and SEAD programs used in the insertion of the personnel aircraft mask the penetration of the heavy-drop aircraft.

c. The aircraft dropping ranger personnel can also carry door bundles to deliver bulky or heavy items to the objective area. These door bundles, normally made up of A-7 and A-21 containers, may be color-coded to identify different loads or units. The most common door bundle loads are such items as 90-mm recoilless rifles and ammunition, 60-mm mortar ammunition, Dragon missiles, SHORAD missiles, or more demolition material. These bundles are pushed from the personnel aircraft just before the lead jumper's exit. They may be marked with lights or infrared sources for locating them at night.

d. If the tactical situation requires that a ranger force receive a resupply during the course of an operation, the most common method would be by airdrop. Airdrop could be needed if the operation takes longer than planned, initial supplies are lost, or changes occur in the operational plan. The ranger unit logistics officer and operations officers normally plan for resupply needs and schedule follow up loads of Classes I, II, IV, V, and VIII. There are two different airdrop requests channels for the ranger force to use:

(1) Nonemergency airdrop requests are sent through logistics channels. The ranger LOC sends requests for the airdrop of supplies and equipment to its supporting supply unit, and the supporting unit forwards the requests to the COSCOM or TAACOM MMC. The MMC coordinates with the movement control center (MCC). Jointly they direct the actions of the supply and transportation units. The MCC sends the request to the joint transportation board (JTB). The board validates it and assigns it a priority. The JTB then sends the request to the airlift control center for action. If the items requested are not on hand in the airdrop company, the MMC directs the proper supply activity to prepare the items for shipment to the airdrop unit. The MMC coordinates with the MCC, and the MCC ships the items. The airdrop unit prepares the supplies for the mode of delivery to be used. It also makes sure the supplies are delivered to the air terminal at the right time. (See Figure 8-6.)

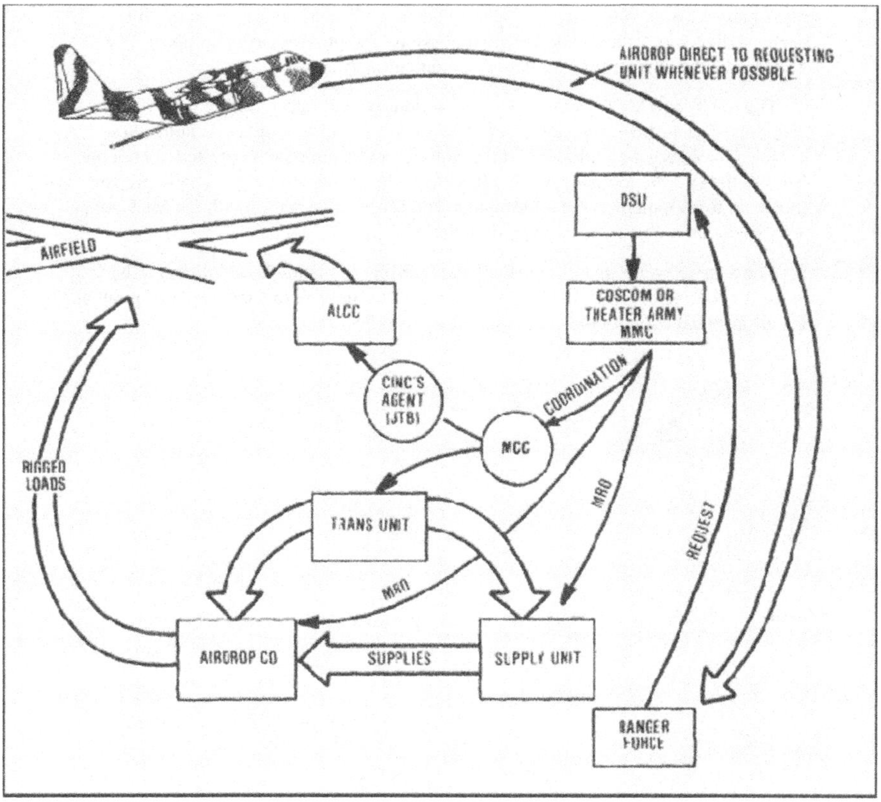

Figure 8-6. Ranger requests for nonemergency airdrop.

(2) Emergency requests for airdrop are passed through command channels in the quickest, most secure way.

(a) The request is usually sent to the command level that has authority to approve it, usually the tactical operations center at corps or above. The Air Force airlift officer at the direct air support center (DASC) coordinates with the Assistant Chief of Staff, Operations (G3) at the tactical operations center to determine whether airlift resources can be used for the mission.

(b) The TOC sends a request for an immediate airlift to the joint force commander's (JFC) designated agent for validation. At the same time, the airlift liaison officer advises the airlift control center (ALCC) that a request is coming through channels so plans can be made. When the ALCC receives the validated request from the JFC agent, it diverts or cancels less essential missions and provides airlift support. The TOC sends the request for supplies or equipment to the MMC.

(c) The MMC and MCC jointly direct the actions of the supply and transportation elements. The airdrop company stocks Classes I, III, and V supplies. It holds these supplies for emergency missions. If the supplies and equipment requested are not on hand in the airdrop company, the MMC directs the proper supply activity to prepare the items for shipment to the airdrop unit. The MMC coordinates with the MCC to ship the items. The airlift is arranged by the ALCC.

(d) The airdrop unit prepares the supplies according to the mode of delivery to be used. It coordinates with the MCC and delivers the supplies to the designated point at the air terminal (usually the cargo aircraft ramp). (See Figure 8-7.)

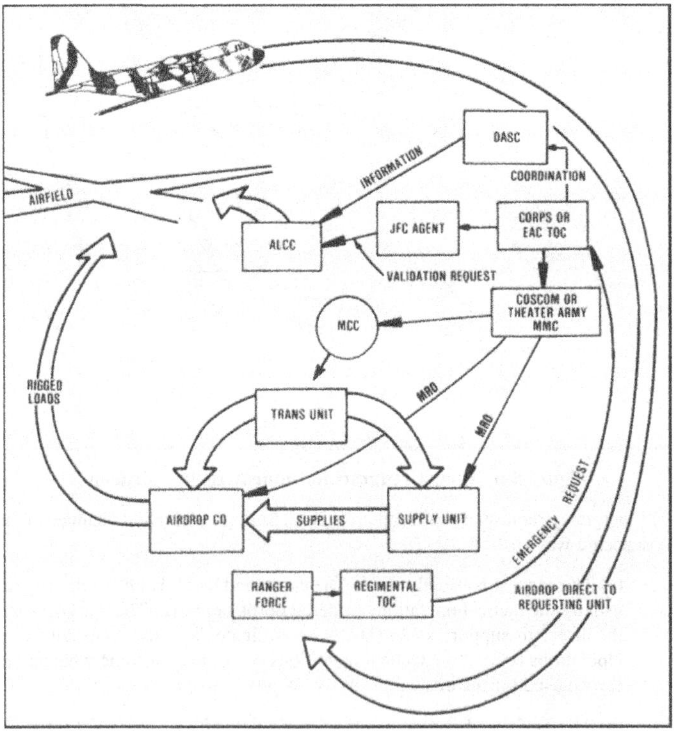

Figure 8-7. Request channels for emergency airdrop.

e. Aircraft delivering airdrop resupply deep behind enemy lines need to take precautions to avoid enemy detection and antiaircraft fire. The USAF is responsible for most of these precautions, such as ECM, suppression of enemy air defense (SEAD), and WILD WEASAL support. These do not concern the ranger commander. The choice of the type of parachute delivery system to use, however, may affect the ground commander's plan. The ranger force commander must specify any tactical restrictions that would require a specific type of delivery system. The ranger LOC and regimental logistics officer must advise the ranger force commander on the type of parachute system to be used. Conflicts must be quickly resolved so as not to interfere with the tactical operations.

(1) The safest way for the airdrop aircraft to penetrate enemy air defense and remain undetected is often by use of very low-level flight. There are four low-level parachute delivery systems that can be used without the aircraft being required to climb to a higher drop altitude and risk detection by enemy ADA.

(a) The low-altitude parachute extraction system (see Figure 8-8) needs a length of level ground with enough clearance for the delivery aircraft to make a very low pass over the extraction zone. The load is extracted by parachute from the aircraft and slides to a stop on a rugged, skid-type pallet. Heavy loads can be delivered using this system.

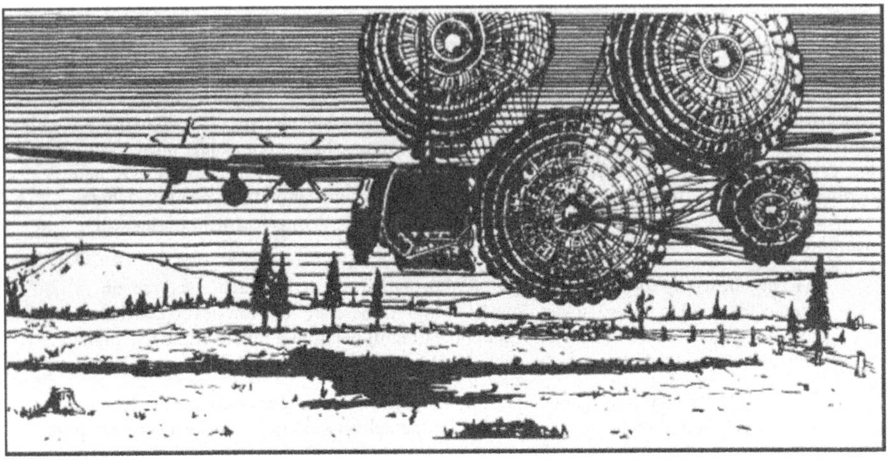

Figure 8-8. Low-altitude parachute extraction system.

(a) The containerized delivery system (CDS) (see Figure 8-9) consists of multiple, individually rigged A-22 containers, each with its own parachute, weighing up to 2,000 pounds each. The system provides single-pass delivery of up to 16 containers by C-130 aircraft and up to 28 containers by C-141 aircraft. The CDS loads may be delivered into drop zones using multiple points of impact to allow for tactical separation. The ranger force commander can direct the use of multiple points of impact, noting the advantages to be gained against the requirement for multiple passes.

Figure 8-9. Containerized delivery system.

(c) The high-speed, low-level, airdrop system (HSLLADS) (see Figure 8-10) is a single A-21 container specially rigged to withstand the parachute opening shock when airdropped from C-130 aircraft at high speed. This system can be used to deliver up to 600 pounds for each container, with a maximum of four containers for each pass, at speeds up to 250 knots.

Figure 8-10. High-speed, low-level, airdrop system.

(d) The CTU-2/A is a high-speed, aerial-delivery container (see Figure 8-11) that can be used to deliver supplies from high-performance aircraft flying at a minimum altitude of 300 feet above ground level and a maximum airspeed of 425 knots. The CTU-2/A is carried on the bomb racks of fighter or bomber-type aircraft. Upon release, a pilot parachute deploys the main chute and the container descends slowly.

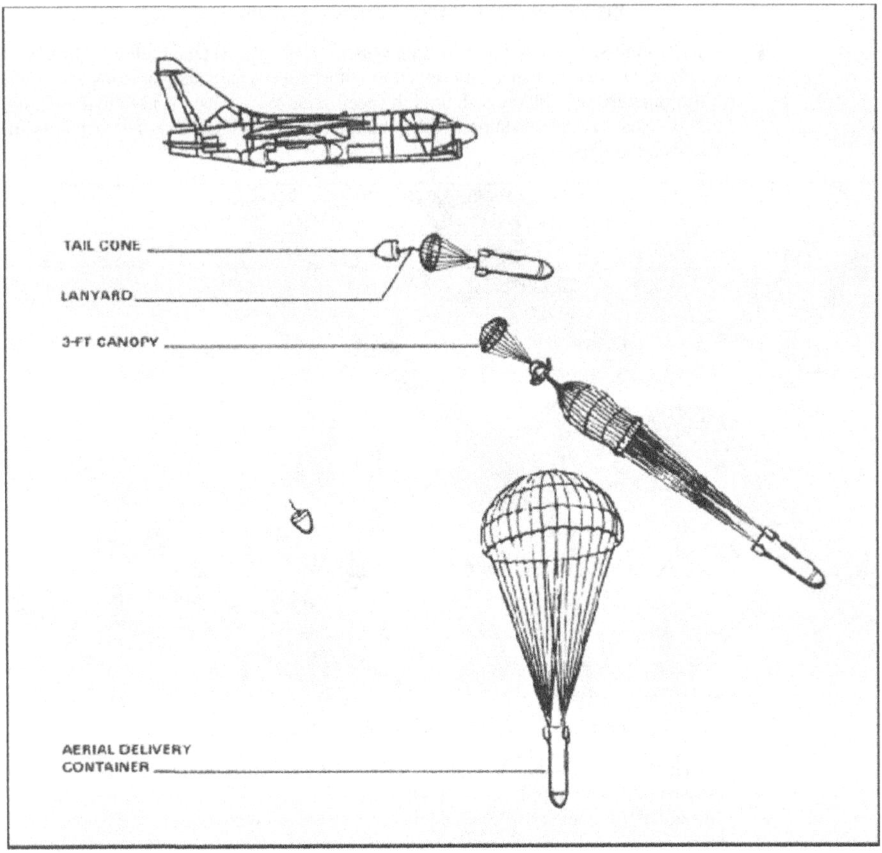

Figure 8-11. CTU-2/A high-speed, aerial-delivery container.

The CTU-2/A can be used to deliver up to 500 pounds of supplies such as weapons, water, food, or munitions. The container can be destroyed by burning. The main advantage of this system is that it can be delivered by high-performance aircraft deep behind enemy lines and in a dense air defense environment. The accuracy of this system is equal to that of a conventional bomb strike.

(2) Sometimes it may be better to drop resupply loads to the ranger force from a high altitude. The HAARS (see Figure 8-12) permits containerized unit loads weighing from 200 to 2,000 pounds to be delivered from aircraft at speeds up to 150 knots from up to 25,000 feet AGL. The HAARS consists of a cargo parachute, an airdrop container, an altitude

sensor, and a pilot chute. The pilot chute gives the descending bundle a speed slightly greater than an accompanying HALO parachutist. The system provides for the steady free-fall descent of loads from altitudes between 2,000 and 25,000 feet to an altitude at which a barometric sensor actuates deployment of the main parachute. This allows safe and accurate delivery of loads onto unprepared drop zones. The HAARS can deliver a payload to within 260 meters from a target impact point from a 10,000-foot altitude, with a proportional degree of accuracy from 25,000 feet. Ranger units can use the system for resupply of battalions and smaller units with rations, ammunitions, and medical supplies, breaking down the containerized material into man-packed loads.

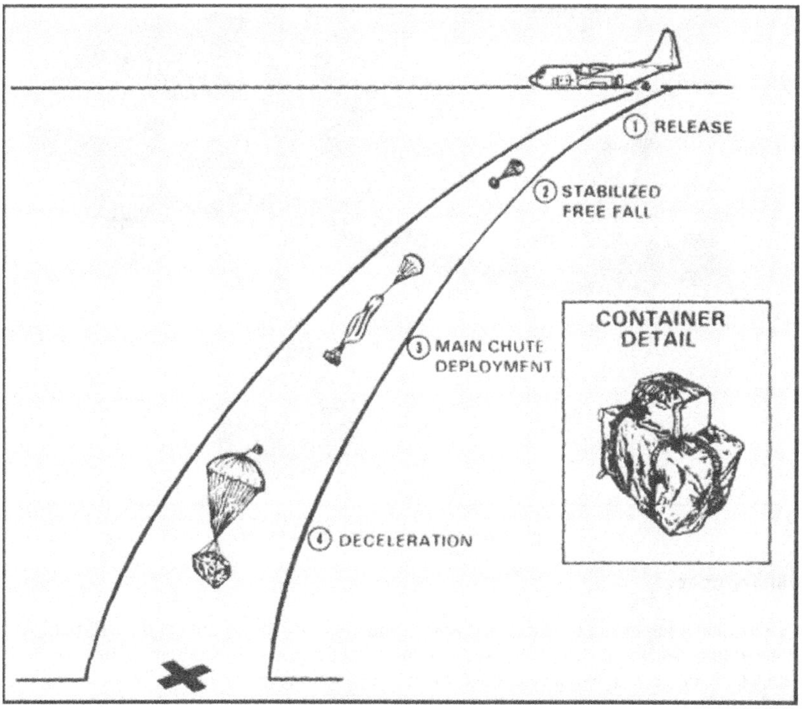

Figure 8-12. High-altitude, airdrop resupply system.

(3) The ranger force may need aircraft loads delivered during periods of poor visibility. The adverse weather aerial delivery system allows aircraft to drop resupply bundles accurately in adverse weather or darkness. These AWADS operations can be conducted safely and effectively with a minimum 300-foot ceiling AGL and a minimum visibility of less than half a mile. The minimum-drop altitude is 500 feet above the highest obstruction. These minimum ceiling or visibility restrictions do not apply to the Combat Talon aircraft.

f. Resupply to the ranger force can be provided by Army helicopters (see Figure 8-13). Normally this is done by SOA units. The danger to helicopters during long-range, low-level penetration of enemy rear areas requires that resupply be conducted when there is limited visibility.

Figure 8-13. Resupply bt helicopter.

8-4. Military police.

a. The ranger regiment needs military police support to provide security at the REMAB. Although ranger units can provide their own security and local defense, using military police for this purpose allows the ranger unit to concentrate its manpower and training on mission preparation. A military police security company is normally designated to provide support to 1st SOCOM elements located in the communications zone. Military police provide security patrols of the REMAB, convoy security, and vehicular search. They provide entry and exit control of the sensitive compartmented information facility (SCIF).

b. A preemptive strike by enemy special operations forces is a threat to the ranger regiment while located in the REMAB; either by conventional attack or by terrorist-type bombing. The preparation and location of the headquarters, the type of buildings, the location of landing zones, the traffic flow, and the size of the local population all influence the size of the military police unit. A military police platoon, augmented with bomb and guard dog teams, should be enough security at the REMAB. These security forces would not normally deploy from the REMAB to an ISB with the ranger unit.

8-5. Counterintelligence.

Counterintelligence support to the ranger regiment is normally provided by a CI team from the military intelligence group at TA level. The CI team provides a liaison and informational flow from the CI agencies operating within the theater. It also acts as advisor to the ranger unit commander on OPSEC while performing CI surveillance operations in the vicinity of the REMAB. The CI support may also be provided through 1st SOCOM channels. It is through 1st SOCOM that liaison with national intelligence and counterintelligence agencies is achieved.

8-6. Personnel service support.

The personnel service support needs of a ranger unit are much like that of any other unit of the same size. To receive the support his unit needs, the ranger personnel officer must meet with the proper agencies within the theater of operations.

a. **Strength accounting.** Within the ranger regiment, strength accounting is conducted through normal personnel channels. Regimental strength figures are passed quickly to the TAACOM or COSCOM. These figures are crucial to the theater commander for planning future operations. When matched with casualty reports, they provide the information needed for replacements.

b. **Replacement operations.** Ranger units need exceptionally well-trained and physically conditioned personnel as replacements. Ranger operations require that replacements train with the unit before they can become an effective member of the team. There are five sources of replacements for ranger units:

(1) **In-theater volunteers.** Use of in-theater volunteers, preferably with experience in ranger units, is the quickest way to replace personnel. These volunteers have a knowledge of the basic skills required of rangers and have some operational experience. The ranger regimental personnel officer coordinates with the theater personnel officer to screen and select those volunteers with the required MOSs and grades. Volunteers from other special operations force components, or from other theaters of war, are identified by DA and screened for assignment to the ranger regiment. Even though these volunteers have individual training and experience, it takes several weeks to become fully integrated into the ranger unit's operations and to achieve the needed skill level.

(2) **Reorganization of units.** This method of reconstitution is used when two or more ranger units have high personnel losses. The remaining members of these units are reorganized into a single unit. This method provides the ranger regimental commander with enough ready-trained, physically fit personnel to form a full-strength unit. Extensive unit training must be conducted before such a reorganization regains the unity and esprit of its component parts. The high standards found among the subelements of the ranger regiment makes this a good method of reconstituting a ranger unit. The units that have lost all their personnel may be reconstituted later, using some other method, or left unreconstituted,

depending on the situation.

(3) **Hospital returnees.** These personnel, former members of the ranger regiment who were injured, evacuated, treated, and have now recovered, are valuable assets. They provide the unit with a major reconstitution ability. These personnel should be returned quickly to their original battalion. This enhances unity and increases morale and esprit. Returnees may not be physically able to return to full duty right away. While recuperating, they may be assigned to limited duty as instructors. The ranger unit uses the returnee's combat experience and individual skills to screen and train newly assigned volunteers. Lessons learned in combat can help prepare the unit for future operations.

(4) **Replacement volunteers from CONUS.** These personnel have training in the basic skills and may have completed the ranger course, but they do not have any ranger unit experience. They need several weeks of unit training to become prepared for combat operations.

(5) **Ranger indoctrination program.** During wartime, this program provides screening and training of newly assigned personnel. Volunteers who have neither served in a ranger unit before nor have graduated from the ranger course require extensive individual training before they are ready to begin unit training with a ranger battalion. This method of reconstitution of ranger units requires the longest training period. During this reconstitution phase, a ranger battalion needs extensive use of ranges, maneuver areas, and support by aviation and other special operations-type units. The ranger regimental commander may choose to move the program from CONUS to a location near the regiment's REMAB.

c. **Casualty reporting.** Casualty reports are handled through regimental personnel channels and must be timely and accurate. The reports are used to check strength accounting data, to assist graves registration personnel in the recovery and identification of remains, and to notify next of kin.

d. **Health service.** The health service support needed by the ranger units deployed OCONUS is usually small. Area medical support provided by host country or theater medical command assets usually meets their needs. Evacuation of wounded members of the ranger force begins upon extraction from the objective area. Personnel wounded in action are evacuated through USAF or Army channels to the nearest medical facility that can provide treatment. The ranger force medical officer and personnel officer coordinate with the area medical support facility providing preventive and routine medical care to the unit while at the REMAB. Vaccine needed for innoculation of the ranger force members and any special Class VIII supplies needed for the mission are drawn through normal medical channels.

e. **Financial service.** Financial support varies with the situation. In mid- and high-intensity conflicts, regular paydays are suspended in the theater of operations. As a nondivisional unit, the ranger regiment coordinates with the theater finance center or corps finance group for financial support. In the REMAB, ranger units are provided financial support from the closest finance support unit. The ranger unit personnel officer appoints Class A agents to help with the disbursement of funds. The ranger force may be given a specific amount of local currency or military scrip depending on the length of time in the REMAB and the operational situation. The ranger unit personnel officer also coordinates with the theater finance center to get any third-country currency that may be included as part of an escape-and-evasion kit provided to ranger elements.

f. **Postal service.** Postal service to the ranger unit is through normal channels. Mail for members of the deployed force is forwarded by the rear detachment commander to an established theater Army general support postal activity. This activity forwards the mail to the postal unit designated to support the REMAB site. The ranger unit commander appoints subordinate unit officers as censors. The exact location of the REMAB, and all aspects of the mission and its location, remains classified until released by an official Army spokesman. The ranger unit commander must ensure that the forwarding of mail, or the writing home of personal letters, does not endanger the ranger force mission. All outgoing mail is censored during wartime. For security reasons, outgoing mail may be held for a short time before it is forwarded. Mail from the ranger unit is never sent by host country or other nation's postal services.

g. **Chaplain activities.** Chaplain activities are provided by the chaplain assigned to the ranger battalion. He provides unit coverage through services, rites, and sacraments regardless of his or the unit member's religious affiliation.

h. **Legal service.** Ranger units do not need extensive legal service support. Legal service support is obtained through coordination with the theater Army staff judge advocate.

APPENDIX A

Organization and Equipment Introduction of the Ranger Regiment

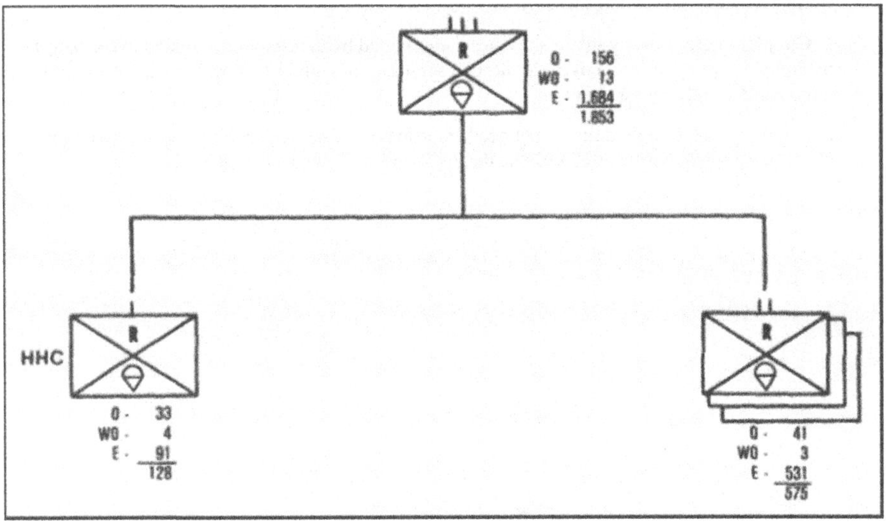

Figure A-1. Ranger regiment.

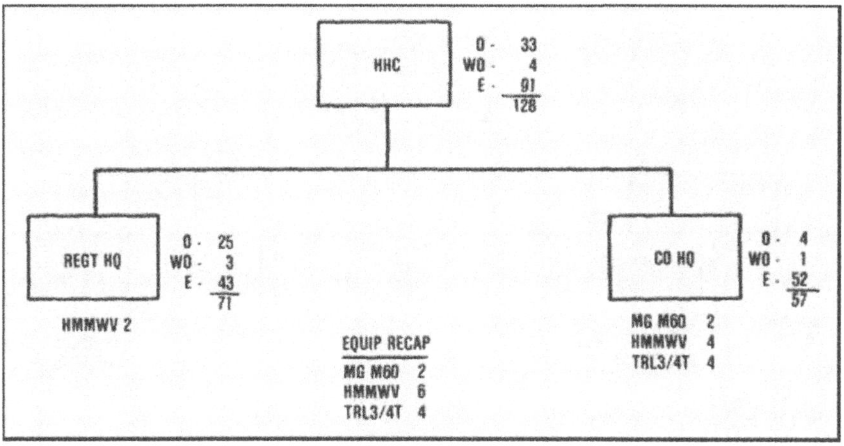

Figure A-2. Ranger regimental HHC.

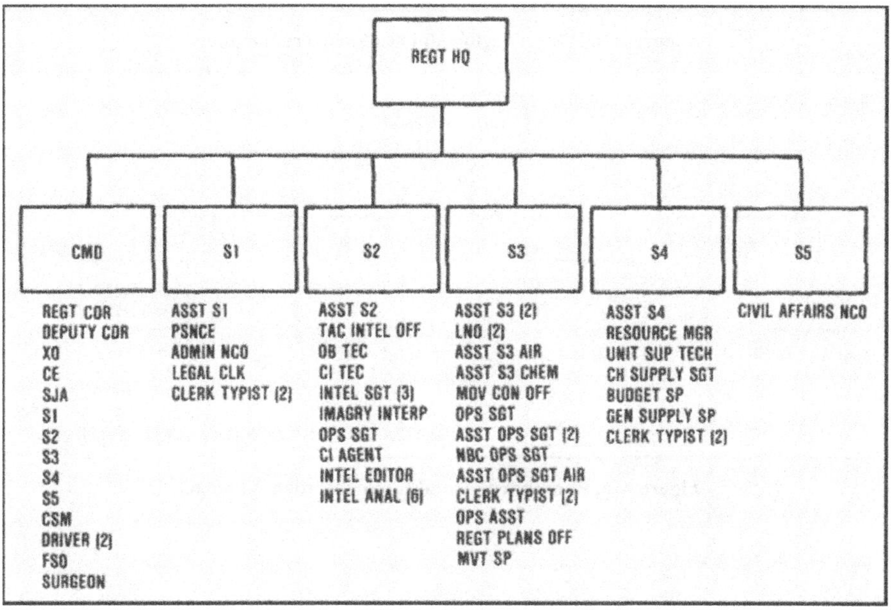

Figure A-3. Ranger regimental staff.

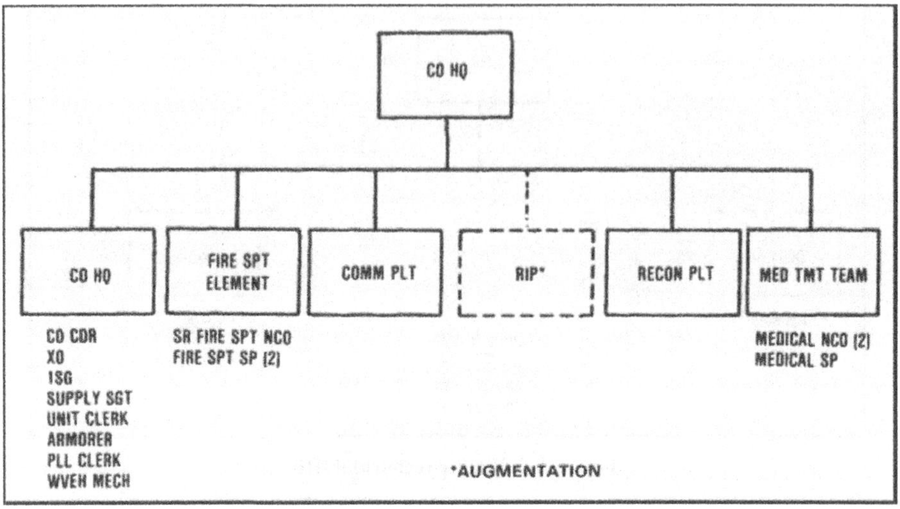

Figure A-4. Ranger regimental headquarters company.

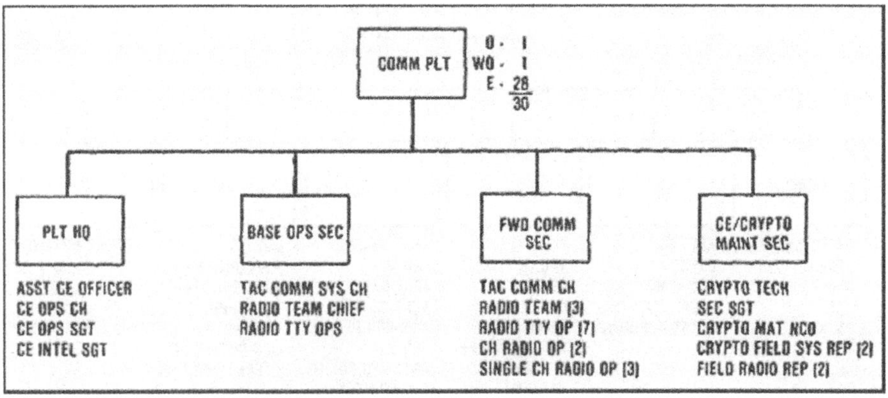

Figure A-5. Ranger regimental communications platoon.

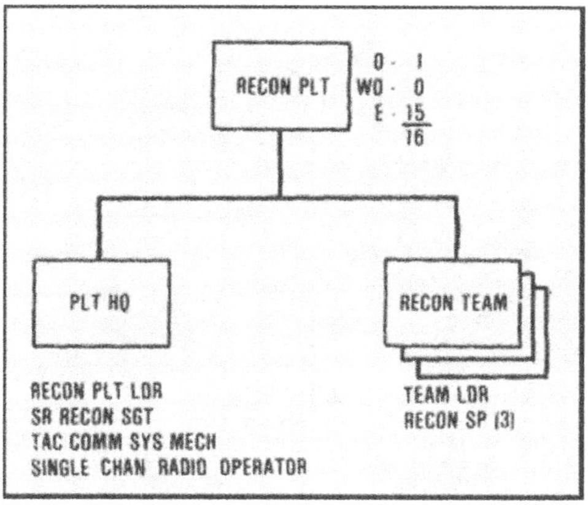

Figure A-6. Ranger regimental reconnaissance platoon.

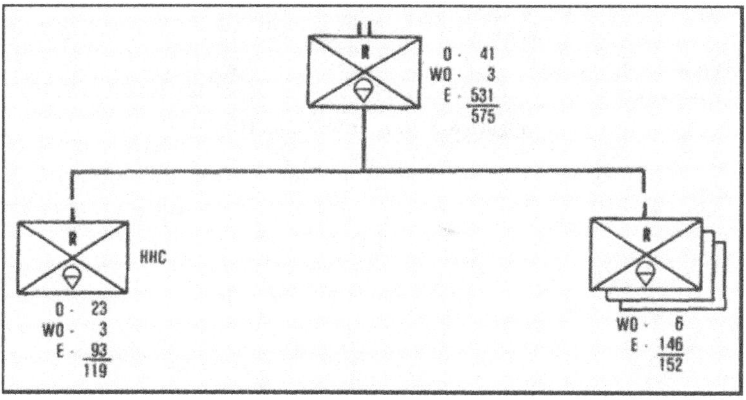

Figure A-7. The ranger battalion.

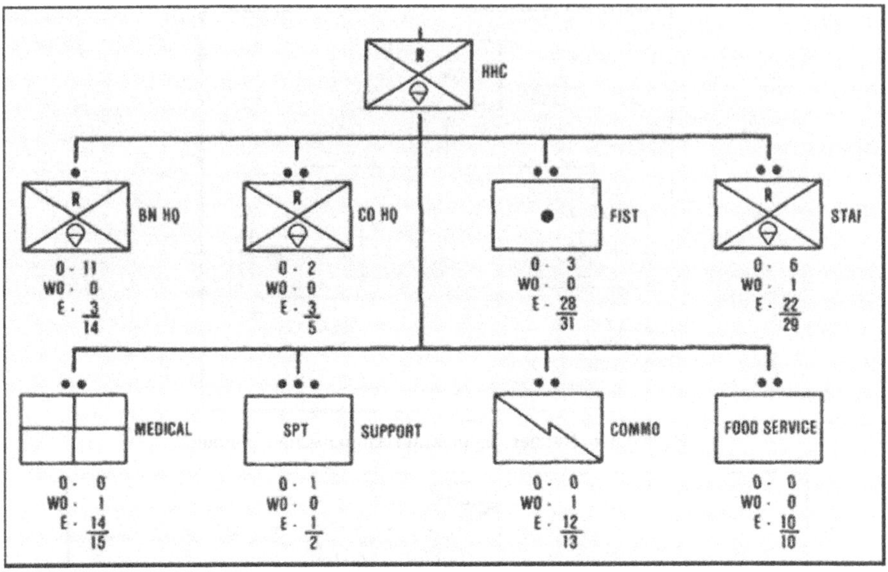

Figure A-8. Ranger battalion HHC.

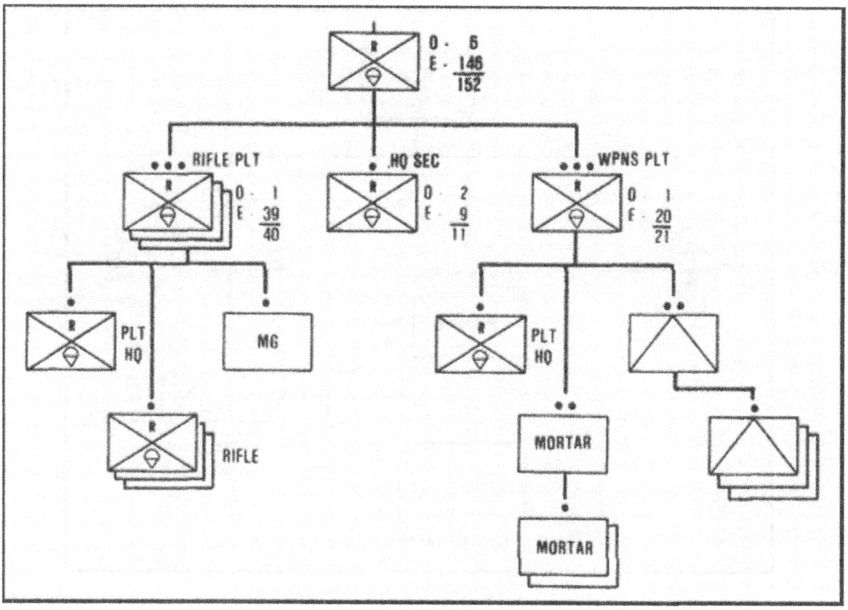

Figure A-9. Ranger rifle company.

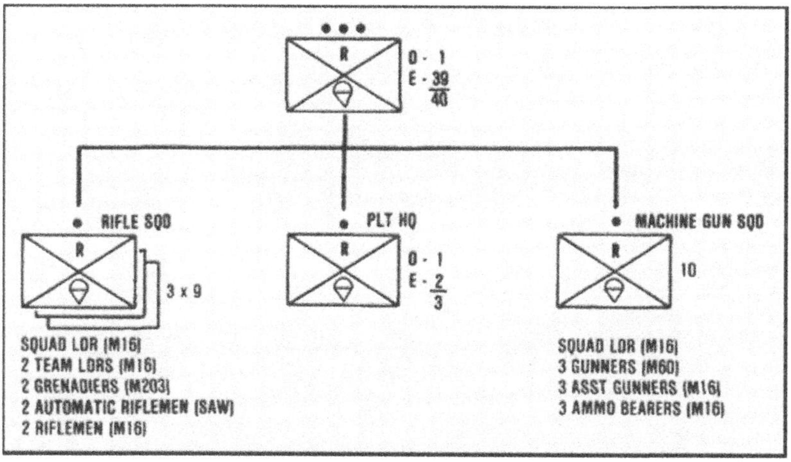

Figure A-10. Ranger rifle platoon.

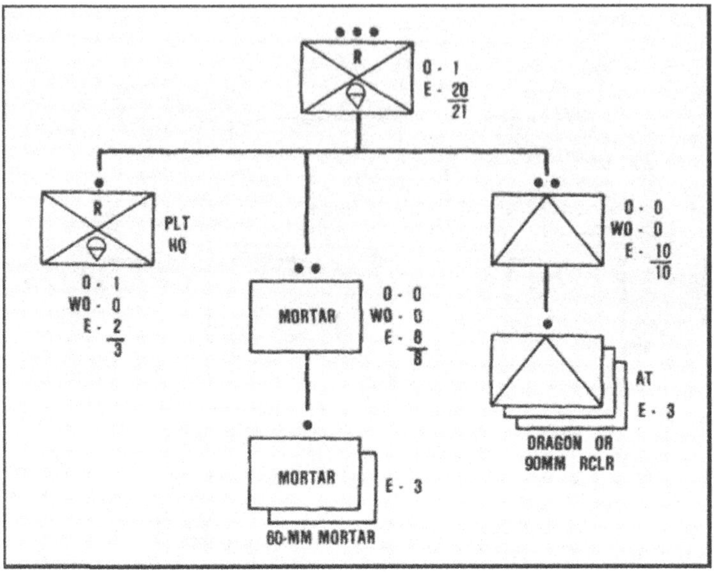

Figure A-11. Ranger weapons platoon.

APPENDIX B

Composition and Equipment of Ranger Regiment Liaison Teams

Team Red	**RANK**
Liaison Officer	MAJ
Senior Intelligence Analyst	SSG
Clerk-Typist	SP4
Tactical Communications System Operator/Mechanic	SGT
Combat Telecommunications Center Operator	SP4
Tactical Communications System Operator/Mechanic	SP4
Team White	
Liaison Officer	CPT
Senior Intelligence Analyst	SSG
Clerk-Typist	SP4
Tactical Communications System Operator/Mechanic	SGT
Combat Telecommunications Center Operator	SP4
Tactical Communications System Operator/Mechanic	SP4

Equipment (for Each Team)

HMMWV with 3/4-ton trailer	5-kw generator
Communications shelter	Tent, GP, small
AN/PCS-3 radio set (3)	AN/VSC-7 net control station (1)
MX 300 (4)	KY 57 with battery case (5)
HYX 57 with battery case (4)	HYP 57 (5)

KYK 13 (2)

KY 65 with battery case (1)

KG 84 (1)

H189 (2)

GRC 193 (1)

VGC-74 (1)

PP 1648 (1)

SB-22 switchboard (1)

KYI 18 (1)

KY 65 WLA (1)

Digital message device (1)

AN/PRC-77 radio (1)

Facsimile transmission device (1)

TA 312 with WD1 (5)

APPENDIX C

Equipment Density

C-1. Communication equipment.

	Regimental Headquarters	Battalions (x3)
AN/VSC-7 net control station	7	2
AN/PSC-3 radio set	23	13
AN/PRC-41 radio set	2	5
AN/PRC-104 radio set	9	26
AN/GRC-193A radio set	4	0
AN/PRC-70 radio set	6	0
AN/PRC-77 radio set	22	50
AN/VRC-46 radio set	2	0
AN/VRC-49 radio set	0	2
AN/PRC-68 radio set	0	66
TSEC/KY-57 speech security device	60	68
TSEC/KY-58 speech security device	6	0
TSEC/KY-65 speech security device	11	0
AO-8990/P digital message device group	15	23
AN/GXC-7A tactical FAX device	7	2
AN/UGC-74A(V)3 communications terminal	7	0
SB-22 switchboard	2	0
SB-3614(V)/TT switchboard	2	1

TSEC/KG-84	5	0
SB-993/GT switchboard	0	12

C-2. Weapons and equipment each battalion.

Infrared aiming light	96	Interrogator set, Stinger missile	3
Rifle, M16	451	Grenade launcher, M203	67
Rocket launcher, M202	13	Machine gun, M60	27
Mortar, 60-mm M224	6	Rifle, sniper, M21	24
Recoilless rifle, 90-mm	18	Viewer, infrared	9
Tracker, Dragon	9	Night vision devices (all types)	322
Truck, 1/4-ton	2	Laser target designators	11
Radio (all types)	164	Tactical satellite radios (all types)	15
Squad automatic weapon	54		
Laser range finder	14		

APPENDIX D

Standing Orders, Rogers' Rangers

1. Don't forget nothing.
2. Have your musket clean as a whistle, hatchet scoured, sixty rounds powder and ball, and be ready to march at a minute's warning.
3. When you're on the march, act the way you would if you was sneaking up on a deer. See the enemy first.
4. Tell the truth about what you see and what you do. There is an army depending on us for correct information. You can lie all you please when you tell other folks about the Rangers, but don't never lie to a Ranger or officer.
5. Don't never take a chance you don't have to.
6. When we're on the march we march single file, far enough apart so one shot can't go through two men.
7. If we strike swamps, or soft ground, we spread out abreast, so it's hard to track us.
8. When we march, we keep moving till dark, so as to give the enemy the least possible chance at us.
9. When we camp, half the party stays awake while the other half sleeps.
10. If we take prisoners, we keep 'em separate till we have had time to examine them, so they can't cook up a story between 'em.
11. Don't ever march home the same way. Take a different route so you won't be ambushed.
12. No matter whether we travel in big parties or little ones, each party has to keep a scout 20 yards ahead, 20 yards on each flank, and 20 yards in the rear so the main body can't be surprised and wiped out.
13. Every night you'll be told where to meet if surrounded by a superior force.
14. Don't sit down to eat without posting sentries.
15. Don't sleep beyond dawn. Dawn's when the French and Indians attack.
16. Don't cross a river by a regular ford.

17. If somebody's trailing you, make a circle, come back onto your own tracks, and ambush the folks that aim to ambush you.

18. Don't stand up when the enemy's coming against you. Kneel down, lie down, hide behind a tree.

19. Let the enemy come till he's almost close enough to touch. Then let him have it and jump out and finish him up with your hatchet.

MAJ Robert Rogers

APPENDIX E

The Ranger Creed

R Recognizing that I volunteered as a Ranger, fully knowing the hazards of my chosen profession, I will always endeavor to uphold the prestige, honor, and high "esprit de corps" of the Ranger Regiment.

A Acknowledging the fact that a Ranger is a more elite soldier who arrives at the cutting edge of a battle by land, sea, or air, I accept the fact that as a Ranger, my country expects me to move farther, faster, and fight harder than any other soldier.

N Never shall I fail my comrades. I will always keep myself mentally alert, physically strong, and morally straight and I will shoulder more than my share of the task, whatever it may be. One hundred percent and then some.

G Gallantly will I show the world that I am a specially selected and well-trained soldier. My courtesy to superiors, my neatness of dress, and my care of equipment shall set the example for others to follow.

E Energetically will I meet the enemies of my country. I shall defeat them on the field of battle, for I am better trained and will fight with all my might. Surrender is not a Ranger word. I will never leave a fallen comrade to fall into the hands of the enemy and under no circumstances will I ever embarrass my country.

R Readily will I display the intestinal fortitude required to fight on to the Ranger objective and complete the mission, though I be the lone survivor.

RANGERS LEAD THE WAY

APPENDIX G

Training

G-1. Application and guidance.

a. To prepare for combat, ranger units train on missions derived from their analysis of operational and contingency plans and the ranger mission training plans (MTPs). Training guidance is in doctrinal manuals, field circulars, and soldier's training publications. Training must ensure that the ranger unit is skilled in all basic light infantry skills and in ranger-unique skills. In the final analysis, good training depends upon the leadership and teamwork of ranger unit officers and NCOs.

b. Everything rangers do should prepare them to perform well in combat. Ranger units train IAW the battalion training management system (BTMS) concentrating on basic skills and on live-fire, hands-on training.

c. An effective training program breeds high morale and few discipline-related problems. Some qualities of ranger units that have good training programs are as follows:

(1) Leaders who can solve problems.

(2) A thorough and continuous analysis of the unit's mission.

(3) Validated training objectives.

(4) Attention to the basics (move, shoot, and communicate).

(5) Multiechelon sustainment training.

(6) Standardized procedures.

(7) Physical and mental strength.

(8) Live fire as an integral part of training.

(9) Safety as a basis to all operations.

(10) The use of night and adverse weather and difficult terrain as combat force multipliers,

d. When all training is planned with realism, rangers will be in training situations that require the same attention as in battle. Realism and stress in training result in faster learning and better retention of skills. Training objectives must be clear with measurable standards to determine proficiency. Rangers and ranger units must train to the standards that will be expected in combat.

e. Examples of unit training goals include:

(1) Developing each member of the chain of command to perform his duties with competence, confidence, and pride.

(2) Maintaining small unit integrity during all training and support activities. Accomplishing tasks as units.

(3) Ensuring that all tasks for training are keyed to mission needs and weaknesses. Assets--ranger's time, money, and materiel--are expended only after thorough planning.

(4) Training to standard.

(5) Developing and maintaining discipline and esprit to ensure the abilities of each ranger and all units are achieved.

(6) Emphasizing use of the ranger mission training plan (MTP) for each echelon of the organization.

(7) Emphasizing the basics.

(8) Maintaining a high level of physical fitness.

(9) Working together within the ranger regiment and other elements of the special operations forces community.

(10) Developing a unit atmosphere of quick reaction under stress. Encouraging and rewarding individual initiative. Accomplishing unit missions IAW the commander's intent.

G-2. Philosophy.

a. To be effective, training should be physically and mentally challenging. Each ranger will then have a sense of purpose, accomplishment, and satisfaction. Training is evaluated by the results rather than hours of training. Units that are well trained today need reinforcement training to sustain readiness. Training management must be continuous. Training managers must use the time available to improve training efficiency. Cyclic training will not produce the desired results. The "train-to-correct deficiencies" method is needed. The training managers must analyze weaknesses, conduct prescriptive and reinforcement training, and reevaluate their unit's training status. The process that is used to develop and maintain a continuous training and evaluation program is shown in Figure G-1.

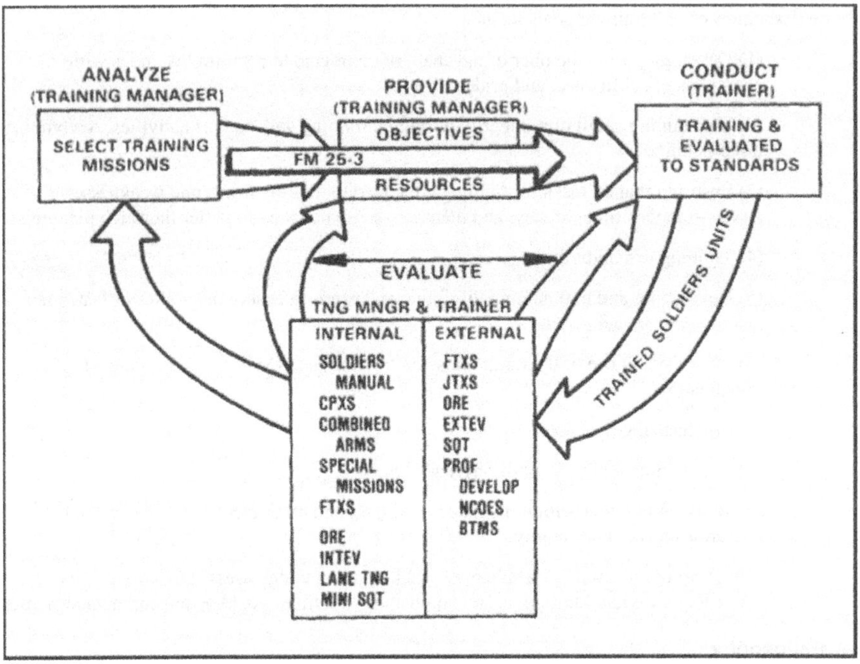

Figure G-1. The Army training and evaluation program.

b. Individual and collective training must be integrated within a unit training program. Training of the individual ranger, leaders, and units at each level must occur at the same time. The ranger commander or leader plans and conducts training for every member of his unit.

c. Ranger commanders analyze operational and contingency plans to determine unit missions and translate the associated tasks into training objectives. These objectives form the basis of the unit training program. The regimental headquarters provides training guidance in a formal regimental training plan. Commanders at lower levels program, supervise, and conduct training within their units.

d. All members of the chain of command are responsible for training. The NCO leader or first-line supervisor conducts individual and small unit collective task training. The officer plans individual training in his unit and directs and conducts collective training.

e. Centralized development of training objectives and decentralized training builds well-trained units and outstanding leaders. Authority and responsibility for the detailed planning, scheduling, conducting, evaluating, and supervising of training are delegated to the lowest element capable of managing this training. Higher headquarters is responsible for establishing objectives, monitoring and evaluating training, and issuing general guidelines.

f. Multiechelon training and evaluation means using all assets and time available to improve combat readiness. The multiechelon structure of the ranger MTP provides a basis for units to be continually trained and evaluated by internal evaluation. Each leader must know his mission tasks and train to accomplish those tasks to the set standards.

G-3. Mission analysis.

a. Mission analysis is continuous for a good training program. The selection of training missions or tasks results from analysis of the unit's overall mission. (See Figure G-1 above.) Tasks and subtasks are then placed in order of priority.

b. The trainer must identify countertasks for the opposing force (OPFOR). The countertasks should be such that an OPFOR that meets the standards for its countertask will be successful. Both sides have a chance to win.

c. The trainer should then develop combat drills and situational training exercises (STXs), if not furnished by the proponent school.

G-4. Unit training.

a. **Responsibility.** Ranger commanders at all levels are responsible for all training and management described herein and current Army training doctrine. This includes individual and collective training at all levels. Units must train to the standards specified in training objectives.

b. **Management.** The training management model for commanders is:

- Set objectives.
- Provide resources.
- Coach subordinates.
- Measure results.
- Provide feedback.

G-5. Individual training.

There are three elements basic to a well-trained individual ranger. He must be physically fit, be proficient in the use of his weapon, and have the psychological commitment to fight the enemy.

a. Individual training is the responsibility of the first-line supervisor under the supervision of the chain of command. It is conducted using performance-oriented techniques under battlefield conditions. New skills must be developed and practiced, and acquired skills must be reinforced throughout the training year. A high level of proficiency must be maintained through continuous training and practice.

b. Trainers should use peer instruction. Commanders ensure needed resources are available to conduct challenging individual training. Ranger units schedule and conduct this training and evaluate individual proficiency.

c. Flexibility in planning and scheduling is necessary. The goal of all individual training must provide for reinforcement training to offset the effects of the forgetting curve. (See Figure G-2).

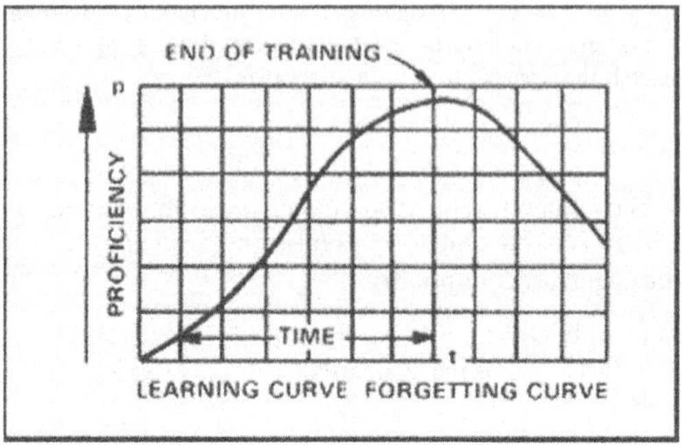

Figure G-2. The learning curve and forgetting curve.

d. The command sergeants major, unit first sergeants, and staff section sergeants are the keys to an effective training program. These senior NCOs must counsel, guide, and critique junior NCOs in training preparation and execution. Fire team leaders, section chiefs, and squad leaders coordinate and present individual training.

e. The NCO supervisors must ensure that the ranger is trained in the skills required for the specific military occupational specialty (MOS) and in those skills used daily. Not all training can be acquired in an office. For example, all rangers must maintain a high skill level with their weapons, be trained in NBC subjects, and be capable of performing contingency missions.

G-6. Collective training.

a. **Guidance.** Collective training trains crews, teams, squads, platoons, companies, and battalions to do their missions. The ranger MTP is used to plan training and evaluate the unit's weaknesses and strengths. If a ranger commander receives new training requirements or missions, they should become part of the collective training program. The ranger MTP describes the combat tasks to be performed by every echelon of the ranger battalion. It is a plan for organizing, conducting, and evaluating the year-round unit training program.

b. **Responsibility.** The ranger unit commander is responsible for the conduct of collective training and for all other training in his unit. Practical exercises emphasize the techniques of terrain use, movement, and effective employment of weapon systems. Training under stress will be conducted to move the unit to a higher plane of combat readiness.

c. **Drills.** Drills are written for small units--for example, fire team and squad, crew and section--and describe an action that is shorter than an entire STX. Drill training objectives cover both individual and collective skills (critical collective tasks) and list them together for effective group action. Drills must be mastered in order to react to rapidly changing situations in combat. The trainer can interrupt the execution of a drill to provide necessary training on individual or collective skills.

d. **Situational training exercise.**

(1) The STXs are short, mission-oriented tactical exercises in which a group of closely related collective tasks are trained. They are like drills but are more complex and flexible. They usually involve a larger unit than does a drill. They integrate leader training, leader tasks, drills, and separate individual tasks into a realistic setting. Trainers may execute STXs as written or tailor them to meet specific unit requirements. An STX is not standardized, but it must have a firm doctrinal base.

(2) Due to their design and purpose, STXs apply mainly to platoon- and company-size units. This does not preclude the development of STXs or similar exercises for unit staffs and headquarters. They combine "how-to-train" specifics for selected collective tasks with the "what-to-train" specifics found in doctrinal manuals and other documents.

G-7. Performance-oriented training.

a. Performance-oriented training emphasizes hands-on performance. Training is conducted to ensure a level of proficiency, rather than established number of hours of required training.

b. Performance-oriented training has precise training objectives and makes good use of resources. Training objectives serve as the basis for preparing, conducting, and evaluating all training.

G-8. Evaluation.

a. A training program must include an evaluation of training (planned and conducted). An evaluation plan must:

(1) Provide immediate feedback to the trainer.

(2) Recognize effective trainers.

(3) Evaluate all units in the conduct of training to provide for a comparison of training programs.

(4) Be flexible to unit needs; provide general and specific evaluations based on the level of training within the unit.

(5) Be in effect at all times.

b. Major considerations for training evaluations include:

(1) Ranger unit commanders establish and maintain an effective internal and external evaluation program.

(2) The director of each staff element is responsible for training evaluations within his area of staff responsibility with emphasis on low-density MOS groups.

(3) The operations officer of the ranger organization develops and validates training and evaluation plans.

c. Evaluation of training must be done vertically from the major task (mission) down through subtasks (collective tasks) to the leader and individual tasks. In selection of individual tasks for evaluation, priority should be given to critical individual tasks that impact on the success of the collective task or mission.

d. The critique and after-action review, after each major training event, are the most important aspects of training. The after-action review takes place immediately after completion of a training task (individual or collective). The individual(s) self-analyzes and critiques the training event. Then after a formal critique of the training, the individuals) trains to standard (not to time).

G-9. Field training.

a. **Considerations.** Ranger commanders schedule training under field conditions to permit maximum training opportunities. Field training emphasis is on mastery of a series of individual (soldier and leader) tasks and collective training tasks. These tasks are repeated until established standards are met. The goal of ranger field training is to ensure realistic, demanding, and meaningful training.

(1) When training is conducted under conditions of moderate stress, most combat critical skills are learned quicker and retained longer. (See Figure G-3 on the following page.) Stress can be incorporated in a training situation by physical deprivation (sleep, food), by live-fire exercises, by operation against an OPFOR under simulated battlefield conditions for an extended period, by the intensity of activity including demands for decision making, or by physically demanding activities. Ranger units train under stress so that they will react quickly in combat.

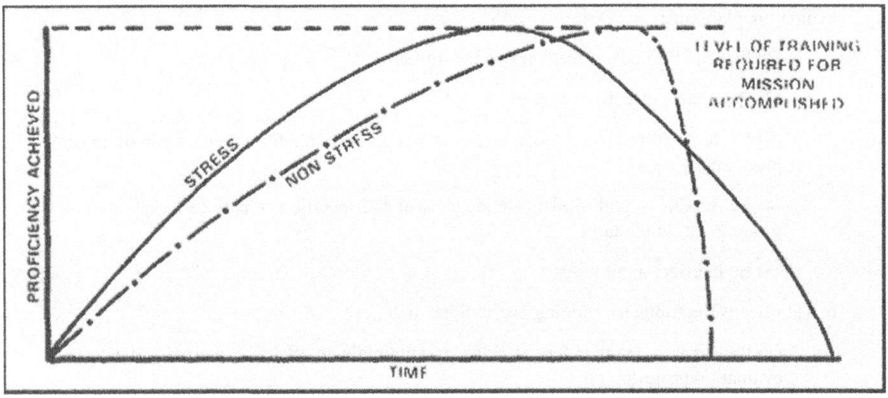

Figure G-3. The Army training and evaluation program.

(2) Antagonistic training situations are two-sided exercises conducted by opposing elements. Each element has a training objective (task and standards). The opposing force has a corresponding countertask. This training may be progressive--for example, starting with a

crawl-through, then a walk-through, and progressing to both elements operating at a fast pace. The use of multiple integrated laser equipment system (MILES) greatly enhances task and countertask training. The units should also be considered for a live-fire exercise against a realistic OPFOR target array when the basics have been mastered.

b. **Methods of training.**

(1) **Considerations.** The chance for success in combat can be increased by becoming proficient in training situations that depict realistic combat conditions. (See Figure G-4.) Meeting the standards of a realistic live-fire training exercise assures the commander that his unit could perform well in combat. Figure G-4 shows the relationship between realistic training and the challenge to the trainer.

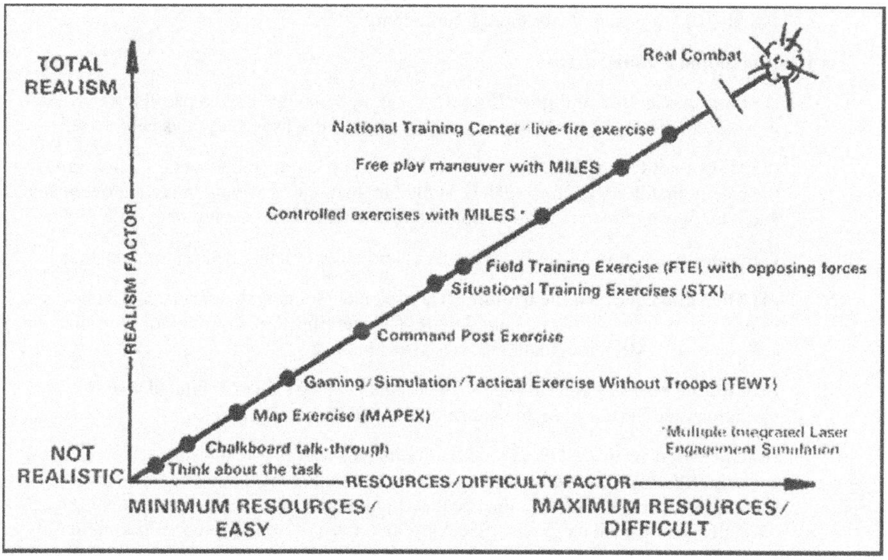

Figure G-4. Training proficiency scale.

(2) **Ranger cohesiveness training.** Ranger units promote teamwork by undergoing tough, stressful, often hazardous training. This challenges the unit, which gains in cohesiveness when it meets these challenges. This includes serving together in live-fire exercises, athletic events, precision parades, high-standard inspections, tough physical training, and combat.

(3) **Field training exercise.** An FTX is training normally conducted under simulated, realistic combat conditions. This training normally exercises command and control systems and the administrative and logistical needs of sustained combat. Stress training is part of this exercise, focusing on live-fire and MILES drills and STXS.

(4) **Combined arms training.** Combined arms training should include all means of support such as tactical air, armed helicopters, and indirect fire with emphasis on fire support coordination and airspace control.

(5) **Live-fire exercises.** Including live-fire exercises in drills and STXs offers a chance to evaluate units in their response to simulated combat missions. Ranger units train under pressure and always anticipate being selected to conduct a live-fire exercise. Live-fire exercises develop confidence and esprit, and reinforce unit discipline.

(6) **Opposing force.** Committing one unit against another adds a competitive spirit to a field training exercise (FTX). The use of MILES and squad combat operations exercises (simulation) (SCOPES) provides realism to two-sided field exercises. Trainers should use OPFOR equipment and training devices available, as well as making their own devices. (See FM 30-103, Appendix R, for more information.)

c. **Key functional training areas.**

(1) **Land navigation and patrolling.** Day and night navigation and patrolling exercises must be emphasized. Orienteering improves these skills, as well as physical fitness.

(2) **Movement techniques.** Infantry combat has two major components--to move and to hit the enemy first. Ranger units must be skilled in the use of traveling, traveling overmatch, and bounding overmatch. Units in contact must be skilled in moving under fire.

(3) **Fighting** positions. Training includes the building of concealed fighting positions.

(4) **Mine and countermine training.** Training includes how to defend against hostile mines, booby traps, and other demolitions. The identification, installation, neutralization, and removal of US mines and devices is also stressed.

(5) **Limited visibility operations.** Ranger units must be skilled in limited visibility operations and the use of night vision and thermal devices.

(6) **Operation security.** Operation security includes deception, physical security, signal security (SIGSEC), and information security. It is taught in all training and at all levels. Although all areas of OPSEC must be measured in training, techniques of camouflage and SIGSEC need special emphasis. (See AR 530-1 and TC 100-1 for more information.)

G-10. Professional development.

Ranger units at battalion and regimental levels conduct many officer and NCO professional development programs. This ensures that all officers and NCOs are skilled in their field, confident, and wellrounded. Professional readings and book reports, tactics seminars, and tactical exercises without troops are good development methods.

G-11. Physical training.

a. This program ensures that each ranger maintains the physical ability and stamina needed to do his mission. Other objectives are to promote unit cohesiveness and discipline. These are met by combining individual testing, formation, running, athletic competition, weight control, proper diet, teamwork drills, and freedom from drug and alcohol abuse. Mental fitness and toughness come from experience and success. Good physical conditioning increases a unit's chances of success.

b. Physical training (PT) programs are designed to include the elements of overload, progression, balance, variety, and regularity.

c. Ranger unit physical fitness standards are:

(1) Score 80 points for each Army physical readiness test (APRT) event and do six chinups.

(2) Pass the ranger swimming test.

(3) Complete an 8-kilometer run in 40 minutes.

(4) Complete a 12-kilometer road march in three hours (with rucksack, helmet, and weapon).

(5) Meet the Army height and weight standards.

G-12. Environmental and specialized training.

a. Rangers conduct training in environments as close to real combat missions as possible. This conditions rangers to survive and fight in a specific environment. Such training includes combat-critical tasks and technique training for that environment.

b. Each ranger battalion's environmental and specialized training includes at least the following:

(1) Jungle (once every 12 months).

(2) Desert (once every 12 months).

(3) Urban (once every 6 months).

(4) Extreme cold (once every 18 months).

(5) Mountain (once every 12 months).

(6) Amphibious (once every 18 months).

G-13. Ranger indoctrination program.

a. The RIP is a training and selection program. It is run by specially selected NCOs from the regimental headquarters. It has as its mission the training and indoctrination of newly assigned soldiers. It physically toughens and mentally sharpens the newly assigned soldier. It teaches basic skills and techniques. The RIP identifies and eliminates in three weeks any applicant who does not show dedication, motivation, physical fitness, and emotional stability.

b. An indoctrination program for newly assigned ranger-qualified officers and NCOs prepares them to assume duty as leaders within the ranger regiment. These personnel monitor and participate in RIP training, as well as undergoing individual ranger leadership training. The RIP provides them an overview of the regiment's history, its mission, and its goals. (See Appendixes C,

D, and E for further information about the history and goals of the ranger regiment.) The RIP also provides a review of the skills required in a ranger unit. Refresher training in airborne and pathfinder operations, use of indirect fires, and light infantry doctrine is conducted. During RIP all newly assigned officers and NCOs must meet the ranger physical fitness standard.

c. The RIP cadre also conducts a preranger program to prepare soldiers to attend the US Army Ranger School. It is a physically demanding course that teaches troop-leading procedures, operation orders, and small unit patrol actions.

Glossary

The following acronyms, abbreviations, and definitions are provided to aid the reader in understanding this publication.

Section I

ACRONYMS AND ABBREVIATIONS

AAT	Army assault team
ABCCC	airborne battlefield command and control center
ACL	allowable cargo load
ADA	air defense artillery
AFFOR	the USAF element of a joint task force
AGL	above ground level
AKZ	armor kill zone
ALCC	airlift control center
ALCE	airlift control element
AM	amplitude modulation
AMC	air mission commander
APERS	antipersonnel
APRT	Army physical readiness test
ARFOR	the Army element of a joint task force
ARSOC	Army special operations command
ARSOF	Army Special Operations Forces
ASG	area support group
ASPC	all-source production center

ATGM	antitank guided missile
AWADS	adverse weather aerial delivery system
BCE	battle coordination element
BTMS	battalion training management system
CA	civil affairs
CAS	close air support
CCT	combat control team
CDS	container delivery system
CE	communications-electronics
CI	counterintelligence
CINC	commander in chief
COMINT	communications intelligence
COMSEC	communications security
CONUS	continental United States
COSCOM	corps support command
CP	command post
CS	combat support; a chemical agent
CSS	combat service support
DACG	departure airfield control group
DACO	departure airfield control officer
DASC	direct air support center
DMDG	digital message device group
DOD	Department of Defense
DS	direct support
DSU	direct support unit
DZ	drop zone
EAC	echelons above corps

EACIC	Echelons Above Corps Intelligence Center
ECCM	electronic counter-countermeasure
ECM	electronic countermeasures
E&E	escape and evasion
EEFI	essential elements of friendly information
EOC	emergency operations center
EW	electronic warfare
FARP	forward area rearm and refuel point
FASCAM	family of scatterable mines
FAX	facsimile transmission
FDC	fire direction center
FEBA	forward edge of the battle area
FFL	friendly front lines
FIST	fire support team
FLOT	forward line of own troops
FM	field manual; frequency modulation
FO	forward observer
FORSCOM	US Army Forces Command
FPF	final protective fire
FPL	final protective line
FRAGO	fragmentary order
FSA	fire support area
FSCOORD	fire support coordination or coordinator
FSE	fire support element
FSO	fire support officer
FSS	fire support station
FTX	field training exercise

G3	Assistant Chief of Staff, Operations
GS	general support
GSR	ground surveillance radar
HAARS	high-altitude, airdrop resupply system
HAHO	high-altitude, high-opening
HALO	high-altitude, low-opening
HE	high explosive
HF	high frequency
HHC	headquarters and headquarters company
HSLLADS	high-speed, low-level airdrop system
HUMINT	human intelligence
IAW	in accordance with
IFF	identification-friend-or-foe
IP	initial point
IRP	initial rally point
ISB	intermediate staging base
JACC/CP	joint airborne communications center/command post
JCS	Joint Chiefs of Staff
JCSE	joint communications support element
JFC	joint force commander
JSOC	joint special operations command
JTB	joint transportation board
JTF	joint task force
kw	kilowatt
LAPES	low altitude parachute extraction system
LAW	light antitank weapon
LNO	liaison officer

LOC	logistical operations center
LRRP	long-range reconnaissance patrol
LRSU	long-range surveillance unit
LTD	laser target designator
LZ	landing zone
MAC	Military Airlift Command
MANPADS	man-portable air defense system
MCC	movement control center
METT-T	mission, enemy, terrain, troops and time available
MI	military intelligence
MILES	multiple integrated laser engagement system
mm	millimeter
MMC	Materiel Management Center
MMEE	minimum mission-essential equipment
MOS	military occupational specialty
MOUT	military operations on urbanized terrain
MRLS	multiple rocket launch system
MTP	mission training plan
NBC	nuclear, biological, chemical
NCA	National Command Authority
NCO	noncommissioned officer
NOD	night observation device
OCOKA	observation, concealment, obstacles, key terrain, and avenues of approach
OCONUS	outside continental United States
OPCOM	operational command
OPCON	operational control
OPFOR	opposing force

OPORD	operation order
OPSEC	operations security
ORP	objective rally point
PAC	personnel administration center
PDF	primary direction of fire
PIR	priority intelligence requirements
PLL	prescribed load list
POM	preparation for oversea movement
PSYOP	psychological operations
PT	physical training
PZ	pickup zone
QRE	quick-reaction element
RATELO	radiotelephone operator
REMAB	remote marshalling base
RFL	restrictive fire line
RIP	ranger indoctrination program
RP	rally point; release point
RPG	rocket-propelled grenade
RRF	ranger ready force
RRP	reentry rally point
RSE	ranger support element
RSTA	reconnaissance, surveillance, and target acquisition
S1	adjutant
S2	intelligence officer
S3	operations and training officer
S4	supply officer
S5	civil affairs officer

SAR search and rescue

SATCOM satellite communications)

SCIF sensitive compartmented information facility

SCOPES squad combat operations' exercises (simulation)

SCUBA self-contained underwater breathing apparatus

SEAD suppression of enemy air defense

SEMA special electronic mission aircraft

SERE survival, evasion, resistance, and escape

SF special forces

SHORAD short-range air defense

SIGSEC signal security

SKE station-keeping equipment

SLAR side-looking airborne radar

SOA special operations aviation

SOCB special operations communications battalion

SOCOM special operations command (usually, 1st SOCOM)

SOP standing operating procedures

SOSB special operations support battalion

SOTF special operations task force

SPL support platoon leader

STX situational training exercise

TA theater Army

TAACOM Theater Army Area Command

TAC Tactical Air Command

TACP tactical air control party

TACSAT tactical satellite terminal

TAMMC Theater Army Materiel Management Center

TASG	theater Army support group
TDA	table of distribution and allowances
TOA	transfer of authority
TOC	tactical operations center
TOE	table(s) of organization and equipment
TRP	target reference point
TTY	teletype
UHF	ultrahigh frequency
US	United States
USAF	United States Air Force
USMC	United States Marine Corps
USN	United States Navy
VHF	very high frequency
WP	white phosphorus
XO	executive officer

Section II

DEFINITIONS

ADVANCE FORCE. An operation requiring the immediate deployment of US forces with the intent to negate a threat before the deployment of follow-on forces is required.

AIRLAND BATTLE DOCTRINE. An approach to military operations that realizes the full potential of US forces. Two notions (extending the battlefield and the ability to integrate conventional, nuclear, chemical, and electronics means) are combined to describe a battlefield where the enemy is attacked to the full depth of his formation.

AREA OF INFLUENCE. A geographical area wherein a commander is directly capable of influencing operations by maneuver or fire support systems normally under his command or control.

AREA OF INTEREST. That area of concern to the commander, including the area of influence, areas adjacent thereto, and extending into enemy territory to the objectives of current or planned operations. This area also includes areas occupied by enemy forces who would jeopardize the accomplishment of the mission.

AREA OF OPERATIONS. That portion of an area of conflict necessary for military operations. Areas of operations are geographical areas assigned to commanders for which they have responsibility and in which they have authority to conduct military operations.

CHALK. One specific aircraft load.

CLANDESTINE OPERATIONS. Activities to accomplish intelligence, counterintelligence, and other similar activities sponsored or conducted by governmental departments or agencies, in such a way as to ensure secrecy or concealment. (They differ from covert operations in that emphasis is placed on concealment of the operation rather than on concealment of the identity of the sponsor.)

COMMUNICATIONS INTELLIGENCE (COMINT). The interception and processing of foreign communications passed by radio, wire, or other electromagnetic means, and by the processing of foreign encrypted communications, however transmitted. Interception comprises search, intercept, operator identification, signal analysis, traffic analysis, crypto-analysis, study of plain text, the fusion of these processes, and reporting the results.

COMMUNICATIONS SECURITY. The protection resulting from all measures designed to deny to unauthorized persons information of value that might be derived from the possession and study of telecommunications, or to mislead unauthorized persons in their interpretations of the results of such study. COMSEC includes crypto-security, physical security, and transmission security.

COUNTERINTELLIGENCE. That aspect of intelligence activities, both offensive and defensive, designed to detect and neutralize or destroy the effectiveness of hostile foreign intelligence activities and to protect information against espionage, personnel against subversion, and installations and materiel against sabotage.

COVERT OPERATIONS. Operations that are so planned and executed as to conceal the identity of or permit plausible denial by the sponsor. They differ from clandestine operations in that emphasis is placed on concealment of identity of the sponsor rather than on concealment of the operation.

DECEPTION. Those measures designed to mislead the enemy by manipulation, distortion, or falsification of evidence to induce him to react in a manner prejudicial to his interests.

DEMONSTRATION. A show of force in an area where a decision is not sought. It is similar to a feint but does not make contact with the enemy.

DRILL. A drill focuses on a "chunk" or "slice" of battle and is targeted at small units. Drills are critical collective tasks that require a high degree of proficiency from a small unit. Generally, small units "drill" collective tasks that require rapid responses by unit members in the absence of detailed orders from unit leaders.

ELECTRONIC COUNTERMEASURES. That division of electronic warfare involving actions taken to prevent or reduce an enemy's effective use of the electromagnetic spectrum.

ELECTRONIC INTELLIGENCE (ELINT). The technical and intelligence information derived from foreign noncommunications electromagnetic radiations emanating from other than atomic detonations or radioactive sources.

ELECTRONIC WARFARE. Military action involving the use of electromagnetic energy to determine, exploit, reduce, or prevent hostile use of the electromagnetic spectrum and action that retains friendly use of the electromagnetic spectrum.

ELECTRONIC WARFARE SUPPORT MEASURES (ESM). That division of EW involving actions taken to search for, intercept, identify, and locate sources of radiated electromagnetic energy for the purpose of immediate threat recognition. ESM provides a source of information needed for immediate action involving electronic countermeasures, electronic counter-countermeasures, avoidance, targeting, and other tactical employment of forces.

ESPIONAGE. Actions directed toward the acquisition of information through clandestine operations.

EVACUATION/SAFEGUARD. The use of US forces to evacuate US citizens or allies from a hostile or potentially hostile area, or to safeguard US personnel or property.

EXFILTRATION. The removal of personnel or units from areas under enemy control by stealth, deception, surprise, or clandestine means. Exfiltration is a method of extracting ranger units.

EXTRACTION. The removal of personnel or units from areas under enemy control by any means, to include exfiltration, airland, airmobile, or amphibious.

FEINT. A diversionary supporting attack conducted to draw the enemy's attention from the main effort. It is normally executed by brigades and small units. Feints are usually shallow, limited objective attacks that go in before or during the main attack.

GUERRILLA WARFARE. Military and paramilitary operations conducted in enemy-held or hostile territory by irregular and predominantly indigenous forces.

HIGH-INTENSITY CONFLICT. A war between the US and any other nation or nations in which both belligerents use the most modern technology and all resources in intelligence, mobility, firepower (including NBC), command, communications and control, and service support.

HUMAN INTELLIGENCE. Intelligence derived from information provided by human resources.

INFILTRATION. The movement through or into an area of territory occupied by either friendly or enemy troops or organizations. The movement is made either by small groups or by individuals, at extended or irregular intervals. When used in conjunction with the enemy, it implies that contact is avoided. Infiltration is a subset of insertion.

INSERTION. Placement of troops and equipment into an operational area by any means.

INTERDICT. To isolate or seal off an area by military means; to prevent, hinder, or delay the use of an area or route by enemy forces.

INTERDICT LINE OF COMMUNICATIONS. An attack to seal off an area or to deny use of a route or approach.

JOINT TASK FORCE. A JTF may be constituted and designated by the Secretary of Defense or by the commander of a unified command, specified command, or an existing JTF. Normally, it performs missions having specific, limited objectives or missions of short duration. It dissolves when it has achieved its purpose. The joint task force commander is responsible to the commander, known as the establishing authority, who created the JTF. The JTF is composed of elements of two or more services operating under a single JTF commander. The JTF commander has operational control over the entire force and may have direct command of his own staff with representatives from the other services. He exercises logistics coordination or control only as necessary to meet his subordinate commanders' logistics needs.

LIAISON. That personal contact or communication maintained between elements of military forces to ensure mutual understanding and unity of purpose and effort.

LINE OF COMMUNICATIONS. All routes (land, water, and air) that connect an operating military force with one or more bases of operations and along which supplies and military forces move.

LOW-INTENSITY CONFLICT. Limited politico-military operations conducted to achieve political, social, economic, or psychological objectives. It is generally confined to a specific geographic area and is often characterized by operational constraints imposed on ranger units by political and economic considerations. It may involve intermittent combat against regular threat military forces, but it more often involves conflict with irregular, security, or paramilitary forces of a country or group involved in an operation that is counter to the best interests and national security of the United States. It is characterized by relatively low levels of combat and constraints placed on the weaponry, tactics, and movements of the ranger force. It may involve counterterrorist or antiterrorist operations and generally occurs before a pro forma declaration of war.

MID-INTENSITY CONFLICT. War, declared or undeclared, between the US and the regular forces of one or more nations and their respective allies, if any, in which the belligerents use the most modern technology and all resources in intelligence, mobility, firepower (excluding nuclear, chemical, and biological weapons), command, communications and control, and support. They are used for limited objectives under definitive policy limitations as to the extent of destructive power that can be used or the extent of geographical area that might be involved.

MILITARY STRATEGY. Art and science of using the armed forces of a nation to secure the objectives of national policy by application of force or the threat of force. Military strategy sets the fundamental conditions for operations.

NATIONAL COMMAND AUTHORITY. This includes the President and Secretary of Defense or the duly deputized alternates or successors. Commonly referred to as NCA.

OPERATIONAL COMMAND/CONTROL. Operational command (OPCOM) and operational control in joint force terminology both refer to the authority exercised by joint commanders over subordinate service components. Those terms are not interchangeable in joint operations. OPCOM applies to the authority exercised by commanders of unified commands. OPCON is the authority that subordinate joint task force commanders exercise in the conduct of specific operations. The authority that military departments exercise over their respective components is commonly referred to as command less operational command.

OPERATIONAL LEVEL OF WAR. The operational level of war uses available military resources to attain strategic goals within a theater of war. It is the theory of larger unit operations. It also involves planning and conducting campaigns. Campaigns are sustained operations designed to defeat an enemy force in a specified space and time with simultaneous and sequential battles. The disposition of forces, selection of objectives, and actions taken to weaken or to outmaneuver the enemy all set the terms of the next battle and exploit tactical gains. They are all part of the operational level of war. In AirLand Battle doctrine, this level includes the marshalling of forces and logistical support, providing direction to ground and air maneuver, applying conventional and nuclear fires in depth, and employing unconventional and psychological warfare.

OPERATIONS SECURITY. Those measures designed to protect information concerning planned, ongoing, and completed operations from unauthorized disclosure. It includes all actions a command takes to deny the enemy information about friendly units and their operations.

ORDER OF BATTLE (OB). The identification, strength, command structure, and disposition of the personnel, units, and equipment of any military force.

RAID. Usually a small-scale operation, involving a swift penetration of hostile territory to secure information, confuse the enemy, or destroy his installations. It ends with a planned withdrawal upon completion of the assigned mission.

RANGER MISSION. To conduct strike, raid, or special light infantry operations. Execution of ranger missions is normally directed by the National Command Authority. Rangers also support operational-level commanders during operations of limited duration when allocated or assigned by the theater commander.

RECONNAISSANCE. A mission undertaken to obtain, by visual observation or other detection methods, information about the activities and resources of an enemy or potential enemy; or to secure data concerning the meteorological, hydrographical, or geographical characteristics of a particular area.

SEARCH AND RESCUE. The use of aircraft, surface craft, submarines, specialized rescue teams, and equipment to search for and rescue personnel in distress on land or at sea.

SEIZURE. As an operational purpose, the capture of a voluntarily restricted portion of an enemy-controlled territory. Capture of an isolated land mass such as an island is usually categorized as a seizure.

SEIZE KEY FACILITY. An attack on an installation with the intent of holding it for a short period. Normally, rangers would make the assault to seize and secure an installation for follow-on forces.

SERVICE COMPONENT. Each service component commander is responsible for recommending the proper use of his forces and for accomplishing operational tasks assigned by the joint commander. He is also responsible for his service in matters of internal administration and discipline; training in own service doctrine, techniques, and tactics; designation of specific units to meet joint requirements; logistics functions normal to the component; tactical employment of service component forces; and service intelligence matters.

SHOW OF FORCE. A mission carried out to demonstrate US resolve, whereby US forces are deployed to defuse a situation that may be harmful to US interests or national objectives.

SIGNAL INTELLIGENCE (SIGINT). A generic term that includes COMINT and ELINT. Activities concerned with the collection and processing of foreign electromagnetic emissions for the purpose of producing intelligence information.

SITUATIONAL TRAINING EXERCISE. An STX is a short, mission-oriented exercise in which a group of closely-related collective tasks are trained, using a doctrinally preferred method of executing these tasks to established standards. An STX is drill-like in nature, but it is more complex and flexible than a drill. It usually involves a larger unit than a drill.

SPECIAL OPERATIONS. Military operations conducted by specially trained, equipped, and organized DOD forces against strategic or tactical targets in pursuit of national, military, political, economic, or psychological objectives. Special operations may support conventional military operations, or they may occur independently when the use of conventional forces is either inappropriate or infeasible. Sensitive peacetime operations, except for training, are normally authorized by the NCA and conducted under the direction of the NCA or designated commander. Special operations may include unconventional warfare, counter-terrorist operations, collective security, psychological operations, and civil affairs measures.

SPECIAL OPERATIONS FORCES. All civil affairs, psychological operations, ranger (active component), and special forces units within the US Army, US Army Reserve, and National Guard.

SPECIAL OPERATIONS COMMAND. The major subordinate command to FORSCOM exercising command and control of all active component Army special operations forces not placed under the operational command of unified or specified commands.

STRATEGIC LEVEL OF WAR. Military strategy uses the armed forces of a nation to secure the objectives of national policy by applying force or the threat of force. Military strategy sets the fundamental conditions for operations.

STRIKE. An attack that is intended to inflict damage on, seize, or destroy an objective. Sometimes referred to as direct action missions.

SURVEILLANCE. The systematic observation of aerospace, surface or subsurface areas, places, persons, or things by visual, aural, electronic, photographic, or other means.

TACTICAL. Pertaining to the employment of units in combat.

TACTICAL LEVEL OF WAR. Tactics are the specific techniques smaller units use to win battles and engagements that support operational objectives. Tactics employ all available combat, combat support, and combat service support. Tactics involve the movement and positioning of forces on the battlefield in relation to the enemy, the provisioning of fire support, and the logistical support of force before, during, and following engagements with the enemy. At corps and division, operational and tactical levels are not clearly separable, but they are guided by the same principles. An operation designed to defeat any enemy force in an extended area does so through operational maneuver and a series of tactical actions.

TACTICAL COMBAT INTELLIGENCE. That knowledge of the enemy, weather, and geographical features needed by a commander in the planning and conduct of combat operations. It is derived from the interpretation of information on the enemy and the environment.

TARGET ANALYSIS. A detailed and systematic examination of processed intelligence to identify and locate targets.

UNIFIED COMMAND. A unified command is established and designated by direction of the President. The command is organized under a single unified CINC and has a broad, continuing mission. The CINC is responsible to the Secretary of Defense for accomplishing his broad and continuing mission and operates under the strategic and operational direction of the JCS. A unified command is composed of assigned forces of two or more services. These forces, as service components, are under the operational command of the unified commander. The CINC exercises operational command authority through service component commanders. When tasked by higher authority, he may establish a subordinate JTF to conduct specific missions. Under emergency conditions, the CINC has the authority to use all facilities and supplies of assigned forces to accomplish his mission.

References

REQUIRED PUBLICATIONS

Required publications are sources that users must read in order to understand or to comply with this publication.

Field Manuals (FM)

FM 7-8	The Infantry Platoon and Squad (Infantry, Airborne, Air Assault, Ranger)
FM 7-10	The Infantry Rifle Company (Infantry, Airborne, Air Assault, Ranger)
FM 7-20	The Infantry Battalion (Infantry, Airborne, and Air Assault)

RELATED PUBLICATIONS

Related publications are sources of additional information. They are not required in order to understand this publication.

Army Regulations (AR)

AR 350-1	Army Training

Field Manuals (FM)

FM 10-52-1	Commander's Handbook for Water Usage in Desert Operations
FM 10-501	Airdrop of Supplies and Equipment: Rigging Containers
FM 10-512	Airdrop of Supplies and Equipment: Rigging Typical Supply Loads
FM 10-542	Airdrop of Supplies and Equipment: Rigging Loads for Special Operations
FM 10-547	Airdrop of Supplies and Equipment: Rigging the High Speed Aerial Delivery Container CTU-2/A
FM 10-553	Airdrop of Supplies and Equipment: Rigging Ammunition
FM 12-3-4	Echelons Above Corps (EAC) Personnel and Administrative Doctrine

FM 14-6	Comptroller/Finance Services in Theaters of Operations
FM 19-15	Civil Disturbances
FM 20-11-1	Military Diving (Volume 1)
FM 20-11-2	Military Diving (Volume 2)
FM 21-20	Physical Training Program
FM 21-26	Map Reading
FM 21-75	Combat Skills of the Soldier
FM 21-76	Survival, Evasion, and Escape
FM 21-77A(S)	Joint Worldwide Evasion and Escape Manual
FM 22-9	Soldier Performance in Continuous Operations
FM 23-67	Machinegun, 7.62-mm, M60
FM 25-1	Training
FM 25-2	Unit Training Management
FM 25-3	Training in Units
FM 25-4	How to Conduct Training Exercises
FM 25-5	Training for Mobilization and War
FM 25-7	Training Ranges
FM 26-2	Management of Stress in Army Operations
FM 29-47	Main Supply and Service Company, Forward Supply and Service Detachments, Supply and Service Battalion, Airborne Division
FM 29-51	Division Supply and Field Service Operations
FM 30-5	Combat Intelligence
FM 30-17	Counterintelligence Operations
FM 31-18	Long-Range Reconnaissance Ranger Company
FM 31-19	Special Forces Military Free-Fall Parachuting
FM 1-24	Special Forces Air Operations

FM 31-25	Special Forces Waterborne Operations
FM 31-70	Basic Cold Weather Manual
FM 34-1	Intelligence and Electronic Warfare Operations
FM 34-60A	Counterintelligence Operations (U)
FM 55-9	Unit Air Movement Planning
FM 57-38	Pathfinder Operations
FM 57-220	Basic Parachuting Techniques and Training
FM 63-3J	Combat Service Support Operations--Corps
FM 63-4	Combat Service Support Operations--Theater Army Area Command
FM 63-5	Combat Service Support Operations--Theater Army
FM 90-3	Desert Operations
FM 90-4	Airmobile Operations
FM 90-5	Jungle Operations
FM 90-6	Mountain Operations
FM 90-8	Counterguerilla Operations
FM 90-10	Military Operations on Urbanized Terrain (MOUT)
FM 90-10-1	An Infantryman's Guide to Urban Combat
FM 100-5	Operations
FM 100-20	Low Intensity Conflict
FM 100-26	The Air-Ground Operations System
FM 101-5	Staff Organization and Operations
FM 101-10-1	Staff Officers' Field Manual: Organization, Technical and Logistical Data (U)

Firing Table (FT)

FT 60-P-1	Firing Tables for Mortar, 60-mm: M224, Firing Cartridge, HE, M720, Cartridge, HE, M49A4, Cartridge, TP, M5OA3, Cartridge, WP, M302AI, Cartridge, Illuminating, M83A3

Soldier's Training Publication (STP)

STP 7-11B24-SM	Soldier's Manual MOS 11B, Infantryman, Skill Level 2/3/4

Technical Manuals (TM)

TM 10-1670-267-12&P	Operator's and Organizational Maintenance Manual Including Repair Parts and Special Tools List for High Altitude Airdrop Resupply System: 500 Pound Capacity

Training Circulars (TC)

TC 23-14	Sniper Training and Employment
TC 23-90	Mortar Training
TC 24-1	Communications-Electronics Operation Instructions, the CEOI
TC 31-20-5(C)	Special Forces Communications (U)
TC 90-6-1	Military Mountaineering

Other Related Publications

King, Michael J. Dr., *Leavenworth Papers, No. 11*, "Rangers Selected Combat Operations in World War II," Combat Studies Institute, USACGSC, Fort Leavenworth, KS

PROJECTED PUBLICATIONS

Projected publications are sources of additional information that are scheduled for printing but are not yet available. Upon print, they will be distributed automatically via pinpoint distribution. They may not be obtained from the USA AG Publications Center until indexed in DA Pamphlet 310-1.

Field Manuals (FM)

FM 7-70	Light Infantry Squad/Platoon
FM 7-71	Light Infantry Company
FM 7-72	Light Infantry Battalion
FM 7-73	Light Infantry Brigade

FM 23-36 60-mm Mortar, M224